DATE DUE			

biological aspects of
inorganic chemistry

BIOLOGICAL ASPECTS OF INORGANIC CHEMISTRY

edited by

The Bioinorganic Group

A. W. Addison
W. R. Cullen
D. Dolphin
B. R. James

Department of Chemistry
The University of British Columbia
Vancouver, Canada

A Wiley-Interscience Publication
John Wiley & Sons
New York · London · Sydney · Toronto

Library of Congress Cataloging in Publication Data:
Main entry under title:

Biological aspects of inorganic chemistry.

   "A Wiley-Interscience publication."
   Includes index.
   1. Biological chemistry. 2. Chemistry, Inorganic.
3. Bioenergetics. I. Addison, A. W. II. University
of British Columbia. Dept. of Chemistry. Bioinorganic
Group. [DNLM: 1. Biochemistry. 2. Chemistry.
QD151.2 B615]
QH345.B536   574.1'92   76-44225
ISBN 0-471-02147-4

Printed in the United States of America

10 9 8 7 6 5 4 3 2 1

preface

In recent years there has been a dramatic increase in interest in that area of science which is found in the interface between inorganic chemistry and biochemistry. This field, generally known as bioinorganic chemistry, has recently seen a remarkable growth. Such growth is evidenced by the establishment of a new journal named *Bioinorganic Chemistry*, through the establishment of research laboratories such as the Unit of Nitrogen Fixation at the University of Sussex, England, by the numerous sections at conferences dealing with such topics, and by the appearance of a number of text and review books on the subject.

The development of this discipline within Canada was invigorated by the award of a negotiated development grant from the National Research Council of Canada to establish the Bioinorganic Group within the chemistry department at the University of British Columbia.

As part of our endeavors in this area the group organized a symposium on the biological aspects of inorganic chemistry which was held at UBC June 20 - 25, 1976. The following 11 papers are symposium addresses given at the conference. Two others given by Dr. J. P. Collman (Stanford University) and Dr. T. R. Parsons (University of British Columbia) were not presented in manuscript form.

A. W. Addison
W. R. Cullen
D. Dolphin
B. R. James

Vancouver, Canada
September 1976

contents

...continued

biological aspects of
inorganic chemistry

on the coupling of oxidation to phosphorylation in biological systems

JUI H. WANG

Bioenergetics Laboratory, Acheson Hall,
State University of New York, Buffalo, New York 14214

1. INTRODUCTION

To maintain life, all living systems must feed on free energy from their surroundings either directly in the form of light or in the form of chemical free energy converted from light and stored in organic compounds used as food. During respiration the free energy stored in food is released through controlled oxidation in mitochondria. A part of this free energy is utilized to synthesize ATP from ADP and inorganic orthophosphate (P_i) for maintaining us in the living state, i.e., for growth and reproduction according to genetic codes and for adaptation to the environment as required by circumstantial factors. The remainder of the liberated free energy is returned to our surroundings in the form of heat, with consequent increase in total entropy, so that those vital processes can take place at satisfactory rates.

The standard free energy change for hydrolysis of ATP at pH 7 and 25°C is about -7.3 kcal/mole. The actual free energy of hydrolysis of ATP to ADP and P_i at physiological concentrations of ATP, ADP, P_i and Mg^{2+} is between -12 and -14 kcal/mole (1,2). Consequently, the condensation of ADP and P_i to form ATP and water is a thermodynamically highly unfavorable reaction which cannot occur unless it is coupled with a free energy supplying process. In both respiration and photosynthesis, the free energy is supplied by an oxidation-reduction or electron transfer process.

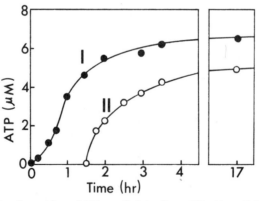

Fig. 1. *Generation of ATP coupled to the oxidization of ferrohemochrome by air.*
Curve I: ADP (0.13 mM), diimidazolium hydrogen phosphate (1.3 mM) and diimidazole-ferrohemochrome (0.4 mM) in dimethylacetamide solutions were mixed and exposed to air at zero time at 25°C.
Curve II: ADP added 1.5 hrs. after the other components were mixed and exposed to air. The sigmoid shape of Curve I suggests that an intermediate is formed which subsequently reacts with ADP to form ATP. The simpler shape of Curve II shows that the intermediate was already formed before the addition of ADP.

But by what molecular mechanism is the free energy liberated in an oxidation process utilized to drive a phosphorylation reaction in which ADP and P_i condense to form ATP and water? There is not yet an universally accepted answer.

2. CHEMICAL REACTIONS FOR COUPLING OXIDATION TO PHOSPHORYLATION

Believing in the ultimate unity of science, we searched for relevant model chemical reactions to guide our thinking on the oxidative phosphorylation problem. It was found that when ferro-hemochrome was oxidized by air in N,N-dimethylacetamide solution containing adenosine-5'-monophosphate (AMP), P_i and imidazole, ATP was formed *(3)*. A typical experiment is illustrated in Fig. 1. Additional experiments with ^{14}C-labeled imidazole and [32]P_i showed that in this model reaction 1-phosphoimidazole was produced first which subsequently transferred its phosphoryl group to AMP or ADP to form ADP or ATP, respectively *(4)*.

$$\text{N} \overset{\curvearrowright}{\underset{\smile}{\text{N}}} \text{–}\overset{\overset{O}{\|}}{\underset{\underset{O^-}{|}}{\text{P}}}\text{–O}^- + \text{AMP} \longrightarrow \text{HN} \overset{\curvearrowright}{\underset{\smile}{\text{N}}} + \text{ADP}$$

1-phosphoimidazole

$$\text{N} \overset{\curvearrowright}{\underset{\smile}{\text{N}}} \text{–}\overset{\overset{O}{\|}}{\underset{\underset{O^-}{|}}{\text{P}}}\text{–O}^- + \text{ADP} \longrightarrow \text{HN} \overset{\curvearrowright}{\underset{\smile}{\text{N}}} + \text{ATP}$$

But how was 1-phosphoimidazole itself formed through the oxidation of ferrohemochrome by O_2? Inspection of the following potential diagram at pH 7 shows that O_2 is a very good 2-electron acceptor but rather poor 1-electron acceptor.

$$\begin{array}{ccccccccc}
& -0.2\,V & & 0.8\,V & & 0.4\,V & & 2.3\,V & \\
O_2 & \rule{1cm}{0.4pt} & O_2^- & \rule{1cm}{0.4pt} & H_2O_2 & \rule{1cm}{0.4pt} & HO+H_2O & \rule{1cm}{0.4pt} & 2H_2O \\
& 0.27\,V & & & & & 1.35\,V & & \\
& & & & 0.815\,V & & & &
\end{array}$$

Therefore as soon as O_2 has accepted one electron from ferroheme, it has a strong tendency to extract another electron from imidazole to form the imidazolyl radical, $C_3H_3N_2$, i.e.,

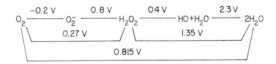

This imidazolyl radical could then react with P_i to form an unstable phosphoimidazolyl radical $(C_3H_5N_2PO_4^-)$, which could subsequently be reduced by another ferrohemochrome molecule to
produce 1-phosphoimidazole and water *(5)*.

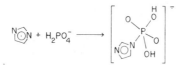

The formation of the trigonal-bipyramidal intermediate compound 1-orthophosphoimidazole $(C_3H_5N_2PO_4)^{2-}$ through a direct
nucleophilic attack at the P atom by imidazole is not only slow
but thermodynamically unfavorable. However, since radical reactions generally require a much lower activation free energy, the
trigonal-bipyramidal phosphoimidazolyl radical can be formed much
more readily through radical addition to the P=O double bond as
illustrated below.

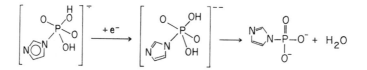

In the subsequent step driven by the oxidation-reduction free
energy change, this phosphoimidazolyl radical is reduced to the
unstable 1-orthophosphoimidazole which then spontaneously elimi-
nates H_2O to form 1-phosphoimidazole. In this way, oxidation can
be coupled to phosphorylation.
    Alternatively, one could imagine that in this model reaction
O_2 was first reduced by ferrohemochrome to the superoxide ion O_2^-,
which subsequently reacted with imidazole to form the hydroper-
oxide ion HO_2^- and imidazolyl radical. The steady-state concentration of O_2^- could be negligibly low if the second step is fast.
Fig. 2 shows the ESR spectrum of a frozen dimethylacetamide (DMAC)
solution containing [17]O-enriched superoxide ion O_2^- at $-170^\circ C$.
The sample was prepared by condensing a small amount of potassium
vapor on a cold glass surface under vacuum and subsequently oxidizing the freshly deposited film of potassium with an excess of
[17]O-enriched oxygen gas. The peaks at $g_\parallel$ = 2.095 and $g_\perp$ = 2.006
are due to [16]O_2^-, and the six smaller peaks are due to hyperfine

splitting of the $g_\perp$-signal of $^{17}O^{16}O^-$ by the magnetic moment of its ^{17}O nucleus *(6)*.

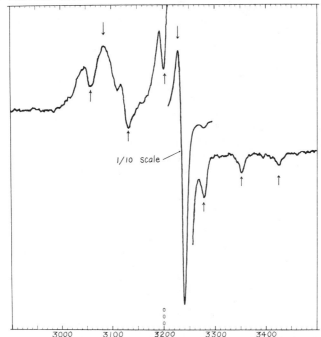

Fig. 2. ESR spectrum of ^{17}O-labeled O_2^- radical in anhydrous DMAC at -170°C. The two downward arrows indicate the resonance peaks of $^{16}O^{16}O^-$ for $g_\perp = 2.006$ at ten times reduced scale, and for $g_\parallel = 2.095$ at full scale, respectively. The six upward arrows indicate the resonance peaks of $^{17}O^{16}O^-$ due to the interaction between the electron magnetic moment ($g_\perp = 2.006$) with the nuclear magnetic moment of ^{17}O (nuclear spin = $(5/2)\hbar$, nuclear magnetic moment = -1.893 nuclear magneton). The six hyperfine peaks are evenly spaced at 74 gauss intervals. The observed large hyperfine splitting indicate appreciable s-character of the unpaired electron in O_2^-. Mass-spectrometric analysis of the oxygen gas used in preparing O_2^- gave the following isotope composition: $^{16}O^{16}O$, 20.89%; $^{16}O^{17}O$, 18.40%; $^{17}O^{17}O$, 4.03%; $^{16}O^{18}O$, 31.17%; $^{17}O^{18}O$, 13.67%; $^{18}O^{18}O$, 11.63%.

In anhydrous DMAC solution O_2^- is fairly stable. The ESR spectrum remained unchanged when the sample was thawed at slightly above the melting temperature for 10 minutes and then refrozen at -170°C. However, when imidazole was added to the sample at slightly above its melting temperature and then quickly refrozen, the ESR spectrum of O_2^- disappeared and was replaced by a different ESR signal.

A much cleaner method of generating imidazolyl radical and 1-phosphoimidazolyl radical is to illuminate an aqueous phosphate buffer at pH 7 containing imidazole, hematoporphyrin (HP) and a catalytic amount of O_2 with orange or yellow light *(7)*. The possible reaction steps are:

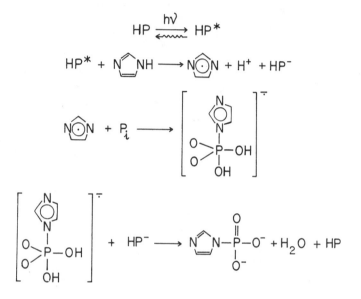

Summarizing the above steps, we obtain the following overall net reaction:

Fig. 3 shows an experiment in which over 50% of the imidazole was photophosphorylated in 24 hours in aqueous phosphate buffer at pH 7 and 4°C. The absorption spectrum of the hematoporphyrin in the reaction mixture remained unaltered at the end of the experiment.

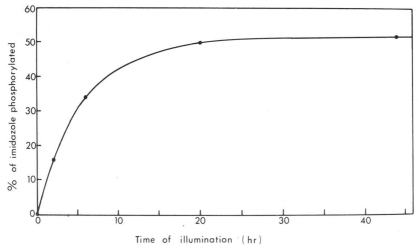

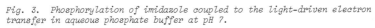

Fig. 3. *Phosphorylation of imidazole coupled to the light-driven electron transfer in aqueous phosphate buffer at pH 7.*

TABLE 1: *PHOSPHORYLATION OF IMIDAZOLE UNDER DIFFERENT PARTIAL PRESSURES OF OXYGEN AT 4°*

Oxygen in Ampule[a]		Imidazole Phosphorylated[b]	
Initial Pressure (Hg)	Total Amount (mole)	%	Total Amount (mole)
150 mm	8.6×10^{-5}	43.0	4.3×10^{-7}
100 μ	5.5×10^{-8}	9.4	9.4×10^{-8}
1 μ	5.5×10^{-10}	5.8	5.8×10^{-8}
0.6 μ	3.3×10^{-10}	4.6	4.6×10^{-8}

[a]*The capacity of each sealed ampule is about 10 ml which includes 9 ml of gaseous space and 1.0 ml of a solution of $1.0 \times 10^{-3}M$ imidazole and $1.0 \times 10^{-4}M$ hematoporphyrin in 0.1 M phosphate buffer at pH 7.0.*

[b]*All sealed samples were illuminated by means of a 500W projector lamp through a yellow filter for 14 hrs. After illumination, the mixtures were spotted separately on DEAE-cellulose TLC plates, developed with a mixture of 2-propanol + concentrated ammonium hydroxide + water (volume ratio 8:1:1), and assayed by scintillation counting of ^{14}C.*

It was observed that a catalytic amount of O_2 is required in the above photophosphorylation reaction *(8)*, although there was no net reduction of molecular oxygen. The effect of molecular oxygen on the phosphorylation rate is summarized in Table 1 *(9)*. The last line in Table 1 shows that the molar ratio of imidazole phosphorylated to the total number of moles of O_2 sealed in the reaction ampule is greater than 100. Thus oxygen seems to play a catalytic role.

Presumably the rate of overall net phosphorylation reaction is proportional to the steady-state concentration of imidazolyl radical which is severely limited by the dissipative reaction

$$N \overgroup{\odot N} + HP^- + H^+ \longrightarrow HN \overgroup{\quad N} + HP$$

Through removal of HP^- as indicated by the reaction

$$HP^- + O_2 \rightleftharpoons HP + O_2^-,$$

molecular oxygen could retard the above dissipative reaction and increase the steady-state concentration of imidazolyl and phospho-imidazolyl radicals. Being a weaker reductant than HP^-, the superoxide radical O_2^- may allow the reaction between imidazolyl radical and P_i to proceed to a greater extent before reducing the phosphoimidazolyl radical to the final products as indicated below:

$$\left[\begin{array}{c} \text{imidazolyl-P(OH)} \end{array} \right]^{-} + O_2^- \longrightarrow N \overgroup{\quad N} - \overset{O}{\underset{O^-}{\overset{\|}{P}}} - O^- + H_2O + O_2$$

By illuminating the frozen mixture at $-150^\circ C$ for 20 minutes and examining the ESR spectra, it was found the imidazolyl radicals are indeed probably present in the mixture *(5,6)*. The experimental data are summarized in Fig. 4A.

The single peak at $g = 2.0036$ in Spectrum A of Fig. 4 may be due to the superposition of the electron spin resonance of several radical species such as the imidazolyl radical, the reduced hematoporphyrin radical (HP^-) and possibly the superoxide radical (O_2^-). In a homogeneous solution, we expect the ESR spectrum of imidazolyl radical to be much broader than Spectrum A on account of contact interactions of the unpaired electron with the magnetic moments of the hydrogen and nitrogen nuclei. However, if the

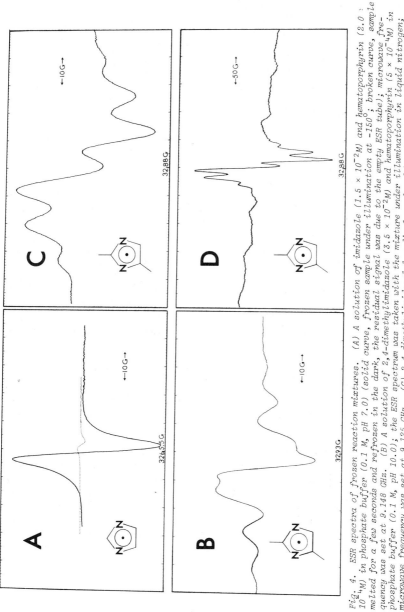

Fig. 4. ESR spectra of frozen reaction mixtures. (A) A solution of imidazole (1.5 × 10⁻²M) and hematoporphyrin (2.0 : 10⁻⁴M) in phosphate buffer (0.1 M, pH 7.0) (solid curve, frozen sample under illumination at -150°; broken curve, sample melted for a few seconds and refrozen in the dark, the residual signal was due to the empty ESR tube); microwave frequency was set at 9.148 GHz. (B) A solution of 2,4-dimethylimidazole (3.5 × 10⁻²M) and hematoporphyrin (5 × 10⁻⁴M) in phosphate buffer (0.1 M, pH 10.0); the ESR spectrum was taken with the mixture under illumination in liquid nitrogen; microwave frequency was set at 9.125 GHz. (C) 2,4-dimethylimidazolyl radical produced by photo-oxidation of 2,4-dimethylimidazole by KMnO₄ at -196°C. (D) Same sample as in (C) but at 500-gauss scan.

9

imidazole molecules aggregate in the frozen sample, it could
appear as a single peak due to possible rapid hydrogen atom or
electron exchange within the aggregate.

In the hope of minimizing aggregation in the frozen sample,
the experiment was repeated with the less symmetrical 2,4-di-
methylimidazole and Spectrum B in Fig. 4 was obtained. Spectrum
B also contains a superimposed single ESR peak which is probably
due to HP$^-$ (g = 2.0026). Spectrum C was obtained with a sample
prepared by rapidly injecting aqueous solutions of potassium per-
manganate and 2,4-dimethylimidazole into an ESR tube, then immediately immersing in liquid nitrogen and subsequently illuminating
the frozen mixture for 2 hours. It seems almost certain that
Spectrum C is due to 2,4-dimethylimidazolyl radical since 2,4-
dimethylimidazole was the only organic component in the system.
A 500-gauss scan of the same sample gave Spectrum D. The smaller
but broader peaks on both sides of the main ESR signal at approximately 80 gauss separation from each other are due to manganous
ions produced by the photo-redox reaction.

After the frozen sample of hematoporphyrin and imidazole in
phosphate buffer was illuminated for many hours, side bands began
to appear on both sides of the single ESR band of Fig. 4A. The
generation of these side bands can be accelerated by preillumina-
ting the liquid sample with yellow light above 0°. As the concentration of phosphorylated imidazole was gradually increased
by preillumination, as determined by chromatographic analysis, so
did the intensities of these side bands. Spectrum A in Fig. 5
was obtained with such a sample after 48 hours of preillumination.
Instead of generating it by prolonged photophosphorylation, the
same spectrum could be produced by freezing a solution of hematoporphyrin and chemically synthesized 1-phosphoimidazole in aqueous
buffer and illuminating for only a few minutes as shown by Fig.
5B. This last observation suggests that Spectrum A in Fig. 5 is
probably due to the phosphoimidazolyl radical discussed above.

Because of their relative simplicity, model oxidative phosphorylation reactions can be studied more thoroughly and hence
more easily understood. However, no matter how interesting their
chemistry is, model oxidative phosphorylation systems are at best
only suggestive of what might happen in mitochondria and chloroplasts. The most direct way to elucidate biological mechanisms
is still to go to the biological systems. Let us now consider
the thermodynamics of energy transduction in living systems.

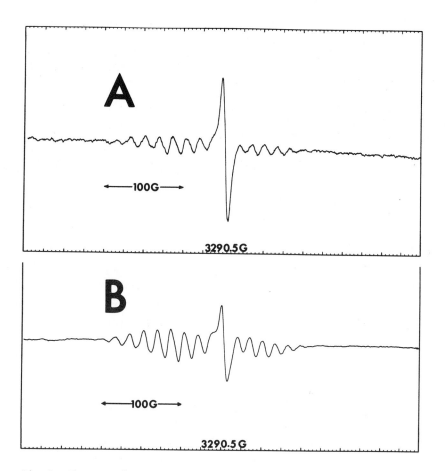

Fig. 5. Electron spin resonance spectra of radicals derived from 1-phosphoimida-zole. All samples contain 2.0×10^{-4}M hematoporphyrin in 0.1 M phosphate buffer (pH 7.0). Special features of each experiment: (A) The mixture containing 1.5 $\times 10^{-2}$M imidazole was illuminated by yellow light in an evacuated and sealed tube for 48 hrs. at room temperature, subsequently frozen in liquid nitrogen and again illuminated for a few minutes, then the light was turned off, and the electron spin resonance spectrum was taken in the dark at -196°. (B) The mixture contain-ing 9.0×10^{-3}M 1-phosphoimidazole was sealed in an evaluated tube, frozen in liquid nitrogen, and illuminated by yellow light for a few minutes, then the electron spin resonance spectrum was taken in the dark at -196°.

11

3.   THERMODYNAMICS OF FREE ENERGY STORAGE IN LIVING SYSTEMS

The free energy liberated by respiratory or photosynthetic electron transport is often first converted to and stored in a certain form prior to the formation of ATP *(10-12)*. The various forms of this stored free energy suggested in the literature include energy-rich intermediates *(10)*, concentration gradients *(13-15)*, electric potential energy *(16-19)*, and changes in macro-molecular conformation and membrane structure *(20-22)*.

In order to develop a thermodynamic treatment of this pheno-menon, let us consider a general energy storage process repre-sented by

$$\sum_i n_i A_i \rightarrow \sum_j n_j A_j \tag{1}$$

in which n_i moles of molecular species A_i, n_{i+1} moles of A_{i+1} etc., change to n_j moles of A_j, n_{j+1} moles of A_{j+1}, etc. The summation $\sum_i$ is over all reactants and the summation $\sum_j$ is over all products of this process. At constant temperature T and pressure P, the increase of free energy due to an infinitesimal change in the above process is given by

$$dG = \int_0^V (\sum_j \mu_j \, dc_j - \sum_i \mu_i \, dc_i) \, dV \tag{2}$$

where μ_i, μ_j represent the chemical potentials, and c_i, c_j repre-sent the concentrations of A_i and A_j respectively in volume element dV, and the integral extends over the whole volume V of the system under consideration.

The electrochemical potential μ_i of reactant molecule A_i of charge number Z_i at electric potential ψ_i may be written as

$$\mu_i = \mu^o + Z_i F \psi_i + RT \ln c_i + RT \ln \gamma_i \tag{3}$$

where γ_i is the activity coefficient of A_i in volume element dV,

F the Faraday, R the gas constant and μ^{o}_{i} the standard chemical potential which is characteristic of A_i, but for a given set of arbitrarily chosen standard conditions is independent of ψ_i and c_i. Likewise, the electrochemical potential μ_j of the product molecular species A_j at electric potential ψ_j, with activity co-efficient, γ_j, in volume element dV may be written as

$$\mu_j + \mu^{o}_{j} + Z_j F\psi_j + RT \ln c_j + RT \ln \gamma_j \tag{4}$$

Substituting Equations 3 and 4 into Equation 2 and integrating over the entire volume V of the system, we get

$$dG = \int_{0}^{V.} \left\{ \left(\sum_j \mu^{o}_{j} \, dc_j - \sum_i \mu^{o}_{i} \, dc_i \right) + F \left(\sum_j Z_j \psi_j dc_j - \sum_i Z_i \psi_i dc_i \right) \right.$$
$$\left. + RT \left[\sum_j (\ln c_j) dc_j - \sum_i (\ln c_i) dc_i \right] + RT \left[\sum_j (\ln \gamma_j) dc_j - \sum_i (\ln \gamma_i) dc_i \right] \right\} dV$$
$$\tag{5}$$

The free energy change due to a chemical process may include contributions from all four terms on the righthand side of Equation 5. This is because in general molecules may react at any concentrations c_i, c_j ....., with activity coefficients γ_i, γ_j.... instead of at their standard states. It was also known since Volta's observations in 1800 and Faraday's experiments in 1833 that chemical reactions can cause changes in electric potential and local concentration. Therefore the experimental detection of concentration as well as electric potential gradients generated by electron transport in chloroplasts and mitochondria does not contradict a general chemical mechanism for coupling oxidation to phosphorylation. A more quantitative study of the problem is necessary before a decision on the correct type of mechanism can be made.

3.1 PROTON GRADIENT MODEL - The failure to isolate high free energy chemical precursors of ATP from mitochondria and chloro-plasts after a quarter of a century of intensive research prompted the formulation of alternative hypotheses of oxidative phosphory-lation. The central assumption of the chemiosmotic or proton

gradient model *(13,14)* is that the first two terms on the righthand side of Equation 5 are negligible and hence the free energy
stored prior to ATP synthesis should be given by

$$dG_s = RT \int_0^V \left\{ \sum_j \ln(c_j \gamma_j) dc_j - \sum_i \ln(c_i \gamma_i) dc_i \right\} dV \qquad (6)$$

Among the numerous experiments conducted to test this model,
the investigations by Jagendorf and his coworkers *(12,23,24)* are
among the most persuasive.

When a slightly buffered suspension of chloroplasts was illuminated, the pH of the medium was found to increase continually
until a steady-state value was reached in about 60 seconds *(23)*.
When chloroplasts suspended in a buffered medium containing N-
methylphenazonium methosulfate at pH 6 were illuminated and subsequently mixed in the dark with an excess of pH 8 buffer containing ADP, P_i and Mg^{2+}, ATP was formed *(11,12)*. By first exposing
chloroplasts to pH 3.8 buffer in the dark, and then incubating
them with a pH 8 buffer containing ADP, P_i and Mg^{2+} in the dark,
a very high yield of ATP was obtained *(12,24)*. These observations
are all consistent with the chemiosmotic hypothesis.

Hind and Jagendorf *(12)* also observed that the free energy
stored by chloroplasts during the light-stage in their two-stage
photophosphorylation experiment decays rapidly in the dark. The
half-lives for this dark decay are 7, 18 and 32 sec. at pH 7.4,
6.7 and 6.0 respectively. However, if the free energy stored
during the light-stage is entirely in the form of proton gradient,
the half-life of decay should be independent of pH.

A simple treatment of the dark decay through diffusional leakage out of a chloroplast gives the following result:

For the decay of the proton concentration gradient, the
half-time is

$$t_{1/2} \approx (\ln 2)/b$$

where

$$b = \frac{S}{\ell} \left(\frac{1}{V_1} + \frac{1}{V_2} \right),$$

V_1 and V_2 represent the volume of liquid inside and outside of the
organelle respectively, D the diffusion coefficient, S and ℓ the
area and mean thickness of the organelle membrane.

For the decay of the pH gradient, the half-time is

$$t_{1/2} \approx (pH_2^o - pH_1^o) / (2b),$$

where pH_1^o and pH_2^o represent pH values outside and inside the organelle respectively at the instant when the light was turned off.
In neither case is $t_{1/2}$ expected to decrease as the pH of the
buffered medium is raised. Therefore this part of their results
seems to contradict the proton gradient model.

Thermodynamic studies of two-stage photophosphorylation also
yielded results which seem to contradict the proton gradient
model. For example, let us examine the efficiency of two-stage
photophosphorylation by chloroplasts suspended in a medium buffered by the base B^- and its conjugate acid BH in the presence of
a much higher concentration of the supporting electrolyte M^+X^-.
For simplicity, let us consider two homogeneous liquid phases of
volumes V_1 and V_2 respectively which are separated by thylakoid
membrane, and adopt the following notation:

$[H^+]_1$, $(H^+)_1$ denote the concentration and activity respectively of H^+ (or H_3O^+) on one side of the chloroplast membrane (Side
1);

$[H^+]_2$, $(H^+)_2$ denote the similar quantities on the other side
of the chloroplast membrane (Side 2);

$[X^-]_1$, $[X^-]_2$, $(X^-)_1$, $(X^-)_2$, $[M^+]_1$, $[M^+]_2$, $(M^+)_1$, $(M^+)_2$ denote
similar quantities for X^- and M^+ respectively.

Illumination of the membrane for a short time dt causes $d\delta$ mole
of H^+ to move from Side 1 to Side 2 of the membrane, with co-
transport of X^-, or in exchange for M^+, to maintain electric
neutrality. Since

$$V_1 d[H^+]_1 = -V_2 d[H^+]_2 = d\delta,$$

the resulting free energy increase according to Equation 6 is
equal to

$$dG_s = RT \left\{ \ln \frac{(H^+)_2 (X^-)_2^{1-n} (M^+)_1^n}{(H^+)_2 (X^-)_1^{1-n} (M^+)_2^n} \right\} d\delta \tag{7}$$

where n<1.

If the ionic strength is maintained approximately constant by
an excess of M^+X^- so that

$$(M^+)_1 \approx (M^+)_2, \ (X^-)_1 \approx (X^-)_2, \text{ and } \frac{(H^+)_2}{(H^+)_1} \approx \frac{[H^+]_2}{[H^+]_1} , \qquad (8)$$

then equation 7 becomes

$$dG_s^\ast \approx RT \left\{ \ln \frac{[H^+]_2}{[H^+]_1} \right\} d\delta \qquad (9)$$

Since in general $[H^+]_2$ and $[H^+]_1$ may change as a result of proton translocation, the free energy increase due to the translocation of δ moles of H^+ from Side 1 to Side 2 of the membrane is given by

$$\Delta G_s = RT \int_0^\delta \ln \left\{ \frac{[H^+]_2}{[H^+]_1} \right\} d\delta \qquad (10)$$

In order to integrate the righthand side of Equation 10, let us define the apparent dissociation constant K_a' of the buffer acid BH as

$$K_a' = \frac{[B^-] \ [H^+]}{[BH]} = K_a \cdot \frac{\gamma_{BH}}{\gamma_{B^-} \cdot \gamma_{H^+}} ,$$

where

$$K_a = \frac{(B^-) \ (H^+)}{(BH)} ,$$

and the activity coefficients γ_{BH}, γ_{B^-}, γ_{H^+} are constants for a system kept at approximately constant ionic strength by the same supporting electrolyte M^+X^-.

If the chloroplasts were equilibrated with the buffer in the dark for at least ½ hour before the illumination, we have

$$\frac{[BH]_1}{[B^-]_1} = r = \frac{[BH]_2}{[B^-]_2}$$

at the beginning of the photoexperiment.

According to the proton gradient model *(13)*, the number of moles of H^+ translocated is proportional to the number of moles of electrons transported. By using intense light for a very short illumination period, it is possible to translocate a sufficiently large number, δ, of moles of H^+ from one side of the chloroplast membrane to the other, with concurrent transport of X^- and M^+ which are present in excess to maintain electrical neutrality, but without appreciable amounts of B^- and BH diffusing across simultaneously. Under such experimental conditions, the concentration of H^+ on the two sides immediately after the light-driven proton translocation are given by

$$[H^+]_2 = K_a' \left\{ \frac{[BH]_2}{[B^-]_2} \right\} = K_a' \left\{ \frac{a_2 r/(1+r) + \delta}{a_2/(1+r) - \delta} \right\} \quad , \qquad \bullet \tag{11}$$

$$[H^+]_1 = K_a' \left\{ \frac{[BH]_1}{[B^-]_1} \right\} = K_a' \left\{ \frac{a_1 r/(1+r) - \delta}{a_1/(1+r) + \delta} \right\} \quad , \tag{12}$$

where a_2 and a_1 represent the total number of moles of buffer, B^- plus BH on Side 2 and Side 1 of the membrane respectively. In obtaining Equations 11 and 12 it is assumed that the buffer as well as the intensity and duration of the illumination are properly chosen so that $\delta < a_2 r/(1+r) < a_1 r/(1+r)$ and $\delta < a_2/(1+r) < a_1/(1+r)$.

Substitution of Equations 11 and 12 into Equation 10 and integration gives *(25)*

$$\Delta G_s \approx \frac{(1+r)^2}{2r} \left(\frac{1}{a_1} + \frac{1}{a_2} \right) RT\delta^2 \tag{13}$$

If there is a threshold of stored free energy, $W = RT\Delta pH$, below which the phosphorylation of ADP cannot take place, Equation 13 should be replaced by

$$G_s \approx \frac{(1+r)^2}{2r} \left(\frac{1}{a_1} + \frac{1}{a_2} \right) RT\delta^2 - W\delta \tag{14}$$

Both equations 13 and 14 show that for a given number, δ, of moles of proton translocated the free energy stored should decrease as the buffer concentration increases according to the proton gradi-

ent models. Since a_1 and a_2 represent the total amounts of buffer, rather than its concentrations, Equation 13 or 14 should
remain valid even when the chloroplasts shrink or swell as a
result of osmotic flow during the illumination, provided that
the light was sufficiently bright and that the illumination period
was sufficiently short so that the number of moles of B^- or BH
diffused across was small as compared to the number of moles of
proton translocated. For longer illumination periods, Equation
13 or 14 should include correction terms due to the diffusion of
B^- and BH, but the above qualitative conclusion concerning buffer
concentration should still hold.

An experiment designed for testing Equation 13 or 14 is to
illuminate chloroplasts suspended in media containing different
concentrations of the same buffer in the absence of P_i and ADP
for a brief period, rapidly inject each illuminated sample into a
strongly buffered solution of P_i and ADP in the dark, and determine the yield of ATP as a function of the buffer concentration
in the light-stage mixture. In order to simplify the theoretical
treatment of data, pH changes due to net oxidation-reduction processes were eliminated by performing the photo-experiments under
the conditions of cyclic electron transport *(11,12)*. In a typical
series of experiments *(25)*, the chloroplast suspension contained
0.5 to 0.6 mg. chlorophyll per ml and the following components:
Phenazine methosulfate (PMS), 1×10^{-4} M; NaCl, 1×10^{-2} M; NaN_3,
1×10^{-3} M; Tricine (N-*tris*[hydroxymethyl]-methyl glycine),
2.0×10^{-4}; maleic acid plus sodium maleate, 0 to 1.0^{-2} M at pH
6.2. The dark-stage solution contained the following components:
ADP, 2.0×10^{-3} M; P_i, 1×10^{-2} M; $MgCl_2$, 1×10^{-2} Tricine,
0.1 M at pH 8.0.

The experimental results in Fig. 6 show that the production
of ATP by such a two-stage photophosphorylation process increases
with both the illumination time and the concentration of buffer
in the light-stage mixture. Fig. 7 shows that with the illumination time fixed at 5 sec., the observed yield of ATP increases rapidly as the buffer concentration in the dark-stage mixture
is raised. Similar results were obtained when malonate, succinate and phosphate respectively were used to replace maleate as
the buffer in the light-stage mixture.

Inasmuch as each experimental point in Fig. 7 was produced
by the absorption of the same number of protons of the same frequency distribution and over the same interval of time, the
number of electrons transported along the cyclic electron transport path must be the same in each case. Consequently the number,
δ, of moles of proton translocated in each experiment must also
be the same according to the proton gradient model which assumes
a stoichiometric relationship between electron transport and

proton translocation *(26)*. However, the data in Fig. 7 show
that for the same number, δ, of moles of proton translocated, the
free energy stored, ΔG_S, increases rapidly as the concentration
of the buffer is raised. Obviously this observation is in a
direction opposite to that predicted by Equation 13 or 14 deduced
from the proton gradient model.

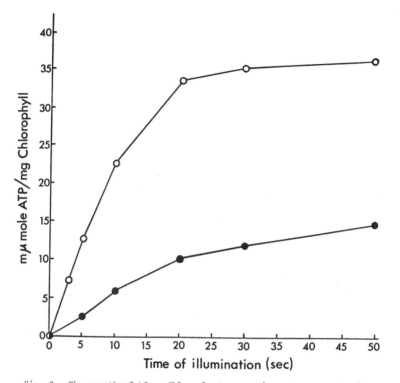

Fig. 6. *The growth of* ΔG_S. *Chloroplast suspension at a concentration
of 0.50 mg chlorophyll per ml was illuminated in the presence of 10 mM
sodium maleate buffer at pH 6.2 (o——o) and in the absence of sodium
maleate at pH 6.2 (●——●). The concentration of NaCl in these suspensions was maintained at 10 mM.*

Attempts have been made to improve the proton gradient model
by postulating that protons are not translocated from Side 1 to
Side 2, but from the exterior to the interior of the thykaloid or
inner mitochondrial membrane. This type of proton translocation
would produce a much larger proton gradient for the same number

of moles of protons translocated. But the improved model also contradicts the observed rapid increase of ΔG_S as the buffer concentration is raised.

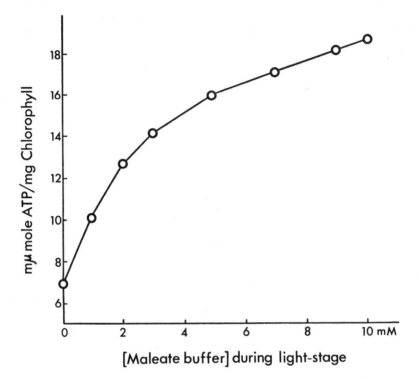

Fig. 7.   The dependence of ΔG_S on the concentration of sodium maleate buffer. Chloroplast suspensions at a concentration of 0.63 mg chlorophyll per ml were illuminated for 5 sec. in the presence of different amounts of sodium maleate buffer at pH 6.2. Complemental amounts of the maleate were added to the dark-stage solutions so that the dark transphosphorylation reactions were carried out at a constant sodium maleate concentration of 5 mM in all experiments. The concentration of added NaCl was 10 mM in all cases.

One may also attempt to reconcile the proton gradient model to the above data by assuming that although the free energy stored during the light-stage of the above experiments was initially in the form of proton gradient, it was rapidly turned over into another form, e.g., a non-phosphorylated chemical intermediate. But such a transformation can only make ΔG_S decrease less rapidly with increasing buffer concentration than that predicted by Equation 13 or 14, it cannot make ΔG_S increase with buffer concentration.

3.2 THE CHARGED MEMBRANE MODEL - This model assumes that the free
energy of electron transport is first used to charge the thykaloid
membrane of chloroplasts or the inner membrane of mitochondria, and
in this way partially stored as electric potential energy. This
stored free energy, ΔG_S, is later utilized to translocate ions and
phosphorylate ADP *(16-19)*. Accordingly, ΔG_S is equal to the work
required to charge a membrane condenser by redistributing the ions
in the chloroplastic or mitochondrial system. The work required
to charge a condenser of constant capacity C to a potential E is
equal to $E^2/2$. But since practically all the electric charges in
a biological system reside on ions and since the capacity of a
membrane condenser also depends on the distribution of ions, we
expect ΔG_S to vary with the concentration of the principal ionic
species, say Na^+ and Cl^-, but to be practically independent of the
concentration of the dilute buffer, $[B^-]+[BH]$, if the latter does
not contribute significantly to the total ionic strength.

As a convenient example, let us consider the free energy ΔG_S
stored by charging an infinite planar membrane, say 65Å thick, by
bringing Na^+ and Cl^- ions to both of its interfaces from the bulk
of the electrolyte solution until the potentials of the two interfaces becomes ψ_0 and $-\psi_0$ respectively. For our present purpose,
it is sufficient to calculate the work ΔG_S required for charging
one of the two interfaces by bringing Na^+ ions to the interface
until the surface potential is ψ_0. To obtain the free energy
stored by the whole membrane, we need only replace the calculated
ΔG_S by $2 \Delta G_S - \Delta G'$, where $\Delta G'$ is the interaction free energy of
the oppositely charged interfaces. For an idealized planar mem-
brane 65Å thick, $\Delta G'$ is comparatively small and practically independent of the salt concentration C_S in the bulk of the solution.

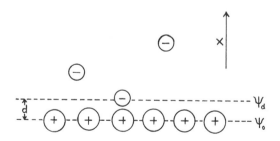

*Fig. 8. The diffuse double layer of hydrated ions with finite
radii.* **d** *is the distance of closest approach of the negatively
charged ion to a positively charged interface.*

Let us consider the infinite planar interface as a diffuse double layer of ions illustrated in Fig. 8. To simplify the problem, let us assume that the concentration C_s of the supporting electrolyte, Na^+ and Cl^-, is much higher than that of the buffer, i.e.,

$$C_s \equiv [Na^+] \approx [Cl^-] \gg [B^-] + [BH].$$

The potential Ψ at any point in the system obeys the Poisson equation

$$\nabla^2 \Psi = - \frac{4\pi\rho}{D} \tag{15}$$

where D is the dielectric constant of the solvent, ρ is the charge density given by

$$\rho = \sum_i Z_i e n_i \tag{16}$$

and the number n_i of ions i per unit volume is related to the average number n_{io} of that ion by the Boltzmann equation

$$n_i = n_{io} \exp \left(- \frac{Z_i e \Psi}{kT} \right) \tag{17}$$

Combination of Equations 15, 16 and 17 gives

$$\nabla^2 \Psi = - \frac{4}{D} \sum_i Z_i e n_{io} \exp \left(- \frac{Z_i e \Psi}{kT} \right) \tag{18}$$

For the infinite planar interface in Figure 8,

$$\frac{\partial \Psi}{\partial y} = 0, \quad \frac{\partial \Psi}{\partial z} = 0,$$

and ∇^2 reduces to d^2/dx^2.

Multiplication of both sides of Equation 18 by $2d\Psi/dx$ gives

$$2 \frac{d\Psi}{dx} \cdot \frac{d^2\Psi}{dx^2} = - \frac{8\pi}{D} \sum_i Z_i e n_{io} \left\{ \exp \left(- \frac{Z_i e \Psi}{kT} \right) \right\} \frac{d\Psi}{dx} \tag{19}$$

Since

$$\Psi \to 0 \text{ and } \frac{d\Psi}{dx} \to 0 \text{ as } x \to \infty \text{ ,}$$

multiplying Equation 19 by dx and integrating from ∞ to x gives

$$\frac{d\Psi}{dx} = \frac{8\pi kT}{D} \Sigma\, n_{io} \left\{ \exp\left(-\frac{Z_i e\Psi}{kT}\right) - 1 \right\} \tag{20}$$

For a simple, symmetrical electrolyte, Equation 20 simplifies to

$$\frac{d\Psi}{dx} = -\sqrt{\frac{8\pi n\ kT}{D}} \left\{ \exp\left(\frac{Ze\Psi}{2kT}\right) - \exp\left(-\frac{Ze\Psi}{2kT}\right) \right\} \tag{21}$$

Because the total surface charge should be opposite in sign but equal in magnitude to the total space charge per unit cross-sectional area, we have

$$\sigma = -\int_d^\infty \rho\, dx = \frac{D}{4\pi} \int_d^\infty \frac{d^2\Psi}{dx^2} \cdot dx = -\frac{D}{4\pi}\left(\frac{d\Psi}{dx}\right)_{x=d} \tag{22}$$

Substitution of Equation 21 into Equation 22 yields

$$\sigma = \sqrt{\frac{DnkT}{2\pi}} \left\{ \exp\left(\frac{Ze\Psi_d}{2kT}\right) - \exp\left(\frac{-Ze\Psi_d}{2kT}\right) \right\} \tag{23}$$

For a dilute solution containing C_S moles of salt per liter and a moderate interface potential Ψ_o, e.g., $C_S < 0.05M$ and $\Psi_o < 1V$, we may use the approximation $\Psi_d \approx \Psi_o$, and write

$$d\sigma \approx \sqrt{\frac{DnZ^2 e^2}{8\ kT}} \left\{ \exp\left(\frac{Ze\Psi_o}{2kT}\right) + \exp\left(-\frac{Ze\Psi_o}{2kT}\right) \right\} d\Psi_o \tag{24}$$

Consequently the free energy ΔG_s stored by charging a unit area of the planar interface to a potential of Ψ_o is equal to

$$\Delta G_s = \int_0^\sigma \Psi_o\, d\sigma$$

$$\approx \sqrt{\frac{DnZ^2e^2}{8\pi kT}} \int_0^{\Psi_o} \left\{ \Psi_o \, \exp\left(\frac{Ze\Psi_o}{2kT}\right) + \Psi_o \, \exp\left(-\frac{Ze\Psi_o}{2kT}\right) \right\} d\Psi_o$$

$$= \sqrt{\frac{2DnkT}{\pi}} \left\{ \Psi_o \, \sinh\left(\frac{Ze\Psi_o}{2kT}\right) - \frac{2kT}{Ze} \left[\cosh\left(\frac{Ze\Psi_o}{2kT}\right) - 1 \right] \right\}$$

$$= \sqrt{\frac{2DC_s RT}{1000\pi}} \left\{ \Psi_o \, \sinh\left(\frac{Ze\Psi_o}{2kT}\right) - \frac{2kT}{Ze} \left[\cosh\left(\frac{Ze\Psi_o}{2kT}\right) - 1 \right] \right\}$$

$$(25)$$

Therefore in the range of experimental data for which Equation 25 is applicable, ΔG_s at a given salt concentration should be proportional to C_s but independent of the smaller concentration of the buffer according to the charged membrane model.

If charging the thylakoid membrane can be regarded as charging a perfect planar condenser with no leakage, we expect the potential Ψ_o at the membrane surface under illumination for a long time to reach its maximum or saturation value as determined by the photo-e.m.f. of the system. For a condenser with a small leakage, we expect the steady-state value of Ψ_o to be lower than the maximum value. But in either case, ΔG_s should be proportional to C_s for $C_s \leq 0.05$ M according to Equation 25, whereas for a given C_s it should be independent of the concentration of the buffer which does not contribute significantly to the total ionic strength.

Contrary to this prediction, experimental results show that ΔG_s increases rapidly as the buffer concentration is raised and that at a given buffer concentration ΔG_s does not increase as C_s changes from 10 to 50 mM *(25)*. These observations show that the above theoretical result based on the charged membrane model is not right. Therefore we can only conclude that most of the free energy ΔG_s produced by photosynthetic energy conversion under the conditions of cyclic electron transport is stored in a form other than that of a charged thylakoid membrane.

Inasmuch as Equation 13 or 14 predicts ΔG_s to decrease with buffer concentration and Equation 25 predicts ΔG_s to be independent of buffer concentration, the observed rapid increase of ΔG_s with buffer concentration also contradicts the hypothesis that a combination of proton gradient and electric field, *viz.*, the electrophysical potential gradient, as the principal driving force for cyclic photophosphorylation.

By measuring directly with microelectrodes, Tedeschi and co-workers *(27)* recently measured membrane potentials of liver mitochondria, 5 to 10 μm in size, from cuprizone-treated mice. The potentials ranged between 10 and 25 mv, positive inside, which are much smaller than that required by the charged membrane theory.

3.3 THE CONFORMATIONAL MODEL (28,29) - According to this model ADP and P_i combine on the surface of inner mitochondrial membrane spontaneously to form ATP and water. The free energy increase due to ATP formation is more than compensated by the free energy released due to the tight binding of ATP to the membrane. Electron transport is assumed to trigger conformation changes in the membrane and cause the release of the tightly bound ATP. If the proton gradient model of oxidative phosphorylation can be regarded as the reversal of the active transport of protons, then the conformational model may be regarded as reversal of muscle contraction. The available quantitative information on this model is presently still not sufficient for a decisive thermodynamic treatment.

4. SELECTIVE LABELING OF PROTEIN MOLECULES WHICH COUPLE OXIDATION TO PHOSPHORYLATION

The sequence of electron-carriers in mammalian respiratory chain is illustrated in Fig. 9. With the help of a group of proteins known as coupling factors, the free energy release by electron transport is utilized in synthesizing ATP from P_i and ADP. A large number of coupling factors have been reported in the literature *(30)*. Among them, the coupling factors F_1 *(31-34)*, CF_1 *(35)*, B *(36,37)* and F_B *(38)* seem to be of the highest purity. Because of the tight coupling between electron transport and phosphorylation, electron transport in the respiratory chain cannot occur rapidly unless there is a sufficiently high concentration of ADP to be phosphorylated. This self-regulatory property of mitochondria, which is often referred to as respiratory control, is important in conserving the stored food for survival. It allows cellular respiration to take place rapidly only when more ATP is needed.

Many kinds of small molecules have been found which uncouple oxidation from phosphorylation, i.e., accelerate respiration without concomitant phosphorylation. Since these uncoupling agents or uncouplers are often effective at very low concentrations (10^{-5} to 10^{-9} M), it is quite possible that they interact with specific macromolecules at the coupling sites. Therefore,

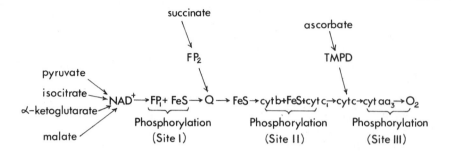

*Fig. 9. The mammalian respiratory chain. The arrows indicate directions of elec-
tron transfer in normal respiration. The abbreviations for electron carriers are
NAD⁺, nicotinamide adenine dinucleotide; FP₁ and FP₂ flavoproteins; FeS, non-heme
iron sulfur proteins; Q, ubiquinone; cyt b, cytochrome b; cyt c₁, and cyt c, cytochromes C₁ and C; TMPD, tetramethylenephenylenediamine; cyt aa₃, cytochrome oxidase.
The sites where electron transport is coupled to phosphorylation are indicated by
brackets.*

by designing and synthesizing radioactive, uncoupler-like molecules with reactive side-chains, we might be able to label
selectively some of the macromolecules at the energy transducing
sites. We hope to identify these macromolecules and their functional groups which participate in the coupling of electron
transport to phosphorylation by means of these selective labeling
agents. Our strategy is to exploit the specificity of local membrane structures and label these macromolecules *in situ*, subsequently pull the labeled molecules out of the membranes for
identification, and use the functional groups of the so identified
molecules to guide us in exploring the primary energy transducing
process.

4.1 SELECTIVE LABELING BY DNBP - The first discovered uncoupler
was 2,4-dinitrophenol (DNP) *(39)*, which is effective in the 10^{-5}
to 10^{-3}M range. By analogy to DNP, the alkylating uncoupler 2,4-
dinitro-5-(bromo-[2-^{14}C]acetoxyethoxy)phenol (DNBP) was synthesized and shown to alkylate selectively a very small number of
sulfhydryl groups in mitochondria which are essential for coupling
electron transport to phosphorylation *(40,41)*.

This *in situ* alkylation was essentially complete in a few
minutes at physiological pH and temperature. In the case of rat
liver mitochondria, only 1 out of about 650 cysteinyl sulfhydryl
groups was so labeled. For bovine heart mitochondria, over 90%
of the labels were covalently attached to only two polypeptide
bands *(42)*, and only 1 out of about 250 cysteinyl sulfhydryl
groups was so labeled. A typical SDS-gel-electrophoresis pattern
of DNBP[^{14}C]- labeled bovine heart mitochondria is shown in Fig.10.

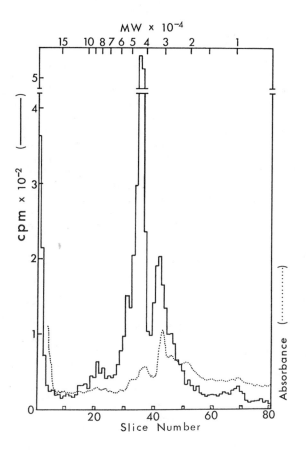

Fig. 10. SDS-gel-electrophoresis pattern of DNBP-^{14}C-labeled polypeptides in bovine heart mitochondria. Concentrations: DNBP, 2×10^{-5}M; sucrose, 0.25 M; Tris buffer, 0,01 M at pH 7.4; BHM, 5 mg protein/ml. Volume of sample = 2 ml. The mixture was incubated at 23°C for 10 minutes and quenched by the addition of 0.2 ml 2-mercaptoethanol. The labeled BHM were washed, centrifuged down and dissolved in 0.4 ml 20% SDS solution with the addition of 0.1 ml of 2-mercaptoethanol, electrophoresed in polyacrylamide gel (10%, 10 mm i.d., 90 mm length), and assayed as previously described (1). Dansylated BSA (MW 65,000), ovalbumin (MW 45,000), chymotrypsinogen (MW 24,000) and cytochrome c (MW 13,000) were used to calibrate the logarithmic molecular weight scale shown on top of the diagram. Dotted curve (.....) represents total protein concentration as scanned at 280 mm.

   Like other uncouplers, DNBP added to mitochondria was found
to decrease the P:O ratio, i.e., the ratio of ATP formed to oxygen reduced. It caused an initial stimulation of respiration in
the absence of ADP. However, in contrast to ordinary uncouplers,
the DNBP-stimulated respiration diminished with time and declined
to a constant rate in 3 to 4 minutes. Even 2,4-dinitro-5-(aceto-
xyethoxy)phenol (DNAP) did not exhibit this peculiar property.
For comparison, a pair of respiration curves are shown in Fig. 11.
Since chromatographic analysis showed that the concentration of
free DNBP molecules in the medium remained practically unchanged
during these respiration measurements, it seems likely that the
observed bending of the DNBP-stimulated respiration curve shown
in Fig. 11 is due to the alkylation of specific sulfhydryl groups
by DNBP. This probable conclusion is also supported by the observation that the rate of DNAP-stimulated respiration did not change
with time during the same experimental period. In other words,
it seems that in the DNBP case, we were monitoring an alkylation
reaction in mitochondrial membrane by a physiological method.

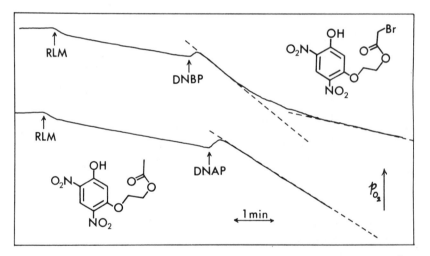

Fig. 11. Respiration of rat liver mitochondria stimulated by DNBP (6.7 × 10⁻⁵M)
and DNAP (6.7 × 10⁻⁵M) respectively. The assay medium contained succinate in pH
7.4 buffer.

4.2 SELECTIVE LABELING BY BDNA – The labeling reagent DNBP suf-
fers the drawback of instability, because its phenolate group is
a sufficiently strong nucleophile to cause slow decomposition of
the compound during storage. After having synthesized and studied
a large number of uncoupler derivatives, we found the compound

2'-bromo-2,4-dinitro[2'-^{14}C]acetanilide (BDNA) to be much stabler and an even better *in situ* alkylating reagent for mitochondrial proteins *(43)*.

The sodium dodecyl sulfate (SDS) gel electrophoresis pattern of bovine heart mitochondria labeled by BDNA showed essentially two radioactive bands of approximate molecular weights 44,000 and 31,000 respectively. Labeling of phosphorylating submitochondrial particles (ETPH) by BDNA gave essentially the same result. In both cases, the labeling was practically complete in a few minutes at room temperature. Labeling by BDNA suppresses respiration in the presence of ADP (State 3 respiration) and oxidative phosphorylation in mitochondria as well as ATP-driven reduction of NAD$^+$ by succinate in ETPH. As in the experiments with DNBP, the radioactive labels were found by amino acid analysis of the acid hydrolysate to be covalently attached to the sulfhydryl groups of cysteine residues.

Apparently an optimum balance between the affinity of the labeling reagent for the hydrophobic mitochondrial coupling sites and the reactivity of the leaving group is required for selective labeling. For example, 2,4-dinitrophenylbromo [2'-^{14}C]acetate effectively suppresses State 3 respiration, but labels mitochondrial proteins almost non-discriminatively. On the other hand, N-bromoacetyl-3'-nitroanthranilanilide ($O_2NC_6H_4NH-CO-C_6H_4NH-COCH_2$ Br) has no effect on State 3 respiration of mitochondria, in spite of its structural similarity to the powerful uncoupler S-13 (5-Cl, 3-t-butyl, 2'-Cl, 4'-NO$_2$-salicylanilide).

In order to examine the effect of BDNA labeling on oxidative phosphorylation without the additional complication due to the presence of excess BDNA molecules in the medium, the labeled mitochondria were washed with bovine serum albumin (BSA) solution prior to oxidative phosphorylation measurements. The data summarized in Table 2 show that with malate + pyruvate as the substrate BSA-washing caused negligible damage, but that the efficiency of oxidative phosphorylation decreased by a factor of 5 as the concentration of BDNA at the preincubation stage (10 minutes at 23°C) increased from 0 to 1×10^{-4}M.

The numbers of nmoles of covalently bound radioactive label per milligram of mitochondrial protein as calculated from the known specific radioactivity of BDNA are also listed in Table 2. It may be noticed that after incubation with 1×10^{-5}M BDNA, altogether only 0.27 nmoles of covalently bound radioactive label was found in subunit bands I and II per mg of mitochondrial protein or about 0.4 label/cytochrome aa$_3$. But the *in situ* labeling of such a small number of functional groups is sufficient to reduce the rate of phosphorylation of ADP coupled to the oxidation of malate by a factor of 2. The total number of sulfhydryl groups

in bovine heart mitochondria, as determined by titration with
5,5'dithiobis(2-nitrobenzoic acid) in the presence of sodium
dodecyl sulfate, was found to be 89 nmole/mg of protein. Therefore after the incubation with 1×10^{-5}M BDNA, only 1 out of 330
mitochondrial sulfhydryl groups was labeled. Thus the labeling
by BDNA seems to exhibit considerable selectivity at both the
protein level and functional group level.

TABLE 2: EFFECT OF BDNA-LABELING ON OXIDATIVE PHOSPHORYLATION
IN BOVINE HEART MITOCHONDRIA

BDNA at preincubation	n moles of label /mg protein		moles ATP formed/min/mg protein		
$(10^{-5}$M)	Band I (MW-44000)	Band II (MW=31000)	malate + pyruvate	β-hydroxy-butyrate	succinate
0 (unwashed)	--	--	0.288	0.200	0.068
0 (BSA-washed)	--	--	0.281	0.078	0.054
1	0.160	0.107	0.139	0.049	0.043
3	0.77	0.53	0.091	0.045	0.039
10	1.38	1.42	0.058	0.050	0.023

It would be interesting to know whether the observed suppression of oxidative phosphorylation is due to the labeling of the
substrate dehydrogenases or electron carriers or coupling factors
or a combination of these effects. In attempting to find the
answer, we studied the effect of BDNA labeling on the biochemical
activities of a number of enzymes and electron carriers which are
known to be present in mitochondria as well as in phosphorylating
submitochondrial particles. The results are summarized in Table
3.

The data in Table 3 show that although the activity of β-
hydroxybutyrate dehydrogenase diminishes as the concentration of
BDNA increases, the activity of malate dehydrogenase is unaffected
by BDNA labeling under the experimental conditions. Therefore,
it may be concluded at least for the case with malate as substrate
that the observed suppression of oxidative phosphorylation is not
due to labeling of the dehydrogenase. Similarly the activity of
succinate-cytochrome c reductase is unaffected by BDNA labeling.
Table 3 also shows that the activity of rotenone-sensitive NADH-

cytochrome c reductase (calculated as the difference between the total and the rotenone-insensitive activities) as well as the activity of cytochrome c oxidase are both unaffected by BDNA labeling. Consequently, we may conclude that neither the observed suppression of phosphorylation coupled to the oxidation of malate nor the observed suppression of ATP-driven reduction of NAD^+ by succinate can be due to the labeling of any of the electron carriers, from NAD^+ to cytochrome oxidase inclusive. Therefore the observed decrease in the rate of oxidative phosphorylation is most likely due to the selective labeling of one or more of the coupling factors with essential sulfhydryl groups by BDNA.

TABLE 3: ACTIVITIES OF RESPIRATORY CHAIN COMPONENTS
IN BDNA-TREATED ETP

Component(s)	BDNA Concentration (10^{-5}M)			
	0	1	3	10
β-hydroxybutyrate dehydrogenase	144	113	24.5	0
malate dehydrogenase (in ETP + supernatant)	2080	2090	2120	2100
NADH-cyt c reductase (total)	560	553	526	532
NADH-cyt c reductase (rotenone-insensitive)	23.4	23.7	25.2	23.0
succinate-cyt c reductase	73.1	76.5	79.9	74.1
cytochrome c oxidase	957	922	976	934

4.3 ISOLATION AND IDENTIFICATION - At least two mitochondrial coupling factors with essential sulfhydryl groups have been reported in the literature. Lam, Sanadi and their coworkers *(36, 37)* obtained a purified protein called Factor B which catalyzed ATP-driven reduction of NAD^+ by succinate in the presence of A-particles with a specific activity of 2-6 μmoles of NADH per minute per mg of the coupling factor per 0.5 mg of A-particles at pH 7.8 and 38°. They also reported a less pure preparation named Factor B' of 20-fold lower specific activity and approximate molecular weight 45000 *(44)*. Since these coupling factors seem to serve as a functional link *(45)* between the electron carriers and the terminal enzyme F_1 of oxidative phosphorylation, their characterization appears to be of basic biochemical significance.

In the course of preparing Factor B by the procedure of Lam *et al.*, we discovered that an extraordinarily efficient coupling factor F_B can be isolated in highly purified form by oxidative precipitation *(46)*. It has a specific activity of more than 1000 µmoles NADH per minute per mg of F_B per 0.5 mg of A-particles. As isolated, F_B is an oligomer of very high molecular weight, but in SDS gel electrophoresis, F_B exhibited a single protein band of subunit molecular weight 44000 ± 1000.

Fraction I   II   IIA   IIB   III   IV

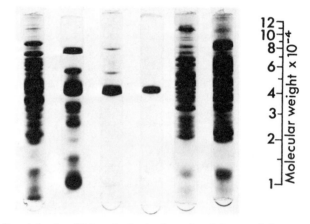

Fig. 12.   *Sodium dodecyl sulfate gel electrophoresis of F_B-containing protein fractions at various stages of purification. I, II, III, and IV refer to the various protein fractions obtained at consecutive stages of purification by fractional precipitation and ion-exchange chromatography. Fraction IIA was obtained from fraction II by oxidative precipitation. Fraction IIB, which contains pure F_B, was obtained from Fraction IIA by oxidative precipitation again. The logarithmic molecular weight scale on the right was determined by the best linear plot determined by dansylated standard proteins.*

Isolated F_B was found to be effectively and irreversibly inhibited by BDNA. The subunit molecular weight of F_B and its irreversible inhibition by BDNA suggest that the *in situ* BDNA-labeled protein factor may be F_B. By subjecting the BDNA-treated mitochondria to the same extraction, ion-exchange chromatography and oxidative precipitation procedure, the radioactive protein of subunit molecular weight 44000 was finally isolated *(43)*. The result of amino acid analyses of this labeled protein and coupling factor F_B are summarized in Table 4. The purity of F_B-

containing protein fractions at different stages of purification
is illustrated in Fig. 12. Values in Table 4 show that except
for carboxymethyl-cysteine (Cm-Cys), the BDNA-labeled protein has
within experimental uncertainties the same amino acid composition
as the coupling factor F_B.

TABLE 4: AMINO ACID COMPOSITION OF PURIFIED BDNA-LABELED
MITOCHONDRIAL PROTEIN OF SUBUNIT MOLECULAR WEIGHT 44000
OF COUPLING FACTOR F_B

| Amino Acid | Residues/subunit | | Amino Acid | Residues/subunit | |
	Labeled Protein	F_B		Labeled Protein	F_B
CM-Cys[a]	2.3	6.0	Met[b]	7	7
Asp	40.9	41.7	Ile	19.6	20.4
Thr	17.1	17.3	Leu	31.5	33.1
Ser	17.8	18.3	Tyr	11.6	11.7
Glu	39.1	39.6	Phe	13.2	13.1
Pro	21.6	22.9	Try	5.3	5.5
Gly	29.3	29.2	Lys	24.4	26.6
Ala	22.1	21.5	His	9.0	9.1
Val	24.1	25.2	Arg	30.5	31.5

[a]*The number of original cysteine residues per subunit was determined as carboxy-methylcysteine by preincubating the protein with iodoacetate for 2 hrs. at 40°C.*

[b]*The number of methionine residues per subunit was calculated to be 7, based on the experimental amino acid composition and the approximate subunit molecular weight of 44,000.*

Except for a minor radioactive product of unknown structure,
the only radioactive amino acid derivative found in the acid
hydrolysate of the labeled protein was carboxymethylcysteine. A
probable mechanism for the selective labeling by BDNA is summarized in Fig. 13. While Cm-Cys is a stable radioactive amino acid
derivative, Compound 2 can be easily hydrolyzed to regenerate the
non-radioactive Cys. By incubating the BDNA-labeled mitochondria
at pH 10 and 22°C for 2 hrs. and subsequently performing SDS gel
electrophoresis, it was indeed found that about one-third of the
radioactive labels was removed from the mitochondrial proteins.
     The close similarity in amino acid composition and the equa-
lity of subunit molecular weights suggests that the above radio-

active protein was probably produced by the *in situ* labeling of
F_B by BDNA. Accordingly we may conclude that the observed suppression of phosphorylation coupled to the oxidation of malate
and of ATP-driven reduction of NAD^+ by succinate can both be
attributed to the *in situ* labeling of F_B by BDNA.

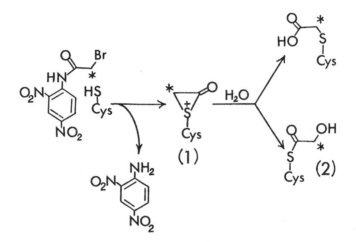

*Fig. 13. A proposed mechanism for the selective labeling of mitochondrial
proteins by BDNA. Cys-SH represents a cysteine residue at the active site.
The asterisk indicates the position of ^{14}C atom.*

In the field of bioenergetics there are many competing theories based on different points of view. However, the discussion
of coupling mechanisms without adequate structural information on
the energy transducing molecules is unlikely to be intellectually
satisfying, because without knowing the chemical identity of the
functional groups which participate in the coupling of electron
transport to phosphorylation we cannot understand the biology of
energy transduction at the truly molecular level no matter which
point of view we may take in formulating our working hypothesis.
Therefore, it may be desirable for us to focus at least as much
attention on structural biochemistry as on phenomenological
observations and general theories of energy-transducing membranes.

ACKNOWLEDGEMENT. The above described experimental work from our
laboratory was supported in part by research grants from the
National Institute of General Medical Sciences and from the National Science Foundation.

REFERENCES

1. R.A. Alberty, *J. Biol. Chem.*,**244**, 3290 (1969).
2. K. Shikama and K.-I Nakamura, *Arch. Biochem. Biophys.*,**157**, 457 (1973).
3. W.S. Brinigar, D.B. Knaff and J.H. Wang, *Biochemistry*,**6**, 36 (1967).
4. T.A. Cooper, W.S. Brinigar and J.H. Wang, *J. Biol. Chem.*,**243**, 5854 (1968).
5. J.H. Wang, *Acc. Chem. Res.*,**3**, 90 (1970).
6. K.P. Huang and J.H. Wang, unpublished work (1967).
7. S.-I Tu and Jui H. Wang, *Biochemistry*,**9**, 4505 (1970).
8. M.J. Bishop, *Biochim. Biophys. Acta*,**267**, 435 (1972).
9. S.-I Tu and Jui H. Wang, unpublished work (1972).
10. E.C. Slater, *Nature (London)*,**192**, 975 (1953).
11. Y.K. Shen and G.M. Shen, *Scientia Sinica*,**11**, 1097 (1962).
12. G. Hind and A.T. Jagendorf, *Proc. Nat. Acad. Sci. U.S.A.*,**49**, 715 (1963).
13. P. Mitchell, *Nature (London)*,**191**, 144 (1961).
14. R.J.P. Williams, *J. Theoret. Biol.*,**1**, 1 (1961).
15. V.P. Skulachev, in "Energy Transducing Mechanisms", E. Racker, Ed., University Park Press, Baltimore, p. 31 (1975).
16. H.H. Grunhagen and H.T. Witt, *Z. Naturforsch*,**25 b**, 373 (1970).
17. W. Junge, *Eur. J. Biochem.*,**14**, 582 (1970).
18. E.A. Lieberman and V.P. Skulachev, *Biochim. Biophys. Acta*,**216**, 30 (1970).
19. W. Junge, *Ber Deutsch. Bot. Ges.*,**88**, 283 (1975).
20. P.D. Boyer, in "Oxidases and Related Redox Systems", T.E. King, H.S. Mason, and M. Morrison, Eds., Wiley, New York, Vol. 2, p. 994 (1965).
21. D.A. Harris and E.C. Slater, *Biochim. Biophys. Acta*,**38**, 335 (1975).
22. G.A. Blodin and D.E. Green, *Chem. Eng. News*, Nov. 10, 26 (1975).
23. A.T. Jagendorf and J.S. Neumann, *J. Biol. Chem.*,**240**, 3210 (1965).
24. A.T. Jagendorf and E. Uribe, *Proc. Nat. Acad. Sci. U.S.A.*, **55**, 1970 (1966).
25. J.H. Wang, C.S. Yang and S.-I Tu, *Biochemistry*,**10**, 4922 (1971).
26. P. Mitchell, in "Regulation of Metabolic Processes in Mitochondria", J.M. Tager, S. Papa, E. Quagliariello, and E.C. Slater, Eds., Elsevier, Amsterdam, p. 65 (1966).
27. B.L. Maloff, S.P. Scordilis and H. Tedeschi, *Biophys. J.*,**16**, No. 2/Part 2, 19a (1976).

28. P.D. Boyer, in "Oxidases and Related Redox Systems", T.E. King, H.S. Mason and M. Morrison, Eds., Wiley, New York, p. 994 (1965).
29. D.A. Harris and E.C. Slater, *Biochim. Biophys. Acta,* **387,** 335 (1975).
30. R.B. Beechey and K.J. Cattell, *Current Topics in Bioenergetics,* **5,** 305 (1973).
31. H.S. Penefsky, *J. Biol. Chem.,* **242,** 5789 (1967).
32. L.L. Horstman and E. Racker, *J. Biol. Chem.,* **245,** 1336 (1970).
33. A.E. Senior and J.C. Brooks, *Arch. Biochem. Biophys.,* **140,** 257 (1970).
34. A.F. Knowles and H.S. Penefsky, *J. Biol. Chem.,* **247,** 6624 (1972).
35. S. Lien and E. Racker, *Methods in Enzymology,* **23,** 547 (1971).
36. K.W. Lam, J.B. Warshaw and D.R. Sanadi, *Arch. Biochem. Biophys.,* **119,** 477 (1967).
37. K.W. Lam, D. Swann and M. Elzinga, *Arch. Biochem. Biophys.,* **130,** 175 (1969).
38. T. Higashiyama, R.C. Steinmeier, B.C. Serrianne, S.L. Knoll and J.H. Wang, *Biochemistry,* **14,** 4117 (1975).
39. R.H. deMeio and E.S.G. Barron, *Proc. Soc. Exp. Biol. Med.,* **32,** 36 (1934).
40. J.H. Wang, O. Yamauchi, Shu-I Tu, K. Wang, D.R. Saunders, L. Copeland and E. Copeland, *Arch. Biochem. Biophys.,* **159,** 785 (1973).
41. L. Copeland, C.J. Deutsch, S.-I Tu and J.H. Wang, *Arch. Biochem. Biophys.,* **160,** 451 (1974).
42. C.C. Chen, T. Higashiyama and J.H. Wang, unpublished work (1974).
43. C.C. Chen, M.C. Yang, H.D. Durst, D.R. Saunders and J.H. Wang, *Biochemistry,* **14,** 4122 (1975).
44. K.W. Lam, M.E. Karunakaran and D.R. Sanadi, *Biochem. Biophys. Res. Commun.,* **39,** 437 (1970).
45. D.R. Sanadi, K.W. Lam and C.K.R. Kurup, *Proc. Nat. Acad. Sci. U.S.A.,* **61,** 277 (1968).
46. T. Higashiyama, R.C. Steinmeier, B.C. Serrianne, S.L. Knoll and J.H. Wang, *Biochemistry,* **14,** 4117 (1975).

recent advances in zinc biochemistry

BERT L. VALLEE

Biophysics Research Laboratory, Dept. of Biological Chemistry, Harvard Medical School, and Division of Medical Biology, Peter Bent Brigham Hospital, Boston, Massachusetts

1. INTRODUCTION

The presence of metals in biological matter in amounts approaching their detection limits has intrigued biologists for
generations, though efforts to ascertain their functional significance have often proved frustrating. The remarkable acceleration of the rate of progress in this field is the result of
conjoint advances in many disciplines. Nutritional and metabolic
experiments can now be monitored both by advanced methods of analysis and through suitable control of contamination. Major progress in isolating and characterizing the composition, structure
and function of metalloenzymes has greatly aided the delineation
of the molecular basis of the biological role of metals. Simultaneously, the emerging knowledge has given new direction to
experiments in biochemistry, physiology, pathology, nutrition
and medicine, and the resultant understanding of metallobiochemi-
stry has given hope that metals may play hitherto unrecognized
roles in disease. The possibility that metals might become
therapeutic agents has motivated much of the past effort in this
field.

Such considerations are directly pertinent to zinc metabolism.
The biological importance of zinc has emerged only during the
last two decades. In succession its biological effects were
viewed as mostly harmful, questionable and now essential. It is
now 100 years ago that zinc was found to be indispensable for the
growth of *Aspergillus niger (1)*, and its presence in plants and
animals was established within a decade *(2,3)*. Thereafter, despite the rapid growth of biological science during the next half-
century, almost no progress was made regarding its biological
role which remained conjectural. Not until 1934 was conclusive
evidence obtained that zinc is essential to normal growth and
development of rodents *(4,5)*.

It is now evident that zinc is essential to normal growth and
development of all living matter, but, remarkably, this universal
requirement was unappreciated until quite recently *(6)*. Its
deficiency results in major abnormalities of composition and
function, though the manifestations are complex and can vary
depending upon the particular species studied *(7-11)* (Table 1).
An increasing number of disease entities are proving to be related to zinc deficiency, both in animals and man *(12-16)*, and this
deficiency during pregnancy results in congenital malformations
of the embryo particularly by affecting growing or proliferating
tissues. The consequences of more subtle metabolic interactions,
as in cirrhosis *(18)*, and the basis of genetic or teratological
defects *(19)* have not been examined widely and offer rich inves-
tigative opportunities.

TABLE 1: ZINC DEFICIENCY IN DIFFERENT PHYLA

	Growth	Development	Changes in Chemical Composition		Decreased Enzymatic Activities
			Decrease	Increase	
Micro-organisms	Retardation	Increase in cellular size	Protein RNA (ribosomal) Pyridine Nucleotides	DNA Amino acids Polyphosphates Phospholipids ATP Organic acids	Alkaline phosphatase Alcohol and D-Lactate dehydrogenase Tryptophan desmolase
Plants	Retardation	Small abnormal leaves Chlorotic mottling Decreased fruit production	Protein Auxin Ethanolamine	Amino acids	Tryptophan desmolase Carbonic anhydrase Aldolase Pyruvic carboxylase
Vertebrates	Retardation	Testicular atrophy Parakeratosis Dermatitis Coarse, sparse hair	Red blood cells Serum proteins	Uric acid	Alkaline phosphatase Pancreatic proteases Malate, lactic alcohol dehydrogenase NADH diaphorase

2. BIOCHEMICAL APPROACHES

Experimental work performed mostly in the past generation has led to the cognizance that metals may be essential to enzymatic reactions (20). During the latter part of the 19th century investigations of cellular respiration and the functional role of iron in oxidative processes first suggested their essential role and pointed the way to the discovery of metalloenzymes. However, the idea that the first transition and IIB groups of metals, other than copper and iron, might have significant roles was unappreciated and biochemical knowledge was almost nonexistent.

The search for an explanation of the physiologic role of metals has emphasized their interaction with proteins, especially enzymes. The resultant metalloenzymes are characterized by their stability constants, whose magnitudes have served as the operational basis for the two extremes of metal-protein interactions, *i.e.*, the very stable metalloenzymes and the more labile metal-enzyme complexes (21), the latter exemplified by sodium, potassium, calcium and magnesium, the most abundant cations of mammalian species.

The transition metals and zinc, because of their electronic structures, tend to form stable complexes with proteins which exhibit characteristic functions, such as the oxygen-carrying capacity of iron in hemoglobin or of copper in hemocyanin, in addition to many enzymes containing these or other metals.

The lack of visible color of zinc proteins accounts for the long delay in their recognition, whereas the red iron (22) and blue copper proteins (23) early called attention to themselves. The isolation, purification and recognition (24) of carbonic anhydrase as a zinc enzyme was a happy accident. Through much of the purification work, the attention of Keilin and Mann (24) was focused on a blue protein, apparently exhibiting carbonic anhydrase activity; hence, it was suspected to be a copper enzyme, but it actually proved to be hematocuprein which is now recognized to be superoxide dismutase. With only minor adjustments of conditions, the analytical method used for copper could also detect zinc, which in turn led to the recognition of carbonic anhydrase as the first zinc metalloenzyme.

Eighty-five years after the initial recognition of the metabolic effects of zinc deficiency (1), the identification of bovine pancreatic carboxypeptidase A as a zinc enzyme occurred in 1954 (25). During the next several years (6,21) additional zinc metalloenzymes were discovered quite rapidly and now, twenty years later, more than 70 zinc enzymes have been identified, the majority within the past decade. It is now known that zinc participates in a wide variety of metabolic processes, among them carbohydrate, lipid, protein and nucleic acid synthesis or degradation. Zinc is essential for the function and/or structure of at least one in each of the six categories of enzymes designated by the Commission on Enzyme Nomenclature of the IUB and present throughout all phyla including several dehydrogenases, aldolases, peptidases, phosphatases, and isomerase, a transphosphorylase and aspartate transcarbamylase *(26)*. Detailed studies of the structure and function of many of these enzymes have become central to present understanding of enzymatic catalysis (Table 2).

TABLE 2: SOME ZINC METALLOENZYMES	
Enzyme	Source
Alcohol dehydrogenase	Yeast; horse, human liver
D-Lactate cytochrome reductase	Yeast
Glyceraldehyde-phosphate dehydrogenase	Beef, pig muscle
Phosphoglucomutase	Yeast
RNA polymerase	*E. coli*
DNA polymerase	*E. coli*
Reverse transcriptase	Avian myeloblastosis virus
Mercaptopyruvate sulfur transferase	*E. coli*
Alkaline phosphatase	*E. coli*
Phospholipase C	*Bacillus cereus*
Leucine aminopeptidase	Pig kidney, lens
Carboxypeptidase A	Beef, human pancreas
Carboxypeptidase B	Beef, pig pancreas
Carboxypeptidase G	*Pseudomonas stutzeri*
Dipeptidase	Pig kidney
Neutral protease	*Bacillus sp.*
Alkaline protease	*Escherichia freundii*
AMP aminohydrolase	Rabbit muscle
Aldolase	Yeast; *Aspergillus niger*
Carbonic anhydrase	Erythrocytes
δ-Aminolevulinic acid dehydratase	Beef liver
Phosphomannose isomerase	Yeast
Pyruvate carboxylase	Yeast

Substantial quantities of firmly bound zinc in RNA *(27)* and DNA *(28)* have revealed additionally important avenues for investigation of its role in biology. Zinc appears to stabilize the secondary and tertiary structure of RNA *(29)* and to play an important role in protein synthesis. To date we have little or no information on the association of zinc with lipids or carbohydrates, glyco- and lipoproteins, but their systematic analysis might turn out to be important.

There are unusual opportunities to study the manner in which the specific function of zinc proteins and enzymes is achieved through interaction of the metal with the protein. By removal and restoration of zinc with concomitant loss and restitution of enzymatic activity, it is possible to achieve the differential chemical labeling of the sites of the protein to which the metal binds. This permits physico-chemical approaches to discern what contribution zinc may make to overall protein structure and stability. Other metals can be substituted successfully in a number of enzymes for the native zinc, with corresponding characteristic alterations of physical properties and/or catalytic function. These, as well as other approaches, have assisted in the delineation of the role of zinc in the function and structure of enzymes *(20)*.

3. CHEMICAL FEATURES OF ZINC ENZYMES

Zinc, a IIB element with a completed d subshell and two additional s electrons, chemically combines in the +2 oxidation state. There is no evidence that it is oxidized or reduced in biological reactions. It generally forms tetrahedral complex ions, but many octahedral complexes are known.

The characteristics of zinc in *e.g.*, simple halo-, cyano- and amino-complexes likely differ from those in metalloenzymes, the stereochemistry being determined largely by ligand size, electrostatic and covalent binding forces. The three-dimensional structure of proteins, heterogeneity of ligands and the degree of vicinal polarity of the metal binding site may jointly generate atypical coordination properties *(30)*. Unusual bond lengths, distorted geometries, and/or an odd number of ligands can generate a metal binding site on the enzyme, which when occupied by a metal, can be thermodynamically more energetic than metal ions when free in solution where they are complexed to water or simple ligands. As a result, in enzymes zinc is thought to be poised for its intended catalytic function in the *entatic state (30)*. In this context the term entasis indicates the existence of a condition of tension or stress in a zinc - or other metal - enzyme prior to combining with substrate.

In effect, the entatic state is thought to originate in the genetic heritage of the cell. The primary structure of the enzyme protein dictates the relative spatial positions of those amino acid side chains destined to serve as ligands when the apoprotein combines with the metal ion. Evidence suggests that the metal ion is not incorporated into the growing, ribosome-bound polypeptide chain *(31)*, until the protein is fully formed. According to this view, the metal does not induce its own coordination site, but its interaction awaits the expression of the genetic message.

The lack of suitable physical-chemical probe properties of

the diamagnetic zinc atom led to a search for means to replace it with paramagnetic metals, *e.g.*, cobalt which could signal information on the nature and environment of the active site *(32)*. The spectra of such cobalt substituted zinc enzymes *(33-36)* differ significantly from those of model Co^{2+} complexes and are thought to reflect the entatic state of the cobalt (and zinc) ions in these enzymes *(30)*.

Identification of the metal binding ligands of metalloenzymes has been difficult by means other than X-ray crystallography. Cysteinyl, histidyl, tyrosyl residues and the carboxyl groups of aspartic and glutamic acids have been implicated most commonly *(20)*. Thus far, metal complex ions in which the mode of metal coordination is known quite precisely have not proven adequate to define metal binding in enzymes, invariably lacking the entatic environment seemingly characteristic of metalloenzymes *(30)*. Moreover, the complexes which are known most thoroughly and have been studied most extensively are bidentate, but present evidence indicates that in metalloenzymes, metals are more likely coordinated to at least three ligands *(20)*. With few exceptions *(37)*, multidentate complex ions, suitable for appropriate comparisons with metalloenzymes, have not been studied. Yet absorption *(38)*, magnetic circular dichroic *(39)* and electron paramagnetic resonance spectra *(40)* combined with kinetic studies of cobalt-substituted and chemically modified zinc metalloenzymes have enlarge understanding both of the possible modes of interaction of zinc with the active sites of zinc metalloenzymes and their potential mechanisms of action.

The molecular details of metalloenzyme action have been elucidated greatly in the past few years *(41)*. Crystal structures for bovine carboxypeptidase A *(42)*, thermolysin *(43)* and horse liver alcohol dehydrogenase *(44)* are now available, and chemical and kinetic studies have defined the role of zinc in substrate binding and catalysis *(41,20)*. In fact, many of the significant features elucidating the mode of action of enzymes, in general, have been defined at the hands of zinc metalloenzymes.

4. ZINC AND NORMAL CELL BIOLOGY

In spite of these tremendous advances establishing in but two decades both the very participation of zinc in enzymatic catalysis and many aspects of its presumable mechanisms, knowledge concerning the roles of this element in cellular metabolism is surprisingly sparse and the reactions to which it becomes limiting have not been recognized, defined or integrated. Studies of the role of zinc in cell metabolism require an organism which can be obtained in homogeneous form, grows rapidly when the trace metal content of the medium is controlled rigorously, and which can be disrupted readily to allow definitive measurements on subcellular organelles. Such an organism, a critical prerequisite for such cell biological studies, has been available only in the last decade: the alga, *Euglena gracilis*, was found to satisfy the necessary criteria *(45)*. It has now been used as a model to study the biochemical and morphological consequences of zinc deprivation *(45-50)*.

A number of striking chemical changes accompany zinc deficiency-induced growth arrest: RNA and protein synthesis is depressed; cellular DNA content doubles; cell volume increases, and peptides, amino acids, nucleotides, polyphosphates and unusual proteins accumulate. Further, the intracellular content of Ca, Mg, Mn, Ni, Co and Fe increases from 6- to 35-fold *(49,51)*. This is particularly interesting since both Mg and Mn activate a number of zinc enzymes, *e.g.*, leucine aminopeptidases, alkaline phosphatases and DNA and RNA polymerases (see below). While the specific biochemical derangements which cause these metabolic and metal imbalances in zinc deficient *E. gracilis* remained unknown, the evidence was impressive that they would relate to nucleic acid metabolism and cellular division. Reports recognizing DNA and RNA polymerases of prokaryotic organisms as zinc metalloenzymes *(52-56)* strongly supported this deduction. Recent experiments with zinc deficient *E. gracilis* , to be summarized briefly, now leave little doubt in this regard and, moreover, detail the effect of the metal on the dynamics of DNA metabolism in the cell cycle. It has become increasingly apparent that many functions of zinc in biology are very similar to those of the vitamins: it serves as a coenzyme.

5.   DNA METABOLISM IN ZINC DEFICIENT *E. GRACILIS*

When zinc in the medium is sufficient (10^{-5} M), the initial growth of *E. gracilis*, lasting 5-6 days, is followed by a logarithmic period which continues up to the 12th-14th days and then becomes stationary without further increase in total numbers of cells. When zinc is growth-limiting (10^{-7} M), there is a marked decrease in cellular proliferation (Fig. 1). Solely on addition of zinc, cell growth is resumed within 24-48 hours and reaches precisely the same level as that of initially zinc sufficient cells. Growth impairment is due entirely to zinc deficiency. The zinc content of 13-day old cells grown in zinc sufficient media is almost 10-fold greater than that of cells of the same age when grown in zinc deficient media, *i.e.*, 58 *vs* 8 μg Zn/10^8 cells.

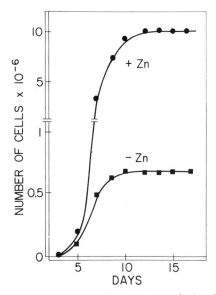

Fig. 1. *Growth of zinc-sufficient (●) and zinc-deficient
(■) E. gracilis grown in the dark. Zinc-sufficient medium
contains 1×10^{-5} M Zn^{2+}; zinc-deficient medium contains
1×10^{-7} M Zn^{2+}. From (49).*

In addition to chemical and enzymatic alterations (Table 1)
the marked morphological changes which are observed are characteristic of the various normal growth phases of *E. gracilis (46,
47).* During the lag period zinc sufficient cells are oval and
become thin and elongated in the course of the logarithmic and
stationary phases; the overall size of the cells decreases throughout this period, probably due to changes in their paramylon content (Fig. 2A). In contrast, cells grown in zinc deficient media
remain round or ovoid throughout the growth period; their size
increases continuously from the lag through the log and stationary
phases (Fig. 2B).

Electron microscopic examination reveals surprisingly little
change either in the size or in the ultrastructure of the principal cellular organelles of zinc deficient cells *(47).* The abundance of ribosomes may be diminished, but that of the mitochondria,
Golgi bodies and endoplasmic reticulum seems unaltered. Importantly, the most striking change is the extraordinary accumulation
of paramylon, the storage form of carbohydrate in these organisms
(Fig. 2).

We have examined the events during the typical eukaryotic cell
cycle in order to define the specific lesions that accompany zinc
deficiency. Although *E. gracilis* is a eukaryote, it undergoes

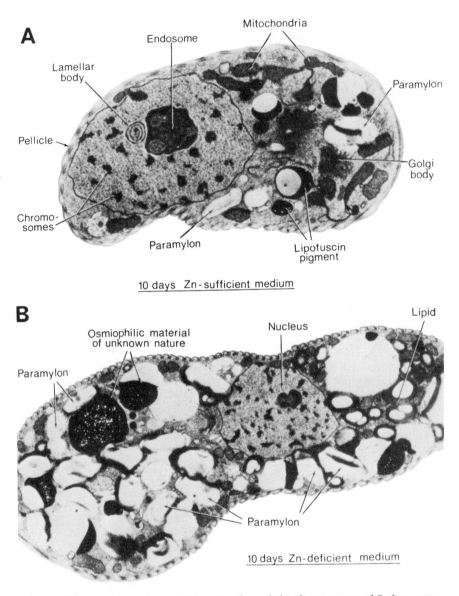

A

Lamellar body

Endosome

Mitochondria

Paramylon

Pellicle

Golgi body

Chromo-somes

Paramylon

Lipofuscin pigment

10 days Zn-sufficient medium

B

Lipid

Osmiophilic material of unknown nature

Nucleus

Paramylon

Paramylon

10 days Zn-deficient medium

Fig. 2. Electron micrographs permitting comparison of the ultrastructure of Euglena grown 10 days with or without zinc, respectively. The zinc-deficient organism is significantly larger, its cytoplasm contains an abundance of paramylon and large masses of dense osmiophilic material presumably rich in lipid. The size difference is actually greater than it appears, since the micrograph of the zinc-deficient Euglena is shown at somewhat lower magnification so that the whole organism can be included. ×8500 and 6500, respectively.

growth phases which are usually considered characteristic of prokaryotic organisms, *i.e.*, lag, log and stationary phases. Yet
their cell division is best characterized in terms of G_1, S, G_2
and mitosis, as described by Howard and Pelc *(57)* (Table 3).

Cell Cycle Stage	Biochemical and Morphological Events	Processes or Cellular Components with Known Zn Requirement
G_1	RNA Synthesis Protein Synthesis Increase in Size	Uridine Incorporation into RNA DNA-dependent RNA Polymerase RNA Stabilization
S	DNA Synthesis	Thymidine Incorporation into DNA DNA-dependent DNA Polymerase DNA Stabilization
G_2	Assembly of Functioning Apparatus for Nuclear and Cell Division	Unknown
Mitosis	Nuclear Division	Dithizone Staining of Nucleoli Spindle Apparatus Chromosomes

TABLE 3: THE ROLE OF ZINC IN THE BIOCHEMICAL AND MORPHOLOGICAL EVENTS OF THE CELL CYCLE

Using these stages, *E. gracilis* cells increase in size and synthesize RNA and protein during G_1. In S, DNA synthesis increases
and is then followed by chromosomal replication. Formation of
the mitotic apparatus occurs during G_2 and is followed by nuclear
and cytoplasmic division.

We have examined the cell cycle with organisms sampled at
various stages in the growth of both zinc sufficient and deficient *E. gracilis* by means of laser excitation cytofluorometry *(48)*.
In this manner, it is possible to detail, delineate and define
those steps of the cell cycle of *E. gracilis* affected specifically
by zinc deprivation.

Aliquots of log phase zinc-sufficient cells were collected
for analysis of DNA content, fixed in ethanol and stained with
propidium diiodide after treatment with ribonuclease to remove
RNA. The stained cells were analyzed in a cytofluorograph provided with a cell sorter which plots the results as histograms of
the numbers of cells *vs.* increasing values of cellular fluorescence, which is a direct measure of DNA content (Fig. 3). The
pattern of cells shown is typical of a log phase population with

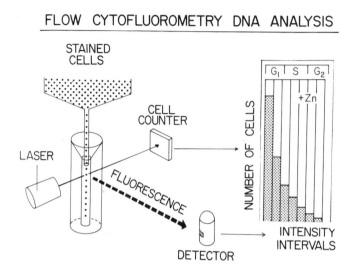

FLOW CYTOFLUOROMETRY DNA ANALYSIS

STAINED
CELLS

CELL
COUNTER

LASER

FLUORESCENCE

DETECTOR

NUMBER OF CELLS

+Zn

INTENSITY
INTERVALS

Fig. 3. Schematic diagram of flow cytofluorometer.

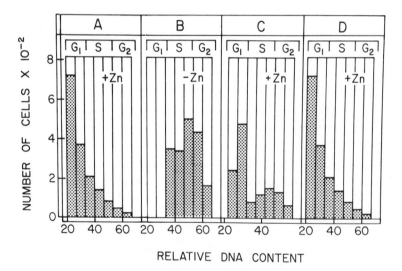

Fig. 4. Comparison of DNA histograms of zinc-sufficient with zinc-deficient E. gracilis. The majority of zinc-sufficient cells are in G_1 with a small fraction in S (Panel A). In contrast, nondividing zinc-deficient cells are mostly in S or G_2 (Panel B). On addition of zinc the number of cells blocked in S or G_2 decreases and histograms typical of dividing log phase cells result (Panels C and D).

48

most cells in G_1 and the remainder distributed between S and G_2.

At early stationary phase, when net cell division ceases, the histogram of the DNA content of zinc sufficient cells indicates that most of the cells are in G_1 with the balance in S (Fig. 4A). The pattern of this histogram contrasts with that obtained for zinc deficient cells (Fig. 4B), also recorded at a time when further growth does not occur. The DNA content of zinc deficient cells is distributed in both the S and G_2 phases of the cycle. This suggests that, among these nonsynchronously growing zinc deprived cells, those that are in S do not continue into G_2, while those in G_2 do not proceed through mitosis. Histograms reflecting the DNA distribution of cells obtained from cultures subsequent to restoration of adequate amounts of zinc demonstrate resumption and progression of cell cycles (Fig. 4C) which return to and restore a pattern typical of dividing log phase cells (Fig. 4D).

Transfer of zinc sufficient cells in early stationary phase, known to be mostly in G_1, to zinc deficient medium reveals further details of the effect of zinc deficiency on the G_1 to S transition. Initially, the number of cells increases by 25%, but growth soon ceases (Fig. 5); it can be restored by addition

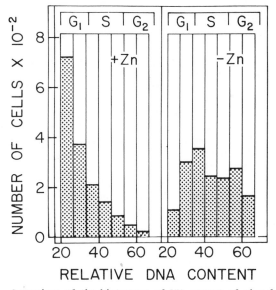

Fig. 5. *Comparison of the histograms of DNA content of zinc deficient with stationary phase zinc sufficient E. gracilis. In stationary phase, the majority of sufficient cells are in G_1, with a small fraction in S. In contrast, nondividing, deficient cells are mostly in S or G_2, with only a fraction in G_1. From (48).*

of zinc which causes a 200% increase in the number of cells within 24 hours (Fig. 4C and D). Hence, zinc deprivation of cells in G_1 blocks their progression into S, which zinc then restores. Clearly, the biochemical processes essential for cells to pass from G_1 into S, from S to G_2 and from G_2 to mitosis require zinc, and its deficiency can block all phases of the growth cycle of *E. gracilis*. Detailed studies of these phenomena have been performed in synchronously dividing organisms [48].

Zinc is an enzymatically essential component of both DNA-dependent DNA and RNA polymerases isolated from prokaryotic organisms [55,56], but its role in the corresponding enzymes from eukaryotic systems has not been explored in similar detail. The reduced incorporation of labeled DNA and RNA precursors into the livers of zinc-deficient rats [58], cultured chick embryos [59] and phytohemagglutinin-stimulated lymphocytes [60] suggests that eukaryotes also require zinc. Further, the rate of incorporation of ^{3}H-uridine into RNA of zinc deficient *E. gracilis* is decreased [47], the element maintains the integrity of their ribosomes [61] and stabilizes both RNA and DNA [28,62]. Zinc is further essential to the function of protein synthesis elongation factor 1 (Light Form EF_L) from rat liver which catalyzes the binding of aminoacyl-t RNA to an RNA-ribosome site through the formation of an aminoacyl-t RNA-EF1-GTP ternary complex [63].

The distinctive red zinc dithizonate complex is the most sensitive means to quantitate this element [64]. By this means the presence of zinc has been demonstrated histochemically in the nucleus, in subnuclear structures and, further, its translocation to and from the nucleolus, spindle and chromosomes at each stage of mitosis has been shown. Zinc dithizonate is present in the nucleolus and spindle during prophase, then in chromosomes during metaphase and anaphase, and reappears in the nucleolus in telophase [65]. This evidence of the participation of zinc both in nucleic acid and protein synthesis as well as in cell division is entirely consistent with the results of cell cytofluorometry already cited.

Potentially, zinc deprivation affects all of these steps (Table 3). In *E. gracilis* zinc might similarly be involved in nucleic-acid metabolism and mitotic activity, and its deficiency might affect multiple processes all of which bear on cell division and nucleic acid function.

It is then reasonable to expect that zinc is essential to systems involved in the polymerization of RNA and/or DNA of *E. gracilis*, constituting a biochemical basis of the above observations - though the metal is apparently indispensable at multiple loci.

Three years ago, as a part of a systematic effort to unravel

these interrelationships, Dr. Kenneth Falchuk began studies of
the nature and role of these enzymes in zinc sufficient *E. gracilis*. Initial observations that RNA polymerase activity in
crude cellular extracts is inhibited by the chelating agent 1,10-
phenanthroline supported the hypothesis that zinc plays a functional role in this enzyme.

6. DNA DEPENDENT RNA POLYMERASES I AND II OF *E. GRACILIS*:
ZINC METALLOENZYMES

Initially, the isolation of DNA dependent RNA polymerases
from *E. gracilis* encountered difficulties quite analogous to
those which beset their purification from other eukaryotic organisms. Examination of a possible role of zinc in the function of
RNA polymerases clearly depends on their purification to obtain
homogeneous enzymes in quantities sufficient for metal analysis.

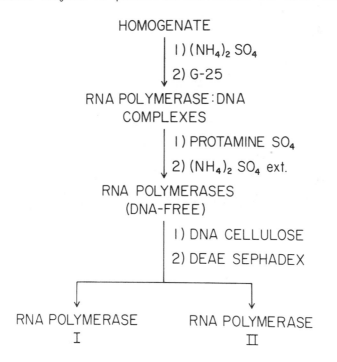

HOMOGENATE

1) $(NH_4)_2 SO_4$

2) G-25

RNA POLYMERASE : DNA
COMPLEXES

1) PROTAMINE SO_4

2) $(NH_4)_2 SO_4$ ext.

RNA POLYMERASES
(DNA-FREE)

1) DNA CELLULOSE

2) DEAE SEPHADEX

RNA POLYMERASE
I

RNA POLYMERASE
II

Fig. 6. Purification of RNA polymerases I and II from zinc sufficient E. gracilis.

While the technical obstacles would have seemed unsurmountable
but three years ago, decisive advances during the last two years
both in methods of isolation of eukaryotic polymerases *(66)* and
the advent of microwave excitation emission spectrometry for
metal analysis *(67)* have now allowed the characterization of
these metalloenzymes. Microwave excitation spectrometry permits
the accurate determination of picogram quantities of metal ions
in microgram quantities of this and other polymerases *(67)* (see
below) and has demonstrated that they are zinc enzymes *(68)*.

Fig. 6 outlines the purification procedure for RNA polymerase
I and II from zinc sufficient *E. gracilis (69,70)*. Cells are
harvested in the log phase of growth and disrupted to solubilize
the enzymes. The homogenate is treated with ammonium sulfate,
the precipitate is dissolved, and then chromatographed on Sephadex G-25. At this stage, the RNA polymerases are bound to DNA.

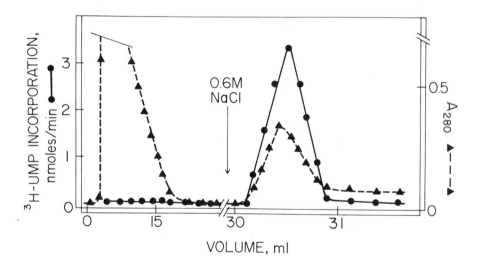

*Fig. 7. Chromatography of E. gracilis RNA polymerases on DNA cellulose.
Three ml of enzyme were solubilized from the protamine sulfate precipitate
(Fraction V), dialyzed against 500 ml of buffer and loaded onto a 0.5 × 3 cm
DNA cellulose column. Proteins which do not bind to DNA were removed with
0.1 M NaCl while the RNA polymerase was eluted with 0.6 M NaCl all at a flow
rate of 2 ml/hr. Two μl fractions were assayed for polymerase activity (•)
and A280 (▲). (68).*

The enzyme DNA complexes are precipitated with protamine sulfate
and the RNA polymerases solubilized by extracting the precipitate
with tenth molar ammonium sulfate.   The enzymes are then free of
DNA and are purified further by affinity chromatography on a DNA
cellulose column which separates two protein fractions.   The
first protein does not bind to DNA and does not exhibit RNA polymerase activity (Fig. 7), the second is eluted with 0.6 molar
sodium chloride and contains all the RNA polymerase activity.
This fraction is further resolved on DEAE Sephadex.   Both protein
fractions contain RNA polymerase activity and are eluted with
0.15 M and with 0.35 M ammonium chloride, respectively (Fig. 8).
The latter accounts for approximately 90% of the total activity.
This ionic strength dependent separation demonstrates the presence of distinct type I and II RNA polymerases in zinc sufficient
E. gracilis (68).

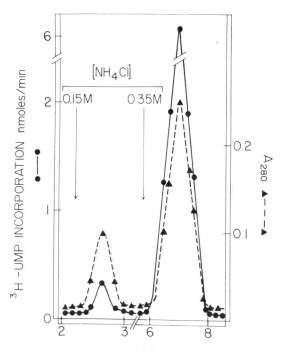

Fig. 8.   *Chromatography of E. gracilis RNA polymerases on DEAE-Sephadex
A-25.   DNA cellulose fractions containing RNA polymerase activity were
pooled and dialyzed in buffer.   One ml was loaded onto a 0.5 × 11 cm
column of DEAE-Sephadex A-25 previously equilibrated with this buffer
and 0.2 ml fractions were collected.   The enzymes were eluted by step
gradients with 0.15 M and 0.35 M ammonium chloride.   Aliquots were assayed for polymerase activity (●) and A280 (▲).   (68).*

Both polymerases are entirely dependent on an exogenous DNA template for activity. The product of their enzymatic reaction is RNA, as evidenced by an absolute substrate requirement for ribonucleotide triphosphates and by digestion of the product by ribonuclease. As with other polymerases, the *E. gracilis* enzymes are inactive in the absence of Mg^{2+} or Mn^{2+}. Both these DNA dependent RNA polymerases are homogeneous on polyacrylamide gels, and their estimated molecular weights, determined on SDS gels, lie between 650,000 and 700,000 for both polymerases. Table 4 summarizes these properties. They are composed of multiple sub-units of varied but unknown molecular weights which remain to be determined to obtain the precise molecular weights of the holo-enzymes.

Property	I	II
TABLE 4: PROPERTIES OF E. GRACILIS RNA POLYMERASES I AND II		
Template Dependence	DNA	DNA
Product	RNA	RNA
Activating Metals	Mg,Mn	Mg,Mn
M.W. (SDS-Page)	650,000	700,000

α-Amanitin differentiates type II RNA polymerases, which are inhibited by this agent, from those of type I, which are not (Fig. 9). The activity of the first DEAE Sephadex fraction is not inhibited by α-amanitin at concentrations up to 100 micro-grams per ml indicating that it is an RNA polymerase I. In con-trast, increasing concentrations of α-amanitin progressively de-crease and at 0.1 μg/ml nearly abolish activity of the second DEAE Sephadex fraction, typical of an RNA polymerase II (Fig. 9).

Both polymerase I and II are inhibited by saturating amounts of chelating agents (Table 5) and 1,10-phenanthroline and EDTA inhibit both their activities completely. Other chelating agents such as 8-hydroxyquinoline and its 5-sulfonic acid, EDTA and α,α'-dipyridyl, reduce the RNA polymerase II activity from 70 to 50%. At saturating amounts the non-chelating analogs of 1,10-phenan-throline, 1,7- or 4,7-phenanthroline, do not inhibit either poly-merase. Hence the inhibition by the 1,10-isomer must be due to chelation of a functional metal atom *(68)*.

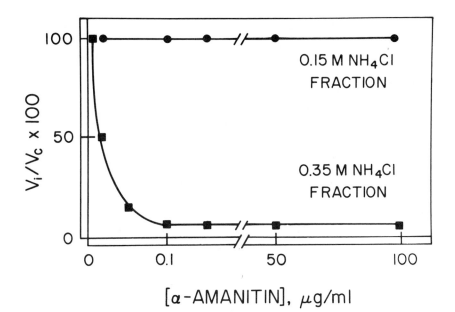

Fig. 9. *Inhibition of RNA polymerases I (●) and II (■) by
α-amanitin. The assay contained 10 µg of enzyme in 0.1 M
Tris-Cl, pH 7.9, preincubated for 15 minutes with inhibitor.*

Agent	$V_i/V_c \times 100$	
	I	II
Chelating:		
1,10-Phenanthroline	0	0
EDTA	0	0
8-Hydroxyquinoline	-	70
8-Hydroxyquinoline 5-Sulfonate	-	70
α,α'-dipyridyl	-	55
Non-Chelating:		
1,7-Phenanthroline	100	100
4,7-Phenanthroline	100	100

TABLE 5: *INHIBITION OF RNA POLYMERASES
I AND II BY CHELATING AGENTS*

The relative sensitivities of polymerase I and II to 1,10-phenanthroline were studied in detail over a range of inhibitor concentrations. Both RNA polymerase I and II are inhibited by this agent but with different pK_I's, 5.2 and 3.4, respectively, further differentiating them (Fig. 10).

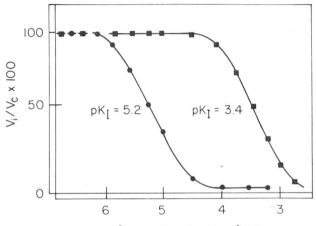

Fig. 10. *Concentration dependent inhibition of RNA polymerases I (●) and II (■) by 1,10-phenanthroline, 10 μg enzyme in 0.1 M Tris-Cl, pH 7.9, 25°, was assayed in the presence of chelator. All activities were measured instantaneously.*

The inhibition of polymerase II by 1,10-phenanthroline is instantaneous (Fig. 11). Addition of 0.5 mM chelating agent, immediately decreases activity to 50%. On incubation up to 60 minutes activity does not change further. Moreover, the inhibition is completely reversible by dilution of the assay mixture with buffer, suggesting the formation of a mixed complex and confirming presence of a metal essential for function. Quantitative measurements of the metal content have established the presence of zinc in both RNA polymerase I and II by means of microwave excitation emission spectrometry. The *E. gracilis* RNA polymerase I and II contain 0.2 μg of Zn per mg of protein. Keeping in mind the as yet provisional nature of the molecular weights of these polymerases which form the basis of the metal/protein ratio, the stoichiometry of both is essentially the same, *i.e.*, 2 g-atom of

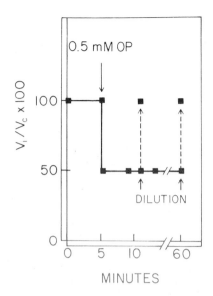

Fig. 11. Reversal of 1,10-phenanthroline inhibition of RNA polymerase II by dilution. The reaction mixture was diluted 1:10 with 0.1 M Tris-Cl, pH 7.9. Other conditions as in Fig. 10.

zinc per mole (Table 6). The sum of Cu, Fe, Mn is less than 0.2 g-atom/mole. Thus, both RNA polymerase I and RNA polymerase II from zinc sufficient *E. gracilis* are zinc metalloenzymes (Table 6). The eukaryotic DNA dependent RNA polymerase I of yeast is also a zinc enzyme containing 2.4 g-atom of Zn per molecular weight 650,000 *(71)*.

The demonstration that *E. gracilis* RNA polymerase I and II are zinc metalloenzymes confirms the essentiality of this element for RNA synthesis, in conjunction with its involvement in other aspects of RNA metabolism *(51,61)*. The isolation of the DNA dependent DNA polymerase of *E. gracilis* and their inhibition by 1,10-phenanthroline *(72,73)* further represent critical steps in deciphering the series of events which may jointly account for some of its requirements in the growth of *E. gracilis* and of other eukaryotic organisms.

In a given zinc enzyme the metal can either be essential for activity, regulate it, stabilize its structure or any combination of these *(74-76)*. While attempts to specify a particular enzymatic (or other) step in zinc deficiency which might limit growth would be conjectural at present, a number of alternatives suggest themselves *(68)*.

TABLE 6: METAL CONTENT OF RNA POLYMERASES I AND II			
AS MEASURED BY MICROWAVE EXCITATION SPECTROSCOPY			
RNA POLYMERASE	PROTEIN mg/ml	Zn μg/mg Protein	g-atom/ mole
I	0.12	0.19	2.0
II	0.10	0.21	2.2
ΣFe, Cu, Mn < 0.2 g-atom/mole			

Metal content is expressed as g-atom per mole of 650,000 and 700,000 for the RNA polymerases I and II, respectively.

Lack of zinc could preclude synthesis of any one or all of these zinc enzymes or result in inactive apoenzyme(s). Further, the possibility must be considered that functional metalloenzymes containing different metals might be generated which could exhibit altered values of K_{cat} or K_m - either for substrate or template. In this regard, the accumulation of Fe, Cr, and Ni which occur in response to zinc deficiency of *E. gracilis (47,49)* could reflect a compensatory mechanism, conceivably designed to overcome a metal-dependent metabolic block. The simultaneous accumulation of Mg and Mn, could similarly serve to regulate such metal substituted RNA polymerases. Certainly, the requirements for both Mg and Mn of the corresponding native *Euglena* enzymes are variable though they are all inhibited by high concentrations of Mn *(69)*.

Whatever the specific mechanism, the zinc deficient state might alter the relative proportions of existent RNA polymerase activities or induce new variants. A number of observations would tend to support such conjectures. Thus, in sea urchin *(77-79)*, liver *(80)*, *Helianthus tuberosus (81)*, and amoebae *(82)*, the activities of RNA polymerase I and II vary as a function of the stage of development. Thus, during normal growth, regulatory mechanisms serve to synthesize or preferentially activate different classes of RNA polymerases; these control processes might be called into play in zinc deficiency and alter the classes of RNA synthesized.

Further, Pogo *et al. (83)*, have shown in experiments with *E. coli* RNA polymerase that the relative preponderance of Mg and Mn can modify the nucleotide sequences of the RNA which is the product of the polymerization reaction, a course of events which could obtain here. One or all of these postulated processes could ultimately lead to the synthesis of unusual RNA sequences and the ensuing synthesis of altered or abnormal proteins *(84-86)*. Zinc deficient *E. gracilis* are known to synthesize peptides and/or proteins whose amino acid composition is unusual *(51)*. Hence, it is conceivable that there could be defects in protein synthesis of zinc deficient *E. gracilis* either during initiation, elongation or termination. The observation that elongation factor I from rat liver contains functional zinc atoms *(63)* is consistent with and could bear importantly on these considerations. At present it is not possible to discriminate between such alternatives. Clearly, critical experiments to delineate between them are both indicated and feasible.

The present data provide evidence for a role of zinc in transcription and translation of *E. gracilis* underlying the essentiality of zinc in cell division. The demonstration that eukaryotic-RNA polymerase II and I *(71)* are zinc enzymes support the hypothesis that the element is essential for RNA metabolism in all phyla *(52-56, 87-89)*. The data further suggest that extension of our studies on zinc deficient *E. gracilis* to other organisms may assist in generalizing the emerging biochemical role of zinc in growth, proliferation and differentiation.

7. ZINC AND NEOPLASTIC DISEASE

In spite of the now extensive biological literature documenting profound effects of zinc on normal growth and development, its possible role in abnormal growth has been examined in much more cursory fashion. The demonstration that a series of reverse transcriptases are zinc metalloenzymes has unexpectedly linked earlier observations of a role of zinc in abnormal growth to present day enzymology.

Reports that tissues of man and other vertebrates contain zinc *(90)*, an element thought to be particularly abundant in certain tumors *(91,93)*, first generated interest in the possible role of this element in the genesis of solid tumors. These and other early reports of abnormal concentrations and possibly altered metabolism of this element in tissues or fluids of tumor-bearing animals would now seem to be primarily of historical interest, considering the methodology then available.

One facet of this early work, however, may be of more lasting significance. The role of zinc in carcinogenesis was explored by injecting zinc chloride or sulfate into the cock testis generating teratomata which can be transplanted *(94,95)*; zinc acetate or zinc sterate proved ineffective *(96)*. These seasonal tumors develop only when the injections are given between January and March, periods of high sexual activity *(95,96)*. In this species spontaneous teratomata are rare.

The relationship of zinc to leukemia has also proven to be of

more enduring importance. A number of years ago we found substantial quantities of zinc in normal human leukocytes *(97)*, later
shown to be associated largely with two zinc proteins whose functional identities remain to be established *(98,99)*.

Such studies of the zinc metabolism of normal and leukemic
leukocytes generated the postulate that disturbance of a zinc
dependent enzyme might be critical to the pathophysiology of mye-
logenous and lymphatic leukemia, to quote: "Since zinc is known
to be present in carbonic anhydrase, it is quite possible that
there is another enzyme system with which it is concerned in mye-
lopoiesis, and that there is some disturbance of this enzyme in
leukemia" *(100,101)*.

This hypothesis has recently found support in studies of Type
C oncogenic RNA viruses, *e.g.*, avian myeloblastosis virus (AMV),
associated with lymphomas and leukemia in a number of species.
The existence of an RNA-dependent DNA polymerase - reverse trans-
criptase - in these RNA tumor viruses *(102,103)* has greatly stimulated study of the initiation, biochemical basis and maintenance
of malignant transformations and of the manner by which viral RNA
is transcribed into a DNA copy. Although a role for zinc in this
process was unknown, the above indications of its importance in
normal and leukemic leukocyte metabolism prompted us to examine
the zinc content of the RNA-dependent DNA polymerase from avian
myeloblastosis, murine, simian, feline and RD-114 RNA tumor
viruses.

As indicated earlier, complexing agents have long been employed to explore the functional role of metals in enzymes by
kinetic methods. However, even if a series of these agents are
employed to study their inhibition of enzymatic activity, of
themselves, their effects do not allow a definitive decision
regarding the specific metal atom which may be involved in a particular instance *(20)*. As in the case of the RNA polymerases the
combination of results obtained with agents which differ in regard
to their selectivity can be suggestive, but ultimately, the identification of a metalloenzyme minimally requires analytical demonstration of the presence of a functional metal atom. The paucity
of the material seemed to present formidable problems in regard
to metal analysis, but the marked inhibition of the enzyme's
activity by metal-binding agents led us to develop and employ an
instrument for microwave excitation spectroscopy and a procedure
capable of quantitative metal determinations at the 10^{-15} g-atom
level (Fig. 12). This has allowed the identification of stoichiometric amounts of zinc in these enzymes which was available
to us in quantities of the order of 10^{-9} moles. The limits of
detection of the procedures are sufficiently low, ultimately to
permit quantitative studies of metals and their metabolism in

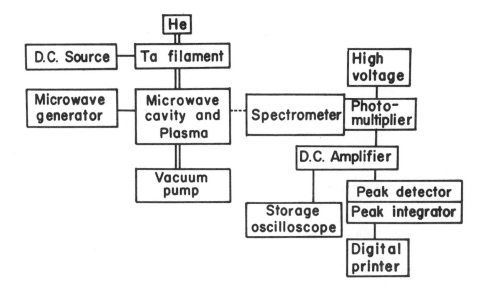

Fig. 12. Block diagram of microwave excitation spectrometer (67).

microgram quantities of enzyme *(52,53,67)*. The enzyme employed exhibited activities comparable to those of the most highly purified preparations then reported.

Using optimal conditions for the assay of the enzyme from the avian myeloblastosis virus *(39)*, the K_m for TTP is 1×10^{-5} M and the V_{max} is 12 pmoles TMP incorporated per minute per µg of protein at 25°.

Metal complexing agents inhibit the AMV polymerase catalyzed reaction. In their absence, the enzyme is completely stable and active for 60 minutes at 25° while in the presence of 1,10-phenanthroline, 1 mM, the enzyme is inhibited instantaneously and reversibly with a K_I of 7×10^{-5} M *(53)*.

The metal complexing properties of 1,10-phenanthroline account for the inhibition: its isomers 1,7- and 4,7-phenanthroline neither bind metals nor do they inhibit the nucleotide polymerization reaction under conditions where 1,10-phenanthroline inhibits it completely. Moreover, a number of other, structurally different metal binding agents, *e.g.*, EDTA and 8-hydroxyquinoline-5-sulfonate also markedly inhibit the polymerase activity instantaneously *(52,53)*. The presence of stoichiometric quantities of metal is, of course, essential to verify that these agents exert their effect by binding to a functional and/or structurally essential metal.

TABLE 7: COMPARISON OF THE CAPABILITIES OF ATOMIC ABSORPTION AND MICROWAVE EMISSION SPECTROMETRY		
	Atomic Absorption	Microwave Emission
Detection Limit (g-atom)	10^{-10}	10^{-15}
Protein (μg)	100	0.1
Volume of Sample (ml)	2	0.005
Effective Range	10^1	10^5

The experimental data pertinent to the characteristics of microwave emission spectrometry are from Ref. 67.

Conventional procedures have not been sufficiently sensitive to allow quantitative metal determinations on very small amounts of enzyme. Table 7 compares the spectral capabilities of atomic absorption, the hitherto conventional procedure, with those of microwave excitation emission spectrometry, the method devised by us. Microwave excitation emission spectrometry increases the detection limits by 5 orders of magnitude and metal analysis can be performed on 0.1 μg protein contained in a 5 μl volume of sample. Thus, precise quantitative metal analyses on the microgram amounts of enzyme available to us for this purpose was feasible by this means. The presence of Zn, Cu or Fe could account for the observed inhibition of the AMV polymerase and these elements and Mn were determined after removal of metal quenching agents and of low molecular weight protein contaminants by gel exclusion chromatography. The elements and protein were measured quantitatively with high precision in 45 μl fractions containing maximal activity when absolute amounts of metal varied from 10^{-11} to 10^{-14} g-atom. The purified enzyme contains zinc in stoichiometric quantities, but Cu, Fe and Mn are virtually absent (Table 8). The zinc content corresponds to from 1.7 to 1.9 g-atom of zinc per mole of enzyme for molecular weights of either 1.6 to 1.8 x 10^5.

Using the same procedures the murine, simian, feline and RD-114 Type C RNA tumor viruses have been shown to be zinc enzymes also. All are inhibited by 1,10-phenanthroline (Table 9) and contain stoichiometric quantities of zinc, assuming specific activities and molecular weights, respectively similar to that of the enzyme from avian myeloblastosis virus (54).

These data extend the role of zinc in enzymes essential to normal nucleic acid metabolism to others presumed to play a role in a leukemic process, confirming a hypothesis of long standing (100,101). In turn, this led us to examine the role of zinc in

TABLE 8: ANALYSIS OF GEL EXCLUSION CHROMATOGRAPHY FRACTIONS (45 µl) EXHIBITING MAXIMAL ACTIVITY*						
Fraction No.	Zn g atom x 10^{11}	Protein µg	Zn/Protein g-atom	Cu	Fe g-atom	Mn
18	0.7	0.7	1.8	< .001	< .01	.06
19	1.5	1.4	1.9	< .001	< .01	.05
20	1.2	1.1	1.9	< .001	< .01	.05

*Calculations are based on a molecular weight of this polymerase of 1.8 × 10^5. Zinc and protein content are expressed as g-atom or µg per 5 µl aliquot, respectively. Analyses were performed in triplicate.

nucleic acid metabolism of leukemic leukocytes.

8. 1,10-PHENANTHROLINE DEPENDENT GROWTH INHIBITION OF HUMAN LYMPHOBLASTS

We have now examined the DNA metabolism of CCRF-CEM lympho-blasts* grown in cell culture, employing instrumentation and pro-cedures described above in regard to the flow cytofluorometric examination of zinc sufficient and deficient *E. gracilis* (Fig. 3). Hydroxyurea results in a block of division of these cells, but their subsequent incubation in fresh media releases it and syn-chronizes their division with a subsequent 100% increase in the number of cells in 12 hours. By releasing the block at discrete

TABLE 9: REVERSE TRANSCRIPTASES: Zn CONTENT AND OP* INHIBITION		
Species	Zn g-atom/mole	OP pK_I
Avian	2	4.2
Murine	1	4.9
Simian	1	4.9
ΣCu, Fe, Mn: 0.06 g-atom/mole		

*OP, 1,10-phenanthroline

*This human lymphoid cell line was established by cultivating cells from the blood of a leukemic pediatric patient.

time intervals the DNA metabolism of synchronized cells can be
followed by means of flow cytofluorimetry performed at specific
stages of the cell cycle. These are identified and defined by
the extent to which ^{3}H thymidine is incorporated and by the
resultant labeling and mitotic indices. In particular, the DNA
content of lymphoblasts in G_2 *(104)* can be ascertained by blocking with podophyllotoxin. Lymphoblasts have not been grown under
zinc deficient conditions thus far, but exposure to 1,10-phenan-
throline can serve.to mimic this condition. This agent forms
stable zinc complexes, both when zinc is free in solution and
when it is firmly incorporated into proteins and enzymes. As
indicated earlier, 1,10-phenanthroline inhibits zinc dependent
reactions and the consequences can closely resemble those of
zinc deprivation, though this agent interacts with other metals
of the first transition group as well, of course.

In normal media, the number of lymphoblasts double within 24
hours but 1,10-phenanthroline, 4 μM, completely arrests their
proliferation and higher concentrations are cytotoxic, actually
decreasing their number by 20% compared with the control. This
contrasts markedly with the consequences of exposure to the non-
chelating isomer, 1,7-phenanthroline, which fails to affect proliferation at any concentration employed, suggesting that 1,10-
phenanthroline exerts its action by chelating a metal essential
to cell division. Both dilution and the addition to the medium
of Zn^{2+}, Cu^{2+} or Fe^{2+} results in the resumption of proliferation
in a manner reminiscent of the reversal of 1,10-phenanthroline
inhibition of specific enzymes *(20)*.

The synchronization of CCRF-CEM lymphoblasts, either blocked
with hydroxyurea or with podophyllotoxin, and their subsequent
release from the block at specified time intervals synchronizes
cells at specific stages of the cell cycle. Subsequent proliferation in the presence of 1,10-phenanthroline can then be correlated to the distribution and content of the DNA of the resultant
cell population and this permits localization of its effects to a
particular stage of cell division. 1,10-Phenanthroline inhibits
cell division of lymphoblasts by blocking their transition from
the G_1 into the S phase. Further, lymphoblasts which are synthesizing DNA in the S phase, are blocked and do not progress into
G_2 *(105,106)*.

Inhibition of ^{3}H thymidine incorporation suggests that the
failure to progress from G_1 into S must involve processes essential for the synthesis of DNA itself. Such inhibition of ^{3}H thymidine incorporation has also been observed with phytohemagglu-
tinin stimulated peripheral lymphocytes incubated with 1,10-phen-
anthroline *(60)*, chicken embryo *(59)* and kidney cortex cells
incubated with EDTA, all suggesting that - both in mammalian and

nonmammalian cells - metal-dependent processes are critical to the transition from G_1 to S and to the progression through S itself. The metabolic properties of both phases are characteristic. In G_1 the activity of enzymes and uptake of substrates essential for synthesis, *e.g.*, amino acids, nucleotides, metals, etc., all increase. DNA synthesis occurs in S. Thus, in these two stages, cells synthesize DNA, RNA and proteins, essential for cell division, and the 1,10-phenanthroline data indicate metal-dependent processes to be most critical *(105,106)*.

Such conclusions are consistent with earlier studies both of histochemical localization of zinc in a number of systems and the consequences of its deficiency. It is essential to the incorporation of labeled DNA and/or RNA precursors into rat embryos *(107)* and livers of zinc deficient rats *(58)*. It has been localized to the nucleoli of sea urchin eggs and, transiently, to the mitotic apparatus of dividing cells *(108)*, suggesting a role in mitosis. Finally, 1,10-phenanthroline, <4 µM, completely *arrests* proliferation both of normal and leukemic CCRF-CEM lymphoblasts but, strikingly, >4 M *destroys* only the leukemic CCRF-CEM cells. The differentiation of these two types of cells, based on their response to 1,10-phenanthroline, may perhaps relate to the known difference of their zinc contents and could suggest novel therapeutic approaches to the disease.

While the present studies do not permit an allocation of these effects either to specific zinc systems or, in fact, to specific biochemical processes involved in the cell cycle, they clearly do indicate that derangements of metal metabolism, likely zinc metabolism, exert their effect in G_1 and S, premitotic phases of the cell cycle of leukemic cells.

9. EPILOG

Many functions of zinc in biology resemble those characteristic of the vitamins: zinc seems to serve as a coenzyme. Both through advanced methods of analysis and progress in isolation and characterizing metalloenzymes decisive nutritional and metabolic studies are now possible. These advances have been potentiated by the advent of cytofluorometric procedures and methods to culture normal and leukemic cells from human blood or bone marrow. Further, populations of cells can be obtained at different stages of maturation, and DNA content can be quantitated *in situ* rapidly and accurately. Hence the cell cycle can be studied as a function of time, cell and metal metabolism *(67)*. Jointly these advances provide the means for studies that can determine the role of zinc in the proliferation and maturation of both normal and malignant cells and to investigate the thesis that derangements in its metabolism affect the pathogenesis of neoplasia.

ACKNOWLEDGEMENT.   This work was supported by Grant-in-Aid GM-15003 from the National Institutes of Health of the Department of Health, Education and Welfare.

REFERENCES

1. J. Raulin, *Ann. Sci. Natl. Botan. et Biol. Vegetale*,**11**, 93 (1869).
2. G. Lechartier and F. Bellamy, *Compt. Rend. Acad. Sci.*,**84**, 687 (1877).
3. F. Raoult and H. Breton, *Compt. Rend. Acad. Sci.*,**85**, 40 (1877).
4. G. Bertrand and R.C. Bhattacherjee, *Compt. Rend. Acad. Sci.*, **198**, 1823 (1934).
5. W.R. Todd, C.A. Elvehjem and E.G. Hart, *Amer. J. Physiol.*, **107**, 146 (1934).
6. B.L. Vallee, *Physiol. Rev.*,**39**, 443 (1959).
7. B.L. Vallee, *J. Chronic Diseases*,**9**, 74 (1959).
8. T.-K. Li and B.L. Vallee, in "Modern Nutrition in Health and Disease", R.S. Goodhart and M.E. Shils, Eds., 5th ed., Lea & Febiger, Philadelphia, p. 372 (1973).
9. E.J. Underwood, "Trace Elements in Human and Animal Nutrition", 3rd ed., Academic Press, New York (1971).
10. A.S. Prasad, (Ed.), "Zinc Metabolism", C.C. Thomas, Springfield, Illinois (1966).
11. W.J. Pories and W.H. Strain (Eds.), "Clinical Applications of Zinc Metabolism", C.C. Thomas, Springfield, Illinois (1974).
12. R.E. Burch, H.K.J. Hahn and J.R. Sullivan, *Clin. Chem.*,**21**, 501 (1975).
13. J.A. Halsted, J.C. Smith, Jr. and M.I. Irwin, *J. Nutr.*,**104**, 347 (1974).
14. E.J. Moynahan, *Lancet*,**2**, 399 (1974).
15. K.H. Neldner and K.M. Hambidge, *New Engl. J. Med.*,**292**, 879 (1975).
16. H.F. Tucker and W.D. Salmon, *Proc. Soc. Exp. Biol. Med.*,**88**, 613 (1955).
17. L.S. Hurley, *Am. J. Nutr.*,**22**, 1332 (1969).
18. B.L. Vallee, W.E.C. Wacker, A.F. Bartholomay and E.D. Robin, *New Engl. J. Med.*,**255**, 403 (1956).
19. L.S. Hurley and R.E. Shrader, in "Neurobiology of the Trace Metals Zinc and Copper", C.C. Pfeiffer, Ed., Academic Press, New York, p. 7 (1972).

20. B.L. Vallee and W.E.C. Wacker, in "The Proteins", H. Neurath, Ed., Vol. 5., 2nd ed., Academic Press, New York (1970).
21. B.L. Vallee, *Advances in Protein Chemistry*, **10**, 317 (1955).
22. W.T. Preyer, "De Haemoglobino Observationes et Experimenta", M. Cohen and Sons, Bonn, p. 27 (1866).
23. E. Harless, *Müller's Arch, f. Anat., Physiol.*, **148** (1847).
24. D. Keilin and T. Mann, *Biochem. J.*, **34**, 1163 (1940).
25. B.L. Vallee and H. Neurath, *J. Biol. Chem.*, **217**, 253 (1955).
26. B.L. Vallee and W.E.C. Wacker, in "CRC Handbook of Biochemistry and Molecular Biology", G. Fasman, Ed., Vol. 2, 3rd edition, p. 276 (1976).
27. W.E.C. Wacker and B.L. Vallee, *J. Biol. Chem.*, **234**, 3257 (1959).
28. Y.A. Shin and G.L. Eichhorn, *J. Am. Chem. Soc.*, **90**, 7323 (1968).
29. K. Fuwa, W.E.C. Wacker, R. Druyan, A.F. Bartholomay and B.L. Vallee, *Proc. Nat. Acad. Sci. U.S.A.*, **46**, 1298 (1960).
30. B.L. Vallee and R.J.P. Williams, *Proc. Nat. Acad. Sci. U.S.A.* **59**, 498 (1968).
31. M.I. Harris and J.E. Coleman, *J. Biol. Chem.*, **243**, 5063 (1968).
32. B.L. Vallee, in "Metal Ions in Biological Systems", S.K. Dhar, Ed., Vol. 40, Plenum Press, New York, p. 1 (1973).
33. B.L. Vallee and S.A. Latt, in "Structure-Function Relationships of Proteolytic Enzymes", P. Desnuelle, H. Neurath and M. Ottesen, Eds., Academic Press, New York, p. 144 (1970).
34. S. Lindskog and P.O. Nyman, *Biochim. Biophys. Acta*, **85**, 141 (1964).
35. B. Holmquist, T.A. Kaden and B.L. Vallee, *Biochemistry*, **14**, 1454 (1975).
36. R.T. Simpson and B.L. Vallee, *Biochemistry*, **7**, 4343 (1968).
37. T. Herskovitz, B.A. Averill, R.H. Holm, J.A. Ibers, W.D. Phillips and J.F. Weiher, *Proc. Nat. Acad. Sci. U.S.A.*, **69**, 2437 (1972).
38. S.A. Latt and B.L. Vallee, *Biochemistry*, **10**, 4263 (1971).
39. T.A. Kaden, B. Holmquist and B.L. Vallee, *Inorg. Chem.*, **13**, 2585 (1974).
40. F.S. Kennedy, H.A.O. Hill, T.A. Kaden and B.L. Vallee, *Biochem. Biophys. Res. Commun.*, **48**, 1533 (1972).
41. J.F. Riordan and B.L. Vallee, *Advan. Exp. Med. Biol.*, **48**, 33 (1974).
42. F.A. Quiocho and W.N. Lipscomb, *Advan. Protein Chem.*, **25**, 1 (1971).
43. P.M. Colman, J.N. Jansonius and B.W. Matthews, *J. Mol. Biol.*, **70**, 701 (1972).
44. C.-I. Brandén, H. Jörnvall, H. Eklund and B. Furugren, in "The Enzymes", P.D. Boyer, Ed., Academic Press, New York, p. 104 (1975).

45. C.A. Price and B.L. Vallee, *Plant Physiol.*,**37**, 428 (1962).
46. K.H. Falchuk and D.W. Fawcett, *Federation Proc.*,**33**, 1475 (1974).
47. K.H. Falchuk, D.W. Fawcett and B.L. Vallee, *J. Cell. Sci.*, **17**, 57 (1975).
48. K.H. Falchuk, A. Krishan and B.L. Vallee, *Biochemistry*,**14**, 3439 (1975).
49. W.E.C. Wacker, *Biochemistry*,**1**, 859 (1962).
50. C.A. Price, in "Zinc Metabolism", A.S. Prasad, Ed., C.C. Thomas, Springfield, Illinois, p. 69 (1966).
51. W.E.C. Wacker, W. Kornicker and L. Pothier, Abstract, Amer. Chem. Soc. 150th Meeting, 88C (1965).
52. D.S. Auld, H. Kawaguchi, D. Livingston and B.L. Vallee, *Biochem. Biophys. Res. Commun.*,**57**, 967 (1974).
53. D.S. Auld, H. Kawaguchi, D. Livingston and B.L. Vallee, *Proc. Nat. Acad. Sci. U.S.A.*,**71**, 2091 (1974).
54. D.S. Auld, H. Kawaguchi, D.M. Livingston and B.L. Vallee, *Biochem. Biophys. Res. Commun.*,**62**, 296 (1975).
55. M.C. Scrutton, C.W. Wu and D.A. Goldthwait, *Proc. Nat. Acad. Sci. U.S.A.*,**68**, 2497 (1971).
56. J.P. Slater, A.S. Mildvan and L.A. Loeb, *Biochem. Biophys. Res. Commun.*,**44**, 37 (1971).
57. A. Howard and S.R. Pelc, Symposium on Chromosome Breakage (supplement to Heredity, Vol. 6) (1953).
58. H.H. Sandstead and R.A. Rinaldi, *J. Cell. Physiol.*,**73**, 81 (1969).
59. H. Rubin, *Proc. Nat. Acad. Sci. U.S.A.*,**69**, 712 (1972).
60. R.O. Williams and L. Loeb, *J. Cell. Biol.*,**58**, 594 (1973).
61. J.A. Prask and D.J. Plocke, *Plant Physiol.*,**48**, 150 (1971).
62. R.A. Tobey and H.A. Crissman, *Can. Res.*,**32**, 2726 (1972).
63. S. Kotsiopoulos and S.C. Mohr, *Biochem. Biophys. Res. Commun.* **67**, 979 (1975).
64. B.L. Vallee and J.G. Gibson, *J. Biol. Chem.*,**176**, 435 (1948).
65. T. Fujii, *Nature*,**174**, 1108 (1954).
66. Cold Spring Harbor Symposia on Quantitative Biology, Vol. 38 (1973).
67. H. Kawaguchi and B.L. Vallee, *Anal. Chem.*,**4** , 1029 (1975).
68. K.H. Falchuk, B. Mazus, D. Ulpino and B.L. Vallee, *Biochemistry*,**15**, 4468 (1976).
69. R.G. Roeder and W.J. Rutter, *Nature*,**224**, 234 (1969).
70. P. Chambon, in "The Enzymes", P.D. Boyer, Ed., Vol. 10, Academic Press, New York, p. 261 (1974).
71. D.S. Auld, I. Atsuya, C. Campino and P. Valenzuela, *Biochem. Biophys. Res. Commun.*,**69**, 548 (1976).
72. A.G. McLennan and H.M. Keir, *Biochem. J.*,**151**, 227 (1975).
73. A.G. McLennan and H.M. Keir, *Biochem. J.*,**151**, 239 (1975).

74. J.P. Rosenbusch and K. Weber, *Proc. Nat. Acad. Sci. U.S.A.*, **68**, 68 (1971).

75. E.R. Stadtman, B.M. Shapiro, A. Ginsburg, D.S. Kingdon and M.D. Denton, *Brookhaven Symp. Biol.*, **21**, 378 (1968).

76. R.A. Anderson, W.F. Bosron, F.S. Kennedy and B.L. Vallee, *Proc. Nat. Acad. Sci. U.S.A.*, **72**, 2989 (1975).

77. M. Nemer and A.A. Infante, *Science*, **150**, 217 (1965).

78. C.P. Emerson and T. Humphreys, *Develop. Biol.*, **23**, 85 (1970).

79. R.G. Roeder and W.J. Rutter, *Biochemistry*, **9**, 2543 (1970).

80. S.P. Blatti, C.J. Ingles, T.J. Lindell, P.W. Morris, R.F. Weaver, F. Weinberg and W.J. Rutter, *Cold Spring Harbor Symposium Quant. Biol.*, **35**, 649 (1970).

81. R.S.S. Fraser, *Eur. J. Biochem.*, **50**, 529 (1975).

82. T. Yagura, M. Yanagisawa and M. Iwabuchi, *Biochem. Biophys. Res. Commun.*, **68**, 183 (1976).

83. A.O. Pogo, V.C. Littau, V.G. Allfrey and A.E. Mirsky, *Proc. Nat. Acad. Sci. U.S.A.*, **57**, 743 (1967).

84. S. Ochoa and R. Mazunder, in "The Enzymes", P.D. Boyer, Ed., Vol. 10, Academic Press, New York, p. 1 (1974).

85. J. Lucas-Lenard and L. Beres, in "The Enzymes", P.D. Boyer, Ed., Vol. 10, Academic Press, New York, p. 53 (1974).

86. W.P. Tate and C.T. Caskey, in "The Enzymes", P.D. Boyer, Ed., Vol. 10, Academic Press, New York, p. 87 (1974).

87. C.F. Springgate, A.S. Mildvan, R. Abramsom, J.L. Engle and L.A. Loeb, *J. Biol. Chem.*, **248**, 5987 (1973).

88. E.J. Coleman, *Biochem. Biophys. Res. Commun.*, **60**, 641 (1974).

89. D.S. Auld, H. Kawaguchi, D. M. Livingston and B.L. Vallee, *Federation Proc.*, **33**, 1483 (1974).

90. C. Delezenne, *Ann. Inst. Pasteur*, **33**, 68 (1919).

91. P. Cristol, "Contribution a l'étude de la physio-pathologie du zinc et en particulier de sa signification dans les tumeurs", Thése Montpellier (1922).

92. P. Cristol, *Bull. Soc. Chim. Biol.*, **5**, 23 (1923).

93. H. Labbé and P. Nepveux, *Le Progrès Med.*, 577 (1927)

94. I. Michailowsky, *Virchow's Archiv. Pathol. Anat.*, **267**, 27 (1928).

95. L.I. Falin and K.E. Gromzewa, *Amer. J. Cancer*, **36**, 233 (1939).

96. J. Guthrie, *Brit. J. Cancer*, **18**, 130 (1964).

97. B.L. Vallee and J.G. Gibson, 2nd, *J. Biol. Chem.*, **176**, 445 (1948).

98. F.L. Hoch and B.L. Vallee, *J. Biol. Chem.*, **195**, 531 (1952).

99. B.L. Vallee, F.L. Hoch and W.L. Hughes, Jr., *Arch. Biochem. Biophys.*, **48**, 347 (1954).

100. J.G. Gibson, 2nd, B.L. Vallee, R.G. Fluharty and J.E. Nelson, *Acta contre Cancer*, **6**, 102 (1950).

101. B.L. Vallee, R.G. Fluharty and J.G. Gibson, 2nd, *Acta Union Intern. Contre Cancer,***6,** 869 (1949).
102. H.M. Temin and S. Mizutani, *Nature,***226,** 1211 (1970).
103. D. Baltimore, *Nature,***22,** 1209 (1970).
104. A. Krishan, *J. Cell. Biol.,***66,** 521 (1975).
105. K.H. Falchuk and A. Krishan, *Federation Proc.,***34,** 530 (1975).
106. K.H. Falchuk and A. Krishan, in preparation.
107. H. Swenerton, R. Shader and L.S. Hurley, *Science,***166,** 1014 (1969).
108. T. Fujii, *Nature,***174,** 1108 (1954).
109. H. Kawaguchi and D.S. Auld, *Clin. Chem.,***21,** 591 (1975).

identification of active sites in iron-sulfur proteins

R.H. HOLM

Department of Chemistry, Stanford University, Stanford, California   94305

(cont'd)

1. INTRODUCTION

Since the initial isolation of purified ferredoxin proteins from *Clostridium pasteurianum* and spinach in 1962 by Mortenson, Valentine, and Carnahan *(1)* and Tagawa and Arnon *(2)*, respectively, nonheme iron-sulfur proteins have passed from obscurity to scientific prosperity in little over a decade. During this period these proteins have been the objects of intensive and extensive biological and physicochemical studies and now rank with myoglobin, hemoglobin, certain cytochromes (especially cytochrome c), and vitamin B_{12} coenzymes in terms of detailed knowledge of the structural and electronic properties of the metal-containing active sites. The field of iron-sulfur proteins has been reviewed in depth with regard to virtually all aspects of these proteins. General reviews are available *(3-12)* and are augmented by others dealing more specifically with biological function *(13, 14)*, protein conformations *(15)*, active site structural and electronic features *(16-19)*, and evolutionary aspects with emphasis on amino acid sequences *(20-24)*. Especially comprehensive coverage is afforded by the three volumes edited by Lovenberg *(25)*.

Iron-sulfur proteins are widely dispersed in nature, having been found in anaerobic, aerobic, and photosynthetic bacteria, algae, fungi, higher plants, and mammals. Indeed, it would appear that they occur in all living organisms. Their physiological function, where known, is that of electron transfer rather than as actual catalysts of chemical transformations. Thus these proteins are the most ubiquitous electron carriers in biology and have been directly associated with, or implicated in, a myriad of metabolic reactions. These include mitochondrial oxidation of $NADH_2$, oxidation of succinate and xanthine, hydroxylation of C—H bonds, the phosphoroclastic reaction, CO_2 and N_2 fixation, sulfite and nitrite reduction, and photosynthetic electron transfer in bacteria and plants, among many others. In view of their central role in cell metabolism, the presence of spectroscopically responsive iron atoms, and their relatively low molecular weight (*ca.* 6,000-20,000 daltons), it is hardly surprising that biologists, chemists, and physicists have found them fascinating molecules for investigation.

An adequate understanding of the biological function and associated physicochemical properties of any protein cannot be achieved without a detailed knowledge of protein structure at all levels and the structure of the active site. Other than amino acid sequences requisite structural details, at least in the solid state, must derive from X-ray diffraction. Fortunately in the case of iron-sulfur proteins rather detailed structures are available for several prototype molecules. These data, together with more precise structural determinations of low molecular weight complexes which serve as synthetic analogues of the Fe-S active sites *(26-27)*, have provided a high degree of structural definition of these sites. Correlation of spectroscopic and magnetic data with active site and analogue structures has provided a fairly secure means of identifying types of sites in low molecular weight proteins. However, in many higher molecular weight Fe-S enzymes, all of which except hydrogenase contain other metals and/or prosthetic groups, spectral properties are more complicated or less easily elicited experimentally and do not allow a definite correlation with active site structure. The purpose of this article is to present and evaluate structure-physical property correlations, and to introduce a new method of active site identification, the core extrusion reaction, which in its present form requires a correlation only between absorption spectra and structures of synthetic analogues. Attention is restricted to the properties of Fe-S sites, the loci of redox behavior; such properties are intimately related to the composition [Fe, sulfide (S*), and/or cysteinate (S-Cys)] and stereochemistry of the site but convey no direct information about the remainder of the protein structure.

2. PROTEIN AND SYNTHETIC ANALOGUE STRUCTURES

2.1 PROTEIN STRUCTURES - In the lower molecular weight iron-sulfur redox proteins there are at present three recognized types of active sites, which in terms of minimal composition are:

$[Fe(S-Cys)_4]$ *(1)*, $[Fe_4S_4^*(S-Cys)_4]$ *(2)*, and $[Fe_2S_2^*(S-Cys)_4]$ *(3)*. Structure **1** has been established by X-ray diffraction studies of oxidized rubredoxin (Rd_{ox}, 6,000 daltons) from

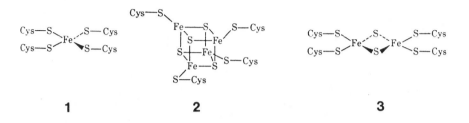

1   **2**   **3**

C. pasteurianum (18,28) and contains a Fe(III)-S_4 coordination unit distorted from tetrahedral symmetry. By similar means the cubane structure **2** of D_{2d} or lower symmetry has been demonstrated in the oxidized and reduced 4-Fe "high-potential" protein ($HP_{ox,red}$, 9,650 daltons) from *Chromatium (29,30)* and the oxidized 8-Fe ferredoxin (Fd_{ox}, 6,000 daltons) from *Peptococcus aerogenes (18,31)*. The binuclear structure *(3)* has not as yet been proven by X-ray methods but is fully consistent with a large body of physiochemical data for oxidized and reduced 2-Fe ferredoxins *(19)*. A substantial number of Rd, Fd, and HP proteins have been sequenced *(20-24,32)* and in each type of active site structure cysteinate sulfur functions as the terminal ligand. Metal bridging is by sulfide such that sites **2** and **3** contain substructural $Fe_4S_4^*$ and $Fe_2S_2^*$ *cores*, the existence and stability of which, as will be seen, lead to biologically significant reactivity properties of these sites. In each type of site iron atoms are coordinated in an approximately tetrahedral arrangement.

2.2 SYNTHETIC ANALOGUE STRUCTURES - Examination of sites **1-3** and the full structural features of Rd, Fd, and HP proteins *(18, 28-31)* reveal that these molecules are fundamentally metal complexes, albeit with elaborate ligand structures. The elegant structural simplicity of these sites together with the substitutional lability of Fe(II,III) suggested to the writer and his co-workers that analogues of **1 - 3** might be inherently stable outside of a protein environment and attainable as thermodynamically favored, soluble products in reaction systems containing Fe(II) or Fe(III), sulfide, and/or thiolate. This has proven to be the case and is the basis of the synthetic analogue approach *(26,27)* to the study of the active sites of iron-sulfur proteins. The syntheses and structural, electronic, and reactivity properties of complexes of the general types $[Fe(SR)_4]^-$ *(33,34)*, $[Fe(SR)_4]^{2-}$

(34), $[Fe_2S_2(SR)_4]^{2-}$ *(35-37)*, $[Fe_4S_4(SR)_4]^{2-}$ *(38-49)*, and
$[Fe_4S_4(SR)_4]^{3-}$ *(50,51)* have been reported. In most of these
complexes simple organic thiolate ligands simulate cysteinate
binding; a number of tetranuclear complexes have also been obtained in which cysteinate derivatives of glycylcysteinyl oligo-
peptides are the terminal ligands *(42,45,49)*. In addition, other
investigators have provided reduced 1-Fe complexes in the form of
$[Fe(SPh)_4]^{2-}$ *(52)* and $[Fe(12\text{-peptide})]^{2-}$ *(53)* [12-peptide = Boc-
Gly-(Cys-Gly-Gly)$_3$-Cys-Gly-NH$_2$)], and the water soluble complex
$[Fe_4S_4(SCH_2CH_2CO_2)_4]^{6-}$ *(54,55)*, isoelectronic with tetranuclear
dianions derived from mononegative thiolates.

Structures of synthetic analogues determined by X-ray diffraction are collected in Fig. 1. The resemblance between
$[Fe(S_2\text{-}o\text{-xyl})_2]^{-,2-}$ (S_2-o-xyl = o-xylyl-α,α'-dithiolate),
$[Fe_4S_4(SR)_4]^{2-}$ (R = Ph, CH$_2$Ph), and $[Fe_2S_2(S_2\text{-}o\text{-xyl})_2]^{2-}$ - $[Fe_2S_2$
(S-p-tolyl)$_4]^{2-}$ and the stereochemistry of sites **1**, **2**, and **3**,
respectively, is apparent. Detailed structural comparisons between 1-Fe and 4-Fe analogues and proteins are available elsewhere *(27)* and leave no doubt that the 8-Fe Fd$_{ox}$ and HP$_{red}$ sites
are essentially congruent geometrically with $[Fe_4S_4(SR)_4]^{2-}$,
especially as regards Fe$_4$S$_4^*$ core structures. While the overall
structures of $[Fe(S_2\text{-}o\text{-xyl})_2]^-$ and the Rd$_{ox}$ site are clearly
related in terms of mean values of Fe-S bond distances and S-Fe-S
bond angles, a similarly high degree of structural correspondence
cannot yet be claimed here. The protein site, at the present
stage of resolution *(18,28)*, is more distorted from idealized T$_d$
symmetry, and is reported to contain an unprecedentedly short
Fe(III)-S bond (2.05 Å), absent in the analogue. Further comparisons must await final refinement of the X-ray structure. Preliminary analyses of extended X-ray absorption fine structure
(EXAFS) spectra from synchrotron radiation experiments with two
Rd$_{ox}$ proteins *(56)* suggest that the Fe-S distances may be more
nearly equal than are the latest published values obtained by
X-ray diffraction. Because of the close resemblance of analogue-
protein spectroscopic and magnetic properties *(vide infra)*,
$[Fe_2S_2(S_2\text{-}o\text{-xyl})_2]^{2-}$ is considered as an adequate minimal representation of the structure of 2-Fe Fd$_{ox}$ sites **3**. In this case an
analogue serves as a confirmatory model for a portion of a protein structure as yet not directly established by X-ray diffraction.

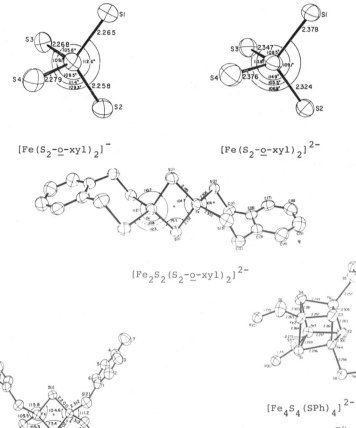

$[Fe(S_2-\underline{o}-xyl)_2]^-$

$[Fe(S_2-\underline{o}-xyl)_2]^{2-}$

$[Fe_2S_2(S_2-\underline{o}-xyl)_2]^{2-}$

$[Fe_2S_2(S-\underline{p}-tol)_4]^{2-}$

$[Fe_4S_4(SPh)_4]^{2-}$

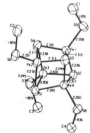

$[Fe_4S_4(SCH_2Ph)_4]^{2-}$

Fig. 1. Structures of the synthetic analogues $[Fe(S_2-o-xyl)_2]^{-,2-}$ (33,34), $[Fe_2S_2-(S_2-o-xyl)_2]^{2-}$ (36), $[Fe_2S_2(S-p-tol)_4]^{2-}$ (36), $[Fe_4S_4(SPh)_4]^{2-}$ (44), and $[Fe_4S_4-(SCH_2Ph)_4]^{2-}$ (39). Portions of the ligands are omitted in the structures of the 1-Fe and 4-Fe complexes.

76

2.3 PROTEIN-ANALOGUE OXIDATION LEVEL EQUIVALENCIES - A requisite
property of any synthetic complex purported to be an analogue of
an active site of iron-sulfur proteins is a redox capacity which
encompasses all known oxidation levels of the corresponding protein site. Analogue oxidation levels are defined by their net
charge, and physicochemical correspondences between analogues and
proteins in fixed oxidation levels unambiguously specify the
redox state of the latter in terms of that of the former. The
analogue electron transfer series of Equations 1 - 3 have been
established by polarographic techniques in nonaqueous and aqueous-
nonaqueous media *(33,34,36,42,49)*. Isoelectronic analogue-protein
relationships indicated in columns (a), (b), and (c) include all
physiologically significant protein levels as well as the (apparently) non-physiological levels in brackets. The latter include
the "super-reduced" form of *Chromatium* HP *(57)* and the "super-
oxidized" forms claimed for *C. acidi-urici* 8-Fe Fd *(58)* and *Azo-
tobacter vinelandii* 8-Fe Fd I *(59)*. The scope of each series can
be conceptualized in terms of terminal reduced and oxidized members containing only Fe(II) and Fe(III), respectively. In this
sense only series (3) is incomplete, lacking $[Fe_4S_4(SR)_4]^0$.

$$[Fe(SR)_4]^{2-} \rightleftharpoons [Fe(SR)_4]^-$$

$$Rd_{red} \rightleftharpoons Rd_{ox} \qquad (1)$$

$$(a) \qquad\qquad (b)$$

$$[Fe_2S_2(SR)_4]^{4-} \rightleftharpoons [Fe_2S_2(SR)_4]^{3-} \rightleftharpoons [Fe_2S_2(SR)_4]^{2-}$$

$$[Fd_{s-red}] \qquad Fd_{red} \rightleftharpoons Fd_{ox}$$

$$2Fe(II) \qquad Fe(II) + Fe(III) \qquad 2Fe(III) \qquad (2)$$

$$(a) \qquad\qquad (b) \qquad\qquad (c)$$

$$[Fe_4S_4(SR)_4]^{4-} \rightleftharpoons [Fe_4S_4(SR)_4]^{3-} \rightleftharpoons [Fe_4S_4(SR)_4]^{2-} \rightleftharpoons [Fe_4S_4(SR)_4]^-$$

$$Fd_{red} \rightleftharpoons Fd_{ox} \longleftrightarrow [Fd_{s-ox}]$$

$$[HP_{s-red}] \longleftrightarrow HP_{red} \rightleftharpoons HP_{ox} \quad (3)$$

$$4Fe(II) \qquad 3Fe(II)+Fe(III) \qquad 2Fe(II)+2Fe(III) \qquad Fe(II)+Fe(III)$$

$$(a) \qquad\qquad (b) \qquad\qquad (c)$$

In view of the apparent instability of the monoanion it is unlikely that the neutral tetramer will be isolated. Of analogues
possessing physiologically significant oxidation levels, $[Fe_2S_2-(SR)_4]^{3-}$ and $[Fe_4S_4(SR)_4]^-$ have not been isolated in substance
and, on the basis of electrochemical and other observations,
appear to have limited stabilities. No attempt has yet been made
to isolate the nonphysiological analogues $[Fe_2S_2(SR)_4]^{4-}$ and
$[Fe_4S_4(SR)_4]^{4-}$. Lastly, it is noted that polarographic half-wave
potentials for analogue redox couples are always more cathodic
than potentials (E_0', E_m) for isoelectronic protein couples in
purely aqueous solution. Recent polarographic investigations of
C. pasteurianum $Fd_{ox,red}$ and $[Fe_4S_4(SR)_4]^{2-,3-}$ potentials (R =
CH_2CH_2OH, Cys(Ac)NHMe) under nominally identical conditions in
$DMSO/H_2O$ solutions containing 0-80% v/v DMSO reveal that potential differences do not exceed 0.19 V and are usually much less
(49). Thus for at least tetranuclear complexes the analogue
designation can be extended with more legitimacy than previously
to cover both scope of oxidation levels and the potentials by
which they are connected.

3. PROTEIN STRUCTURE-PROPERTY RELATIONSHIPS

Iron-sulfur proteins are molecules remarkably responsive to
spectroscopic and magnetic examination, because of the following
features which allow observation of the indicated properties:
(a) molecular chirality and enantiomeric purity - ORD, CD
    spectra;
(b) uv-visible absorption - charge transfer spectra, reso-
    nance Raman spectra of Fe-S(S*) vibrations, ligand field
    spectra (Rd_{red} in near-infrared), MCD spectra;
(c) ^{57}Fe nuclei - Mössbauer spectra;
(d) paramagnetic ground and/or excited states - magnetic
    susceptibilities and antiferromagnetic interactions; ESR,
    ENDOR, and isotropically shifted NMR spectra with hyper-
    fine interactions; magnetically perturbed Mössbauer
    spectra;
(e) electron transfer - redox potentials.
All or nearly all of these properties have been determined for
the lower molecular weight redox proteins in their physiological
oxidation levels. The more commonly measured properties of proteins containing 1-, 2-, and 4-Fe sites are assembled in Tables
1 and 2 *(7-9,15-20,24,25,27-31,57-87)*, and are organized in terms
of the oxidation level equivalencies in Equations 1 - 3. Also
included are mean values of selected structural parameters.

TABLE 1: PROPERTIES OF 1-Fe AND 2-Fe ACTIVE SITES

Property	Rd_{ox} $[Fe(S\text{-}Cys)_4]^-$	Rd_{red} $[Fe(S\text{-}Cys)_4]^{2-}$	Fd_{ox} $[Fe_2S_2^*(S\text{-}Cys)_4]^{2-}$	Fd_{red} $[Fe_2S_2^*(S\text{-}Cys)_4]^{3-}$
Structure				
Fe-S (Å), S-Fe-S (deg)	2.24, 109° (<T_d)[a]	b	b	b
Absorption spectrum				
λ_{max}, nm (ε_{mM})	370-380 (10.9), 490-497 (8.9), 565 (~4), 745 (0.4)[c,d]	311 (10.8), 333 (6.2), 1600 (0.13), ~2700[c,d]	325-333 (12-15), 410-425 (9.0), 455-470 (8.5-9.5), 560, 720 (0.8), 820 (0.3), 920 (0.08)[d,e]	320, 400, 470, 530, 650 (0.60), 820 (0.2), 920 (0.07), 1700 (<0.1)[d,e]
Magnetism				
μ_{Fe} (BM)	5.85 (280-330°K)[f]	5.05 (280-330°K)[f]	1.2, 1.5 (290-300°K)[g]	2.8, 3.1 (298°K)[g]
$-J$ (cm^{-1})	--	--	183, 185[g,j]	≲ 100[g]
NMR spectrum				
$-CH_2S$, ppm	b	2-5[f,h]	-37[i]	-13 to -21[i,k]
Mössbauer spectrum				
δ (mm/sec)[l]	0.25[f,m,n]	0.58, 0.65[f,m,n]	0.18 - 0.30[n,o]	0.22 - 0.29[n,o,p]; 0.54 - 0.60[n,o,q]
ΔE_Q (mm/sec)	0.74 (?)[f]	3.1 - 3.2[f,m]	0.60 - 0.66[o]	0.60 - 0.80[o,p]; 2.7 - 3.0[o,q]
ESR spectrum				
g-values	4.3, 9.4[d,r]	b	b	1.89, 1.94, 2.04[s]
Potential				
E'_0, E_m (V)	-0.04 to -0.06[t]		-0.24 to -0.43[d,u]	

(a) Refs. 18,28; mean values. (b) Undetermined. (c) Refs. 60,61. (d) Numerous literature values. (e) Refs. 62-65. (f) Ref. 66. (g) Ref. 67. (h) Upfield of internal standard, assignment uncertain. (i) Refs. 68,87. (j) Ref. 69. (k) Resonances associated with Fe(III) center. (l) Relative to Fe metal. (m) Ref. 70. (n) Ref. 71. (o) Refs. 19, 72, 73. (p)Fe(III). (q) Fe(II). (r) Additional signals also observed; cf. Ref. 74. (s) Representative values; for a tabulation of data cf. Ref. 75. (t) Refs. 60,76. (u) Refs. 12,27.

TABLE 2: PROPERTIES OF 4-Fe ACTIVE SITES

Property	HP$_{ox}$, Fd$_{s-ox}$ [Fe$_4$S$_4$*(S-Cys)$_4$]$^-$	HP$_{red}$, Fd$_{ox}$ [Fe$_4$S$_4$*(S-Cys)$_4$]$^{2-}$	HP$_{s-red}$, Fd$_{red}$ [Fe$_4$S$_4$*(S-Cys)$_4$]$^{3-}$
Structure			
Fe-S, Fe-S*, Fe-Fe (Å)	2.21, 2.25, 2.73^{a,b} ($\lesssim$D$_{2d}$)	2.22, 2.32, 2.81^{a,b}; 2.18, 2.29, 2.85^{c,d} ($\lesssim$D$_{2d}$)	e
Absorption spectrum			
λ_{max}, nm (ε_{mM})	325, 375, 450^{a,f}	388-400 (15-17);^{g,h} 640, 700, 1040^{a,i}	~340 – 350^{g,j}
Magnetism			
μ_{Fe} (BM)	1.9 (200°K)^{a,k}	~0.1-0.6 (100-150°K)^{a,i} 0.9-1.3 (280-300°K)^{c,l}	1.6-1.7 (275-295°K)^{c,l}
NMR spectrum			
-CH$_2$S, ppm	-23 to -41 (280-320°K)^{a,l,m}	-8 to -19 (270-350°K)^{l,n}	-12 to -47 (280-320°K)^{c,l,m}
Mössbauer spectrum			
δ (mm/sec)^o	0.31^{a,p}	0.42,^{a,p} 0.43^{c,q}	0.59,^{a,r} 0.57^{c,q}
ΔE_Q (mm/sec)	0.80^{a,p}	1.13,^{a,p} 0.91^{c,q}	1.28,^{a,r} 1.25^{c,q}
ESR spectrum			
g-values	2.04, 2.12;^{a,k,s} 2.01^{c,t}	---	1.93, 2.04;^{c,u} 1.88, 1.92, 2.06;^{c,v} 1.88-2.07^{w}
Potentials			
E'$_o$, E$_m$ (V)	-0.42,^{a,t} + 0.35^{a,f}	-0.28 to -0.49^{a,f}	

(a) HP. (b) Ref. 30. (c) Fd. (d) Ref. 31. (e) Undetermined. (f) Ref. 77. (g) Numerous literature values. (h) Ref. 78. (i) Ref. 79. (j) Ref. 65. (k) Ref. 80, S = ½ (4.2-80°K). (l) Ref. 81. (m) Positive and negative temperature coefficients of chemical shift. (n) Positive temperature coefficient of chemical shift. (o) Relative to Fe metal. (p) Ref. 82. (q) Ref. 71. (r) Ref. 83. (s) Additional rhombic set of signals observed. (t) Refs. 58,59. (u) Ref. 57. (v) Ref. 84, 1 4-Fe center/molecule. (w) Ref. 85, 2 4-Fe centers/molecule, ~6 resonances observed. (x) Ref. 27.

Data quoted are representative but not exhaustive; references should be consulted for specification of protein sources. Results from ORD, CD, MCD, resonance Raman, ENDOR, and magnetically perturbed Mössbauer spectra are either too detailed to be economically tabulated, or are of presently uncertain value in structural diagnosis owing to limited measurements, and are not included. However, ENDOR spectroscopy is proving to be of increasing utility as a means of probing electronic, and thus indirectly, geometric structure, and attention is drawn to recent work on this subject *(19,69,86)*.

The data of Tables 1 and 2 serve to show how the collective results of physical measurements can be used to distinguish active site structures **1 - 3** in fixed oxidation levels in proteins containing one or more sites of the same type. Provided these properties are not exceptionally modified by protein milieu, they provide the basis of establishing active site structures and oxidation levels in new proteins short of X-ray diffraction. Of the physical measurements in Tables 1 and 2, bulk magnetism, NMR and Mössbauer spectra have associated sensitivity problems and are less likely to be useful in rapid establishment of active site structures of proteins initially available in limited amounts. Absorption and ESR spectra and redox potentials are readily determined on small quantities and are most commonly used in structure-property relationships.

3.1 ABSORPTION SPECTRA - Representative uv-visible absorption spectra of the three known types of iron-sulfur chromophores are collected in Fig. 2. In the absence of other chromophores (*e.g.*, hemes, flavins) or mixed iron-sulfur sites in the same molecule these spectra are useful active site indicators, especially if proteins are examined in their oxidized forms. The structures of the Rd_{ox} and 2-Fe Fd_{ox} chromophore spectra beyond 400 nm are particularly characteristic.

3.2 ESR SPECTRA - These spectra have been observed for the oxidation levels in columns (b) of series (1) and (2) and in columns (a) and (c) of series (3). Typical spectra at X-band frequencies in the g $\sim$ 1.8-5 region are set out in Fig. 3. ESR spectra of numerous characterized and uncharacterized iron-sulfur proteins and enzymes have been measured under a variety of experimental conditions. Here attention is restricted to a few brief considerations. Provided it is not mistaken for adventitious Fe(III), an intense signal at g $\sim$ 4.3, readily observable at 80 K and below, is the simplest manifestation of a Rd_{ox} site. Among 2-Fe and 4-Fe sites the most ESR-characteristic oxidation level appears to be $[Fe_4S_4(S-Cys)_4]^-$ found in HP_{ox}, which exhibits an

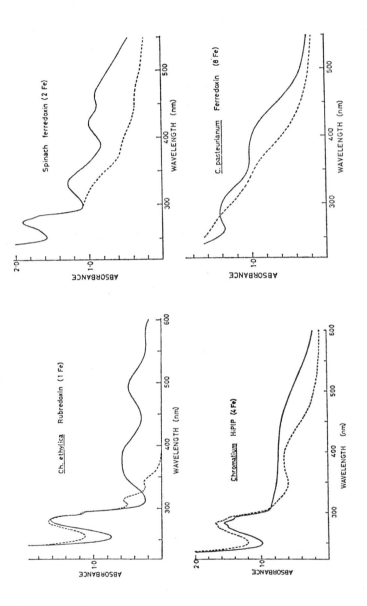

Fig. 2. Uv-visible absorption spectra of 1-Fe, 2-Fe, and 4-Fe chromophores in their oxidized (———) and reduced (— — —) forms. From Hall et al. (12).

axial spectrum with both g-values above 2. The only structurally characterized example of this oxidation level is *Chromatium* HP$_{ox}$ *(29,30)*, whose epr spectrum is in fact more complicated than that shown, exhibiting subsidiary rhombic signals as well *(80)*. Further authenticated examples of this oxidation level in analogues or proteins are required to assess the validity of an ESR-structure correlation.

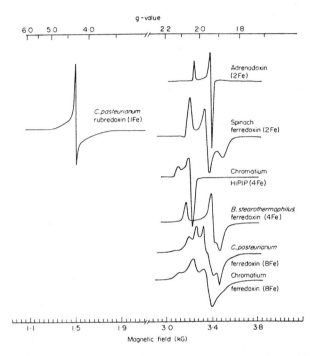

Fig. 3. X-band epr spectra of 1-Fe, 2-Fe, and 4-Fe proteins. For oxidation levels corresponding to these spectra see Tables 1 and 2. From Hall et al. (138).

Sites of the $[Fe_2S_2(S-Cys)_4]^{3-}$ and $[Fe_4S_4(S-Cys)_4]^{3-}$ types cannot be distinguished by their g-values (usually centered at or near 1.94), or g-tensor anisotropies. Indeed, within the 2-Fe class of proteins axial (adrenodoxin, putidaredoxin) and rhombic (spinach and other plant Fds) spectra are observed. An interesting structural manifestation does occur in the spectra of 8-Fe Fd$_{red}$ proteins containing two essentially equivalent 4-Fe sites.
The multicomponent spectrum of *C. pasteurianum* Fd$_{red}$ (Fig. 3) is typical. Modification of this spectrum from the simpler rhombic spectra afforded by single 4-Fe sites (*e.g.*, in *B. stearothermo-*

philus Fd$_{red}$) has been interpreted in terms of weak spin-coupling between spin-doublet sites *(85)*, which in *P. aerogenes* Fd are separated by ∿12 Å *(18,31)*. The appearance of the additional transitions in the g ∿ 1.8-2.2 region is thus an indication of magnetically interacting sites in the same molecule. Inasmuch as similar interactions are conceivable between 2-Fe sites, or 2-Fe and 4-Fe sites, such modified spectra do not necessarily convey the structural type(s) of coupled sites.

Examination of the temperature dependencies of ESR signal intensities of proteins containing 2-Fe and 4-Fe sites of the Fd$_{red}$ types suggests a rough correlation between g-anisotropy and the temperature range of ESR observation at comparable microwave power levels. Cases in point are reduced adrenodoxin and spinach Fd$_{red}$; the spectrum of the former is observable to 150°K whereas that of the latter is well resolved only below 77°K *(87)*, owing to more rapid electron spin relaxation. Spectra of reduced 4-Fe sites are generally well resolved at temperatures not exceeding ∿40-50°K. All of these observations pertain to frozen aqueous solutions in which proteins presumably have their normal tertiary structures. In dimethylsulfoxide (DMSO)/H$_2$O solvent mixtures containing more than ∿60% DMSO by volume it is known that tertiary structure is disrupted *(vide infra)*, with the consequence that g-tensor anisotropy is reduced and axial or near-axial spectra result. Under these conditions extrinsic protein structural and environmental effects on anisotropies and relaxation times appear to be diminished, and these properties intrinsic to each type of isolated magnetic center tend to be more closely approached. These considerations are evident in Cammack's recent investigation of ESR spectra in DMSO/H$_2$O solutions *(88)*, from which the following conclusions emerge: (1) spectra of 2-Fe sites are readily detectable at 77°K and are easily saturated with microwave power at 20°K; (2) spectra of 4-Fe sites are readily detected only below 35°K and are readily saturated at 10°K. Provided additional proteins reveal no serious overlaps of spectral detectability and saturation temperature ranges, this procedure should prove most useful as a structural probe for [Fe$_2$S$_2$-(S-Cys)$_4$]$^{3-}$ and [Fe$_4$S$_4$(S-Cys)$_4$]$^{3-}$ sites in reduced proteins soluble and stable in nonaqueous-aqueous solvent mixtures.

3.3 REDOX POTENTIALS - The data in Tables 1 and 2 indicate that, of the various redox couples in series (1) - (3), potentials for only the 1-/2- reaction of series (1) are unique, being just negative of 0 V. However, Rd potential data are quite limited and of uncertain generality. It is also seen that known potential ranges for the 2-/3- processes of 2-Fe and 4-Fe sites, determined for much larger groups of proteins, overlap appreciably.

Based on the recent report that the 1-/2- potential of one of the two 4-Fe sites in *A. vinelandii* Fd I is -0.42 V *(59)*, potentials in this range cannot necessarily be associated with the 2-/3- reactions of 4-Fe centers. There are no observations which challenge assignment of potentials at *ca.* +0.35 V to the 1-/2- reaction of 4-Fe centers. When comparing potentials of different proteins for any purpose, due attention should be paid to the effects of pH and ionic strength, as revealed by recent careful work on clostridial proteins *(89)*. However, the effects of these variables appear sufficiently small *(89)* that the *ca.* 0.2 V ranges of potentials for the 2-/3- couples of series (2) and (3) must be ascribed mainly to the influences of differing protein structures which modify the inherent potentials of $[Fe_2S_2-(S-Cys)_4]^{2-,3-}$ and $[Fe_4S_4(S-Cys)_4]^{2-,3-}$ couples. In the present context the extent of overlap between the two series renders potentials of the couples of little value in distinguishing 2-Fe and 4-Fe centers.

The foregoing data and considerations make evident the possibility of active site structural determination by means of a combination of physical measurements on the smaller proteins. However, it would clearly be advantageous to have in hand an alternative means of site structural identification incorporating the following features: (1) ready applicability to lower molecular weight proteins and more complex enzymes, which may possess additional prosthetic groups; (2) clear distinction among sites **1 - 3**; (3) sensitivity consistent with the use of *ca.* 10-100 μM solutions or *ca.* 10-100 nmol protein per experiment; (4) rapidity and ease of execution involving, ideally, only one type of conventional physical measurement. Toward this end the core extrusion reaction is under development as a potentially general technique of active site identification. In succeeding sections the chemistry underlying this technique is described, applications are summarized, and an assessment of the scope and limitations of the technique is presented.

4. ACTIVE SITE CORE EXTRUSION REACTIONS

4.1 ANALOGUE REACTIONS WITH RETENTION OF CORE STRUCTURE - The reaction chemistry of synthetic analogues is currently in a developmental stage. At present four types of reactions that proceed with retention of $Fe_2S_2^*$ or $Fe_4S_4^*$ core structures in analogues and/or proteins can be identified and are summarized in Table 3. The protonation reaction *(4)* is a marginal inclusion inasmuch as rapid decomposition follows protonation. Coordinated thiolate can be replaced by water or hydroxide, halide, or

TABLE 3: ANALOGUE AND PROTEIN REACTIONS WITH RETENTION OF $Fe_2S_2^*$ AND $Fe_4S_4^*$ CORES

Reaction Type		References
A. Electron Transfer series (1), (2), (3)		a
B. Fe, S* Exchange[b] (8-Fe Fd_{ox})		90
C. Protonation: H^+ + $[Fe_4S_4^*(SR)_4]^{2-}$ ⇌ $[HFe_4S_4(SR)_4]^-$	(4)	54,55,92
D. Thiolate Ligand Substitution		
1. aqueous solvolysis (L = H_2O, OH^-)[c]		54,55,92
$[Fe_4S_4^*(SR)_4]^{2-}$ + nL ⇌ $[Fe_4S_4^*(SR)_{4-n}L_n]^{Z-}$ + nRS^-	(5)	
2. X^-/RS^- substitution (X = halide)[d]		91
$[Fe_2S_2^*(SR)_4]^{2-}$ + $4R'COX$ → $[Fe_2S_2^*X_4]^{2-}$ + $4R'COSR$	(6)	
$[Fe_4S_4^*(SR)_4]^{2-}$ + $4R'COX$ → $[Fe_4S_4^*X_4]^{2-}$ + $4R'COSR$	(7)	
$[Fe_4S_4^*X_4]^{2-}$ + $4R'S^-$ ⇌ $[Fe_4S_4^*(SR')_4]^{2-}$ + $4X^-$	(8)	
3. $RS^-/R'S^-$ substitution[e]		
$[Fe_4S_4^*(SR)_4]^{2-}$ + $2R'SSR'$ ⇌ $[Fe_4S_4^*(SR')_4]^{2-}$ + $2RSSR$	(9)	44
$[Fe_2S_2^*(SR)_4]^{2-}$ + $4R'SH$ ⇌ $[Fe_2S_2^*(SR')_4]^{2-}$ + $4RSH$	(10)	36
$[Fe_4S_4^*(SR)_4]^{2-}$ + $4R'SH$ ⇌ $[Fe_4S_4^*(SR')_4]^{2-}$ + $4RSH$	(11)	36,40,42,44,45,48,49

a See Tables 1 and 2.

b Demonstrated for 4-Fe protein sites only.

c Analogues and Fd_{ox}.

d Halide and cyanide replacement reactions of thiolate tetramer in aqueous solution have been reported (55).

e Core extrusion references not included; cf. Table 4.

another thiolate under a variety of conditions. The near quantitative yields obtained upon reactions with acyl halides demonstrate that thiolates retain appreciable nucleophilicity when
coordinated. The facility with which reactions (8), (10), and
(11) occur is doubtless due mainly to the tetrahedrally coordinated metal centers (Fig. 1), a factor which generally promotes
lability of coordinated ligands, and the apparent lack of steric
congestion at these centers. Occurrence of reactions (4) - (11)
presages an even richer reaction chemistry based on binuclear and
tetranuclear core units.

Reactions (10) and (11), initially executed with analogues
(36,44), have been observed in nonaqueous [DMSO, N,N-dimethyl-
formamide, acetonitrile, hexamethylphosphoramide (HMPA)] and nonaqueous-aqueous media (*e.g.*, 80% v/v HMPA/H_2O, DMSO/H_2O) and form
the basis for the active site core extrusion method. Reaction
(11) in nonaqueous solvents has been the most thoroughly studied
and exhibits the following important properties: (1) Equilibrium
is attained rapidly. (2) Equilibrium substitution tendencies of
added thiols R'SH tend to parallel their acidities at least to
$pK_a \geq 6.5$, leading to the ligand substitution series R' = alkyl $\leq$
Ac-L-Cys-NHMe $\leq$ aryl. (3) Addition of 4.5-6 equiv. of aryl thiol
(*e.g.*, PhSH) results in complete formation of $[Fe_4S_4(SAr)_4]^{2-}$
from an alkylthiolate tetramer. (4) Initial (and presumably
succeeding) thiolate substitution is bimolecular with rate constants *ca.* $1-10^3$ M^{-1} sec^{-1} *(48)*; kinetic data are fully consistent with the sequence (12) in which proton transfer from R'SH to
coordinated RS$^-$ is followed by departure of weakly ligating RSH
from the metal site and coordination of R'S$^-$.

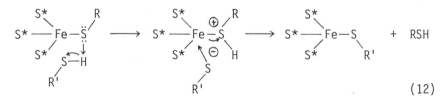

$$(12)$$

(5) No significant amount of core structure decomposition occurs.
(6) Transfer of an $Fe_4S_4^*$ core from an alkyl- to an arylthiolate
coordination environment (or conversion to $[Fe_4S_4Cl_4]^{2-}$ *(91)*)
produces no significant changes in core dimensions, as seen in
Fig. 1. Qualitative observations of reaction (10) indicate that
all properties except (4) apply to it as well; no kinetic studies
of this reaction have been performed. Only small changes occur
in the $Fe_2S_2^*$ core structure found in $[Fe_2S_2(S_2\text{-}o\text{-xyl})_2]^{2-}$ and

$[Fe_2S_2(S\text{-}p\text{-tol})_4]^{2-}$ (Fig. 1). The simplest view of the thiolate substitution reactions is that both rates and equilibrium positions are dominated by the acid-base characteristics of coordinated thiolate and added R'SH, such that substitution of basic thiolate by the stronger thiol acids is both kinetically and thermodynamically favored.

Reactions (10) and (11) at equilibrium may be conveniently monitored by NMR which allows detection of free thiol formed and, in favorable cases, differently substituted tetramers owing to their differences in contact shifts *(40,41,44)*, and by electronic spectra. The latter method is best applied when R = alkyl and R' = aryl because of substantial spectral differences between aryl- and alkylthiolate analogues. Thus for $[Fe_4S_4(SPh)_4]^{2-}$ λ_{max} 450-460 nm ($\varepsilon \sim$17,000-18,000) and for $[Fe_2S_2(SPh)_4]^{2-}$ λ_{max} 480-500 nm ($\varepsilon \sim$11,000-12,000), with the ranges indicating solvent effects. Alkylthiolate tetramer dianions have $\lambda_{max} \sim$ 420 nm *(42)* and $[Fe_2S_2(S_2\text{-}o\text{-xyl})_2]^{2-}$, the only characterized alkylthiolate dimer, has a more complicated spectrum extending to $\sim$700 nm *(36)*, which is readily distinguished from that of $[Fe_2S_2(SPh)_4]^{2-}$.

An important advantage of thiolate substitution reactions is the ready incorporation of various ligand structures around preformed cores. One example is the reactions sequence (13) carried out in DMSO solution *(45)*. Reaction of the *t*-butyl-

$[Fe_4S_4(S\text{-}t\text{Bu})_4]^{2-}$ + 12-peptide $\rightarrow$ 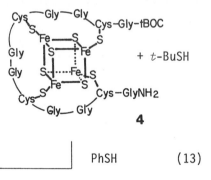 + *t*-BuSH

4

12-peptide + $[Fe_4S_4(SPh)_4]^{2-}$ $\longleftarrow$ $\rceil$ PhSH   (13)

thiolate tetramer (λ_{max} 421 nm) and the glycyl-*L*-cysteinyl oligopeptide with removal of volatile *t*-butylthiol affords complex **4** (λ_{max} 406 nm), or an oligomer of tetramers in which the peptide is linked to more than one core. Treatment of this product with benzenethiol yields $[Fe_4S_4(SPh)_4]^{2-}$ (λ_{max} 458 nm). The two reactions are essentially quantitative and may be monitored spectrophotometrically. The second illustrates removal of an intact $Fe_4S_4^{*}$ core from a peptide environment, and offers a clear precedent for related reactions of proteins.

4.2 ACTIVE SITE CORE EXTRUSION REACTIONS - As pointed out in the
first report *(40)* of reaction (11) and on subsequent occasions
(36,44,45), the existence of facile thiolate substitution offers
the possibilities of protein reconstitution from apoprotein and
preformed cores, and removal of intact cores from proteins as
synthetic analogues, reaction (14). Spectral identification

$$\text{holoprotein + RSH} \rightarrow \left\{ \begin{array}{c} [Fe_2S_2(SR)_4]^{2-} \\ \text{and/or} \\ [Fe_4S_4(SR)_4]^{2-} \end{array} \right\} \text{ + apoprotein} \quad (14)$$

of analogues in turn identifies the active site (**2,3**) provided
$Fe_2S_2^*$ and $Fe_4S_4^*$ cores maintain their structural integrity during
and after the extrusion process. The first published extrusion
reactions were performed with *Spirulina maxima* 2-Fe Fd_{ox} and *C.
pasteurianum* 8-Fe Fd_{ox} *(93)*. A list of proteins subjected to
successful extrusion reactions is provided in Table 4. The
majority of cases are drawn from research in this laboratory;
extensive extrusion studies have also been undertaken by Orme-
Johnson and co-workers *(97,98,100)* but only a part of their results *(97,98)* was published at the time of this writing. All
tabulated extrusions were effected with benzenethiol except for
the reaction of Rd_{ox} with *o*-xylyl-α,α'-dithiol. In this case a
single iron atom rather than a substructural core is removed from
a protein molecule. Designation of reaction (15) as one of extrusion is gratuitous, and this reaction is included here to show

$$Rd_{ox} + 2 \, o\text{-xyl(SH)}_2 \rightarrow [Fe(S_2\text{-}o\text{-xyl})_2]^- + \text{apo-Rd} \quad (15)$$

that ligand substitution reactions apply generally to the three
types of active sites. Prior to consideration of the results in
Table 4, summarized in terms of n, the number of cores extruded
per protein molecule, experimental conditions necessary to the
successful application of the core extrusion technique are examined.

4.2.1 Anaerobicity. Owing to the oxygen sensitivity of all synthetic analogues and many proteins, all operations are best carried out under anaerobic conditions. The apparatus and procedures used in this laboratory have been described *(95)*.

4.2.2 Solvent Medium. In purely aqueous solutions containing

TABLE 4: *ACTIVE SITE CORE EXTRUSIONS*

Protein	n		References	
	$Fe_2S_2^*$	$Fe_4S_4^*$	Extrusion	Protein
1-Fe				
C. pasteurianum Rd_{ox}		[a]	34	9, 18, 60, 61
2-Fe				
Spirulina maxima Fd_{ox}	1.0	--	93	94
Spinach Fd_{ox}	1.0	--	95	[b], 7, 19, 87
Adrenal Fd_{ox}	$+^c$	--	97, 98	5, 99
4-Fe				
B. stearothermophilus Fd_{ox}	--	1.0	95	96
Chromatium HP red	--	1.0	95	29, 30, 77
C. pasteurianum nitrogenase, Fe protein	--	0.93; $+^c$	95; 97	97
8-Fe				
C. pasteurianum Fd_{ox}	--	$2.0^e; +^c$	93, 95; 97, 98	[d]
C. pasteurianum $Fd_{ox} + Fd_{red}$	--	1.9^f	95	[d]
A. vinelandii Fd I	--	1.9	139	59
C. pasteurianum hydrogenase	--	$2.9^f; >1^c$	95; 98	107, 108; 98

[a] $\sim$1.1 g-atom Fe removed in the form of $[Fe(S_2-o-xyl)_2]^-$.
[b] Prototype 2-Fe plant Fd, numerous studies.
[c] n-value unspecified.
[d] Prototype 8-Fe clostridial Fd, numerous studies.
[e] Result of 10 determinations at 5-192 μM (95).
[f] Ferricyanide added to complete extrusion.

no denaturants proteins assume their native tertiary structures and core extrusion proceeds slowly if at all. However, based on the following observations tertiary structure is disrupted and protein is unfolded to a considerable extent in mixed nonaqueous-aqueous solvents: (1) broadened Fd_{ox} NMR contact-shifted signals and spectral changes in the high-field region suggestive of random coil arrangements in $DMSO/H_2O$ solutions exceeding 50-60% v/v in DMSO *(101)*; (2) removal of additional resonances in Fd_{red} ESR spectra *(88)* arising from spin-spin coupling *(85)*; (3) red shifts of the 390 nm Fd_{ox} aqueous absorption *(49,93,95)* and cathodic displacement of Fd_{ox}/Fd_{red} potentials *(49)* such that these two properties approach those of isoelectronic 4-Fe analogues in non-aqueous media *(42,45)*; (4) reduction of HP_{red} to HP_{s-red} by di-thionite in 80% $DMSO/H_2O$ *(57)*. In a mixed solvent medium the protein must be adequately soluble (≥ 10 μM) and stable during the time period of complete extrusion (usually $\leq$ 1 hr.). In this laboratory quantitative extrusions have been accomplished in 60-80% $DMSO/H_2O$ and $HMPA/H_2O$. All extrusion reactions in Table 4 were performed in these media. The 80% HMPA medium, introduced by Erbes *et al.* *(98)*, is at least as effective as $DMSO/H_2O$ mixtures in dissolving a variety of proteins without active site decomposition over a period of several hours. Purification of commercial solvents is advisable; purity criteria and other properties of HMPA have been described *(102)*. Other binary or ternary solvent mixtures may have to be devised to meet solubility and stability properties of individual proteins. Prior to mixing with the nonaqueous component, the aqueous solvent component should be buffered to an alkaline pH to promote stability of both protein and analogue toward solvolysis or acid-induced decomposition and dimer $\rightarrow$ tetramer conversion *(vide infra)*. Aqueous solutions containing 20-50 mM TrisCl at pH 8-9 have proven quite satisfactory.

4.2.3 Extrusion Reagents. All extrusion reactions thus far have been monitored by absorption spectra. Extrusions of oxidized 2-Fe and 4-Fe centers are illustrated in Figs. 4 and 5; only final spectra are shown in the former whereas the time course of the reaction is presented in the latter. These reactions illustrate the following desirable properties of extrusion reagents: (1) formation of analogues with distinctive band shapes and rather intense maxima red-shifted from protein chromophore absorption; (2) lack of appreciable absorption by deprotonated thiolate in the analogue chromophore region; (3) complete extrusion within 1 hr. Properties (1) and (2) afford spectra at complete extrusion which can be so recognized by no appreciable intensity changes upon further thiol addition, identify the site

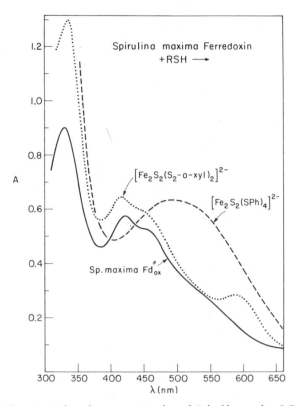

Fig. 4. Active site core extrusion of Spirulina maxima 2-Fe
Fd_{ox} in 80% DMSO/H_2O (20 mM TrisCl, pH 8.5) with benzenethiol
and o-xylyl-α,α-dithiol. From Que et al. (93).

extruded as 2-Fe or 4-Fe, and allow calculation of analogue concentration from previously determined extinction coefficients and
observed spectral absorbancies at ≥ 450 nm with minor or negligible corrections for non-analogue absorptions [apoprotein, excess
thiol(ate)]. Alkylthiols do not satisfy property (1) for extrusion of 4-Fe sites. However, o-xylyl-α,α'-dithiol is useful as
at least a qualitative indicator of 2-Fe sites due to its distinctive band at ∼590 nm (ε ∼5,000, Fig. 4). With this reagent
4-Fe sites are extruded as $[Fe_4S_4(S_2\text{-}o\text{-}xyl)_2]_n^{2n-}$ *(93)*, presumably
an oligomer of tetramers, which lacks a discrete absorption band
in this region. The cases shown in Figs. 4 and 5 together with
the results in Table 4 demonstrate that benzenethiol is an ideal
extrusion reagent for proteins containing 2-Fe or 4-Fe sites and
no other chromophores absorbing in the 400-600 nm region.

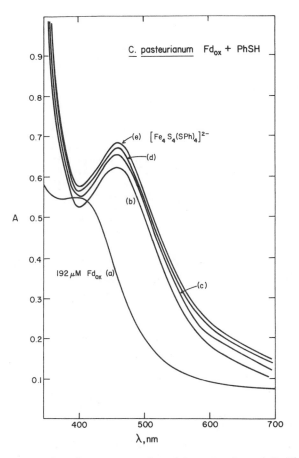

Fig. 5. Active site core extrusion of C. pasteurianum 8-Fe Fd_{ox} in 80% HMPA/H_2O (50 mM TrisCl, pH 8.5). (a) 192 µM Fd_{ox}; (b) 1 ml of solution (a) + 9 µl PhSH 1 min after mixing; solution (b) 6 min (c), 13 min (d), and 25 min (e) after mixing. The PhSH/Fe mol ratio is 57/1. From Gillum et al. (95).

Extinction coefficients of $[Fe_2S_2(SPh)_4]^{2-}$ and $[Fe_4S_4(SPh)_4]^{2-}$ are sufficient to produce absorbancies of 0.3 or greater for *ca.* 20 m*M* solutions of proteins containing one active site when examined in a 1 cm cell. With regard to property (3), extrusions in 80% HMPA/H_2O of a variety of proteins with 4-Fe centers have led to the guideline that complete extrusion is achieved with the mol ratio PhSH/Fe~100/1 at times of *ca.* 20-30 min after thiol addition. In this medium complete extrusion of spinach Fd_{ox} (mol ratio ∿500/1) occurs in *ca*. 90 min.

4.2.4 Oxidation Levels and Dimer → Tetramer Core Conversion.
The extrusion technique will yield reliable results only if core
units of analogues elicited from proteins correspond to the core
substructures of active sites. A complication previously encountered in 2-Fe Fd_{ox} extrusion *(93)* was reaction (16), spontaneous
dimer → tetramer conversion (R = Ph) in 80% $DMSO/H_2O$ solvent
medium. Subsequent observations *(103)* indicate that this reaction
is not as rapid as first described and that its rate is retarded
by increasing pH and by excess thiol, which is always required
for complete extrusion *(95)*. In 80% $HMPA/H_2O$ with aqueous component pH 8.5 and a PhSH/Fe mol ratio of ∿100/1, $[Fe_2S_2(SPh)_4]^{2-}$
completely resists tetramer formation for at least 44 hrs *(95)*.

$$2[Fe_2S_2(SR)_4]^{2-} \rightarrow [Fe_4S_4(SR)_4]^{2-} + RSSR + 2RS^- \qquad (16)$$

These conditions serve as a guide for suppression of reaction (16)
during extrusion. A second dimer → tetramer conversion, reaction
(17), has been proposed to account for the electrochemical behavior observed upon reduction of the R = Ph dimer dianion in aprotic

$$2[Fe_2S_2(SR)_4]^{3-} \rightarrow [Fe_4S_4(SR)_4]^{2-} + 4RS^- \qquad (17)$$

media *(36)*. No dimer trianion has yet been isolated; this reaction is currently under study. Unlike reaction (16) this conversion is not a redox process, but instead is formation of a tetramer core from a dimer core in the same oxidation level [series
(2) and (3)]. If a reduced 2-Fe site extrudes and the analogue
$[Fe_2S_2(SR)_4]^{3-}$ is formed, the possibility exists that unless reaction (17) is suppressed by excess thiol or is eliminated by
addition of an oxidizing agent, incorrect active site identification could result. Extrusion of mixtures initially containing
$\gtrsim$70% spinach Fd_{red} in 80% $HMPA/H_2O$ has yielded $[Fe_2S_2(SPh)_4]^{2-}$ as
the principal product before and after $Fe(CN)_6^{3-}$ addition, which
is required to complete the extrusion *(95)*. Near-quantitative
extrusion of mixtures of oxidized and reduced 4-Fe sites of *C.
pasteurianum* Fd has been achieved in 80% $HMPA/H_2O$ solutions to
which $Fe(CN)_6^{3-}$ was added after reaction with benzenethiol had
reached steady state *(95)*. Lastly, efforts to extrude the $[Fe_4S_4-(S-Cys)_4]^-$ of *Chromatium* HP_{ox} in 80% $HMPA/H_2O$ with benzenethiol
have not afforded spectra attributable to any 2-Fe or 4-Fe ana-
logue chromophore. The spectra resemble those resulting from
solutions of Fe(III) salts and the thiol in the same medium,

indicating that in this oxidation level core integrity is lost upon disruption of ligation with the protein. From these considerations it follows that extrusion reactions are optimally performed on proteins whose active site oxidation levels are adjusted to $[Fe_2S_2(S-Cys)_4]^{2-}$ and/or $[Fe_4S_4(S-Cys)_4]^{2-}$ prior to their introduction into the extrusion medium.

4.2.5 Extraneous Iron. Simple criteria for the singularity of iron-containing chromophores resulting from extrusion are absorbance ratios at two or more wavelengths, which may then be compared with those measured separately for independently prepared analogues. In 80% $HMPA/H_2O$ $A_{458}/A_{550} = 2.18 \pm 0.16$ for $[Fe_4S_4-(SPh)_4]^{2-}$ determined over a wide concentration range, while more limited data for $[Fe_2S_2(SPh)_4]^{2-}$ yield 1.15 ± 0.01 for the same ratio *(95)*. In practice mean values from extrusions of a number of different proteins are 1.90 (4-Fe sites) and 1.12 (2-Fe sites) *(93,95)*. With the more extensively studied 4-Fe site extrusions occasional cases have been encountered where A_{458}/A_{550} ratios were substantially lower than 1.9, a result traced to lack of sufficiently anaerobic conditions or the use of excessively manipulated protein stock solutions. Extrusion with benzenethiol in these cases leads to solutions with perceptible red- or blue-violet casts instead of the red-brown color characteristic of anaerobic extrusions of pure protein samples. Similar colors are developed upon reaction of $FeCl_3$, $[Fe(S_2-o-xyl)_2]^-$, or Rd_{ox} with the thiol, and in the 550 nm region is observed strong absorption, whose intensity decreases with time. Similarly colored and unstable chromophores have been detected in other Fe(III)/thiol systems *(104)*. The increased visible absorption under extrusion conditions is attributed to Fe(III)/PhSH species formed from Fe(III) not incorporated in 4-Fe units and arising from oxidative or other decomposition of protein, and must be avoided if active sites are to be securely identified and extrusions quantitated. As an empirical control, extrusion experiments with $A_{458}/A_{550} \leq 1.8$ are rejected as unsuitable for quantitation of $n(Fe_4S_4^*)$ *(95)*.

4.2.6 Mixed Active Sites. A possibility for any uncharacterized protein of sufficient iron content is the presence of more than one type of site. With coextrusion of oxidized 2-Fe and 4-Fe cores from the same protein in mind, DePamphilis *et al.* *(36)* synthesized pairs of arylthiolate dimers and tetramers including $[Fe_2S_2(SPh)_4]^{2-}/[Fe_4S_4(SPh)_4]^{2-}$ and found their spectral properties to be additive and sufficiently different to allow analysis of mixtures. Subsequently mixtures of spinach Fd_{ox} and *C. pasteurianum* Fd_{ox} were successfully coextruded with benzenethiol in 80% $HMPA/H_2O$ *(95)*. Using $Fe_4S_4^*/Fe_2S_2^*$ core mol ratios of 2.2-0.25

λ_{max} shifted from 460 to 470 nm and A_{458}/A_{550} decreased from 1.64 to 1.23. Experimental spectra could be reconstructed from appropriately weighted sums of the spectra of the two proteins extruded separately under the same conditions. These results indicate the feasibility of detecting and quantitating 2-Fe and 4-Fe sites in the same protein provided there is no interference from extraneous iron.

The active sites of the proteins in Table 4, none of which contains other metals or prosthetic groups, had for the most part been established prior to extrusion experiments by X-ray diffraction (Fd_{ox}, HP_{red}) or by iron, sulfide, and amino acid analysis together with the structure-property relationships of Tables 1 and 2. Consequently, such proteins have been the obvious vehicles for demonstration and quantitation of the extrusion reaction, the first requirement for which is medium-induced disruption of tertiary structure without active site decomposition. While relatively simple, these proteins encompass an interesting range of properties which include molecular weights of 6,000-13,000 daltons, variant isoelectric points, thermophilicity (*B. stearothermophilus* Fd; this feature may be associated with stabilizing salt bridges absent in nonthermophiles *(105)*), active sites located in hydrophobic environments with differing extents of hydrogen bonding to peptide chains *(106)*, and location of a site well within the interior of the protein structure [HP,*(29)*]. The quantitative extrusion of this group of proteins serves as an encouraging forerunner to application of the technique to higher molecular weight proteins and enzymes, whose active site structures are uncertain and may be difficult and laborious to establish even with the armamentarium of physical methods available.

4.3 HYDROGENASE - These enzymes, which have recently been reviewed *(107)*, catalyze reaction (18) when coupled to an endogeneous electron carrier C (NAD^+, cytochrome c_3, Fd) and are key catalysts in microbial metabolism. The most extensively investigated H_2ase is that from *C. pasteurianum*, for which two preparations reported to be purified to homogeneity have been obtained.

$$H_2 + C_{ox} \rightleftharpoons 2H^+ + C_{red} \qquad (18)$$

The enzyme has a molecular weight of $\sim$60,500 daltons and contains no subunits, no metals other than iron, and no prosthetic groups other than those arising from Fe-S*-(S-Cys) interactions. The enzyme isolated by Chen and Mortenson in 1974 *(108)* and later used in extrusion experiments by Gillum *et al.* *(95)* contains $\sim$12 g-atom Fe, 12 g-atom S*, and 12 mols of Cys residues per mol. It is more extensively purified than earlier preparations with a lower iron content (~ 4 Fe) *(109)*. The preparation described by Erbes *et al.* in 1975 *(98)* is reported to contain ~ 4Fe and ~ 4S*/60,000 daltons. This preparation was used in the first published extrusion of a higher molecular weight Fe-S protein. In an important paper these authors first employed the 80% HMPA/H_2O solvent medium and showed that benzenethiol extrusion yielded [Fe_4S_4-$(SPh)_4$]$^{2-}$, indicating the presence of a 4-Fe site in their enzyme preparation.

In order to define the nature of active sites in the Chen-Mortenson preparation of clostridial H_2ase, reported to be substantially richer in Fe and S* content, Gillum *et al.* *(95)* undertook a quantitative extrusion study using the methods developed with smaller proteins (Table 4). One such experiment is depicted in Fig. 6; protein concentration was determined by the Folin-Lowry method. The spectrum in the 80% HMPA/H_2O solution after brief exposure to air and before the addition of benzenethiol is structurally informative compared to the rather ill-defined spectra of the enzyme in water *(95,98,108)*. The maximum near 400 nm and its intensity is consistent with approximately three [Fe_4S_4-$(S-Cys)_4$]$^{2-}$ chromophores, and the lack of structure in the spectrum at longer wave lengths is suggestive of the absence of [$Fe_2S_2(S-Cys)_4$]$^{2-}$ sites resembling those of the Fd_{ox} type (Fig. 2). Addition of thiol generates the [$Fe_4S_4(SPh)_4$]$^{2-}$ spectrum, complete development of which, as in extrusions of Fd_{ox} + Fd_{red} mixtures, is afforded by the addition of ferricyanide. The final spectrum (A_{458}/A_{550} 1.96) is that of [$Fe_4S_4(SPh)_4$]$^{2-}$, with no evidence of additional absorbance by extruded 2-Fe sites or extraneous iron. Initial iron concentration in this experiment was 168 μM based on 11.2 g-atom Fe/60,500 daltons, determined by prior analysis, or 180 μM based on 12 g-atom, the nearest integral multiple of 4. The final iron concentration, independently determined from the [$Fe_4S_4(SPh)_4$]$^{2-}$ spectrum, was 174 μM, a result within experimental uncertainty of the two preceding values. Related extrusion experiments carried out anaerobically at the same and slightly lower concentrations gave similarly excellent agreement between initially calculated iron concentrations and those found by extrusion. From these results it is concluded that this preparation of H_2ase contains only 4-Fe sites and $n(Fe_4S_4^*) \cong 3$. Also consistent with the presence of more than one magnetic site is the appearance of additional ESR lines in the g = 1.8-2.2 region (*cf.* Fig. 3) observed in the reduced forms of H_2ase obtained by Erbes *et al.* *(98)* and by Chen *et al.* *(110)*. The presence of a small proportion of [$Fe_4S_4(S-Cys)_4$]$^{3-}$ sites is considered responsible for the requirement of ferricyanide to achieve full spectral intensity of analogue dianion upon

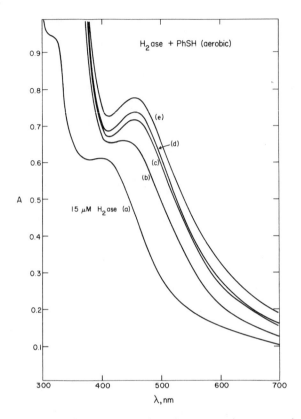

Fig. 6. *Active site core extrusion of C. pasteurianum H_2ase in 80% HMPA/H_2O (50 mM TrisCl, pH 8.5). (a) 1 ml of 15 µM H_2ase solution after exposure to air for 1 min; (b) solution (a) made anaerobic + 2 µl PhSH 2 min after mixing; (c) solution (b) 5 min after mixing; (d) solution (b) 15 min after mixing; (e) solution (d) + 1 µl 29 mM $Na_3Fe(CN)_6$ 5 min after mixing. From Gillum et al. (95) to which reference is made for further experimental details.*

extrusion.

Among various H_2ase preparations from the same and different organisms, that from *C. pasteurianum* containing ∿12 Fe/molecule is unique because of its high iron content. *Chromatium* H_2ase is reported to have been purified to homogeneity and to contain ∿4Fe and ∿4S* per 100,000 daltons, composed of two subunits of ∿50,000 daltons *(111)*. Different preparations from *Desulfovibrio vulgaris* with ∿1-8 g-atom Fe have been reported *(112-115)*. The latest description is that of an enzyme of 89,000 daltons (28,000,

59,000 dalton subunits) containing 7-8 g-atom S* and 7-9 g-atom
Fe *(115)*. Absorption spectra of the oxidized forms of all of
these preparations are quite similar to that of clostridial
H_2ase, indicating that all most likely contain 4-Fe sites.
Establishment of the number and structural types of Fe-S centers
present in hydrogenases is the requisite first step in examining
the mechanism of reaction (18). In the absence of carrier C,
H_2ase catalyzes the isotope exchange reaction (19) in which
heterolytic cleavage of dihydrogen has been frequently proposed
as the first step. An exceptionally thorough study of the catalysis of isotope exchange by *D. vulgaris* H_2ase has been made by
Yagi *et al. (116)*. The establishment of 4-Fe sites by the

$$D_2 + H_2ase \rightleftharpoons H_2ase \cdot D^- + D^+$$
$$D^+ + H_2O \rightleftharpoons H^+ + HDO$$
$$H^+ + H_2ase \cdot D^- \rightleftharpoons HD + H_2ase$$

$$D_2 + H_2O \rightleftharpoons HD + HDO \tag{19}$$

extrusion technique *(95,98)* demonstrates that such a site, alone
or in conjunction with surrounding protein structure, is capable
of activating dihydrogen and presumably stabilizes hydride by
Fe-H coordination. Further mechanistic considerations require
acquisition of the following minimal primary information:
(1) the oxidation level of the activating site(s); (2) the roles
of all 4-Fe sites present. Based on Ru(IV) reduction by H_2 and
D_2-H_2O exchange catalyzed by Ru(III) *(117,118)*, the following
reaction pathway is offered as one possibility for the catalysis
of reaction (18) by the clostridial H_2ase preparation containing
$\sim$12 g-atom Fe. In this scheme a is the activating (catalytic)
center, c is an electron transfer center coupled to the endogeneous 8-Fe Fd electron carrier, and t is a tetrameric (4-Fe)
site. Reactions (20a,b) occur within the enzyme molecule and

$$t_a^{3-} + H_2 \rightleftharpoons t_a^{3-} \cdot H^- + H^+ \tag{20a}$$
$$t_a^{3-} \cdot H^- + 2t_c^{2-} \rightleftharpoons t_a^{3-} + 2t_c^{3-} + H^+ \tag{20b}$$

$$2t_c^{2-} + H_2 \rightleftharpoons 2t_c^{3-} + 2H^+$$
$$Fd_{ox} + 2t_c^{3-} \rightleftharpoons Fd_{red} + 2t_c^{2-} \tag{20c}$$

$$Fd_{ox} + H_2 \rightleftharpoons Fd_{red} + 2H^+ \qquad (21)$$

result in H_2 uptake or evolution autocatalyzed by t_a^{3-}. Reaction
(20c) is the coupling of intramolecular events to the external
carrier yielding the net catalytic reaction (21). It is emphasized that the scheme is entirely *speculative*; alternate pathways
are not difficult to devise. However, it does propose a reasonable role for three 4-Fe centers and is consistent with the currently best defined mechanism of H_2 activation and uptake catalyzed by metal complexes in aqueous solution. With the analogues
$[Fe_4S_4(SR)_4]^{2-,3-}$ in hand *(39,44,51)* it should be possible to
test the above and related reaction schemes with particular regard to the oxidation level (2-,3-) of t_a, stoichiometry of H_2
uptake and evolution, and possibly, those features which differentiate t_a from t_c provided the above or a related scheme requiring such a differentiation proves a viable representation.
Research relating to these matters and the development of a synthetic hydrogenase system is underway in this laboratory.

4.4 SCOPE AND LIMITATIONS – Because the reaction chemistry of
synthetic analogues in general, and the active site core extrusion technique in particular, is still in an exploratory stage,
it is probable that further research will enlarge the reaction
types in Table 3 and require some revision of core extrusion
experimental guidelines in Section 4.2. The latter have been
formulated on the basis of experience in this laboratory, primarily with relatively simple proteins. An assessment of future
applications of the technique, and some attendant difficulties,
can best be made by reference to Table 5. Here is listed a variety of proteins and enzymes ranging in molecular weight from
8,000 to 800,000 and in apparent Fe/S* content from 4 to nearly
40. The existence of Fe-S sites is indicated by the *ca.* 1:1
Fe:S* ratio established analytically and/or by the appearance of
$g \sim 1.94$ type ESR spectra of the reduced forms. The list includes cases in which mixed sites are quite probable, *e.g.*, in
succinate dehydrogenase *(128)* and the nitrogenase FeMo protein
(120). For no protein has the nature of the site(s) been fully
defined. The list is incomplete, especially with regard to enzymes, and reviews and additional tabulations are available *(8,
12,123,129,138)*. In several cases, spinach chloroplast membrane
protein (a possible primary electron acceptor in photosynthesis),
and the hydrogenases already mentioned, the extrusion technique
as described should suffice. Provided it can be unfolded in an
appropriate solvent medium, the same comment applies to pyruvate
dehydrogenase whose thiamine cofactor is not appreciably

TABLE 5: CANDIDATE PROTEINS FOR Fe-S SITE IDENTIFICATION BY THE CORE EXTRUSION TECHNIQUE

Protein	Molecular Weight	g-atom Fe, S*	Other Cofactors	References
Spinach chloroplast membrane protein	8,000	4	--	119
Hydrogenase	60,000-100,000	4-9	--	111,116
Nitrogenase FeMo protein	200,000-270,000	17-28	1-2 Mo	97,120
Trimethylamine dehydrogenase	147,000	4	lumazine (?)	121
Pyruvate dehydrogenase[a]	240,000	6	thiamine	122
Succinate dehydrogenase	97,000	8	FAD	127-129
Xanthine oxidase (milk)	270,000	8	2 Mo, 2 FAD	123,130
Aldehyde oxidase (liver)	280,000	8	2 Mo, 2 FAD	123,131
Glutamate synthase	800,000	38	7 (FAD + FMN)	132
Nitrite reductase[b]	60,000	+	siroheme (Fe)	126,133,134
Sulfite reductase[c]	670,000	$\geq$ 14-15	4 FMN, 4 FAD, 3-4 siroheme (Fe)	124-126
Nitrate reductase[d]				
I[e]	300,000[g]	20	1-2 Mo	135
II[e]	260,000	8	<1 Mo, 4 Fe[f]	136
	180,000	6-7	<1 Mo, ~1 Fe[f]	

[a] C. acidi-urici, 3S* reported.
[b] Spinach, Chlorella.
[c] E. coli B, 20-21 Fe reported.
[d] E. coli.
[e] K. aerogenes.
[f] Nonheme Fe in excess of S* content.
[g] Cf. Ref. 137.

chromophoric in the visible region. Absorption spectra of these proteins in their oxidized forms are highly suggestive of 4-Fe sites. The extrusion technique using benzenethiol should offer confirmatory evidence and afford site quantitation in conjunction with accurate protein concentrations.

For the remaining proteins in Table 5 the situation is not necessarily as straightforward, even if unfolding conditions are found, owing to the presence of other cofactors. These classify into ones which are weakly or non-absorbing in the visible region in the native protein (Mo, Fe in excess of S* content) or others which are very strongly absorbing in this region (flavin, siroheme *(126)*). The contribution of Mo and excess Fe to the absorption spectra resulting from extrusion must be established before definite conclusions can be drawn. Orme-Johnson and Davis *(97)* note that extrusion of the FeMo protein of *C. pasteurianum* nitrogenase yielded a complex mixture whose spectrum could not be analyzed as a sum of 2-Fe and 4-Fe analogue spectra. This situation may arise from the formation of a chromophoric Mo complex and/or the presence of several different types of Fe centers *(120)*, not all of which may correspond in kind to **1**, **2**, or **3**.

In the presence of organic chromophores strongly absorbing in the 400-500 nm region, extrusion with benzenethiol may not prove satisfactory. Barring the inconvenient, if not difficult, procedures of quantitative separation of extruded analogues from remaining protein or selective chemical bleaching of the interfering protein chromophore, determination of analogue spectra requires spectrophotometric difference techniques or an accurate subtraction of protein chromophore from a total spectrum. In either case correction for the organic chromophore (*e.g.*, flavin) in its chemical state under extrusion conditions in the presence of excess thiol must be made. To decrease or eliminate errors associated with this approach certain alternative procedures (1)-(4) are noted. Of these (2)-(4) have not as yet been tested in detail.

(1) Red-shifted analogue spectra. Thiols such as p-$(CH_3)_2$N-C_6H_4SH *(44)* and p-$CH_3OC_6H_4SH$, containing strong electron-releasing groups in the para-position, shift 2-Fe and 4-Fe analogue spectral maxima to the red by *ca.* 20-40 nm compared to benzenethiol. Both are active in 2-Fe and 4-Fe site extrusion, and their use should lead to improved accuracy in obtaining analogue spectra in any subtraction or difference procedure. The use of o-xylyl-α,α'-dithiol as at least a qualitative indicator of 2-Fe sites is re-emphasized here.

(2) Chiral or fluorescent analogues. Thiols which confer these properties on 2-Fe and 4-Fe analogues have not yet been developed. As with absorption spectra, maximally useful CD and

fluorescence spectra of analogues should differentiate the two types of analogues, allow quantitation of mixtures, and not be subject to interference by apo- or residual holoprotein.

(3) Electrochemistry. The basis of this method is the substantial difference in half-wave or peak potentials for the redox couples $[Fe_4S_4(SPh)_4]^{2-,3-}$ (-1.03 V) and $[Fe_2S_2(SPh)_4]^{3-,4-}$ (-1.41 V) as determined in N,N-dimethylformamide *vs.* the saturated calomel electrode *(36)*. The absence of protein and thiol interference and medium breakdown to an applied potential of about -1.6 V is required. For enhanced sensitivity differential pulse polarography should be advantageous; analogue concentrations at the 0.1 mM level or higher are desirable.

(4) NMR. This method has the potential advantage of combining the proven extrusion efficiency of aryl thiols with a conventional spectroscopic technique not subject to interference by competing chromophores or protein in any form. As one possible example of this approach, 3,5-bis(trifluoromethyl)benzenethiol (unreported but synthesizable from the commercial aniline by standard procedures) should extrude 2-Fe and 4-Fe protein cores in the form of analogues which can be independently prepared by known methods *(36,39,44)*. As in the ^{1}H spectra of their non-fluorinated counterparts *(40-42,44)*, the ^{19}F spectra of these analogues should exhibit resonances substantially shifted from that of the free ligand by contact interactions. If under extrusion conditions these shift differences and those between dimer and tetramer are sufficiently large and the ideal situation of slow exchange between free and coordinated ligand obtains, measurements of chemical shifts and integrated intensities *vs.* an added standard will identify and quantitate extruded analogues. The utility of the fast exchange limit, potentially induced by excess thiol added to maximize extrusion, will depend on relative amounts of dimer, tetramer, and thiol in the extrusion mixture and their chemical shift separations. Here extrusion product identification, if possible at all, would rest on the value of the averaged chemical shift compared to those of the separate components. Similar comments apply to the intermediate exchange case. The fluorinated thiol provides a factor of 24 in equivalent nuclei over dimer or tetramer analogue concentration. Thus in an extrusion solution 0.1 mM in the latter, spectrometers with conventional Fourier transform capability should be adequately sensitive.

The considerations offered in this section should prove useful to those who may wish to apply the active site core extrusion technique, which in its most experimentally convenient form requires only the measurement of absorption spectra. Of the desirable features (1)-(4) of a method for Fe-S active site identification listed in Section 3, the core extrusion technique as applied thus far demonstrably satisfies (3) and (4), and is capable of distinguishing and quantitating 2-Fe and 4-Fe sites, (2). Analysis of mixtures containing 1-Fe sites has not been attempted, but such sites when oxidized are the most readily recognized by physical techniques (Table 1). Applicability in the presence of other prosthetic groups or cofactors, (1), remains to be demonstrated, but work in progress in this laboratory with several enzymes containing organic chromophores indicates that this feature can also be realized. Thus the current prognosis for the core extrusion approach is favorable and, together with readily acquired temperature and power saturation characteristics of ESR spectra, its application should prove increasingly valuable as a rapid means of active site identification. Lastly, two points should be borne in mind. First, extrusion directly identifies only the core, but not the ligands terminally coordinated to iron atoms. Second, the sites **1 - 3** are those which have been thus far established in Fe-S proteins. It would be unwise to constrain interpretation of extrusion experiments or any physical measurements to only these sites. Three billion years of evolution may have generated other Fe/S* (±S-Cys) combinations yet to be discovered by the biochemist or synthesized, purposely or serendipitously, by the chemist.

ACKNOWLEDGEMENT. Research in the author's laboratory described herein has been supported by National Institutes of Health Grants GM-22351 and GM-22352. I thank Drs. J. Cambray, W.O. Gillum, C.L. Hill, and L.E. Mortenson for valuable discussions.

REFERENCES

1. L.E. Mortenson, R.C. Valentine and J.E. Carnahan, *Biochem. Biophys. Res. Commun.*,**7**, 448 (1962).
2. K. Tagawa and D.I. Arnon, *Nature,***195**, 537 (1962).
3. B.B. Buchanan, *Struct. Bonding (Berlin)*,**1**, 109 (1966).
4. J.C. Rabinowitz, *Ann. Rev. Biochem.*,**36**, 113 (1967).
5. T. Kimura, *Struct. Bonding (Berlin)*,**5**, 1 (1968).
6. D.O. Hall and M.C.W. Evans, *Nature,* **223**, 1342 (1969).
7. D.O. Hall, R. Cammack and K.K. Rao, *Pure Appl. Chem.*,**34**, 553 (1973).
8. W.H. Orme-Johnson, *Ann. Rev. Biochem.*,**43**, 159 (1973).
9. W. Lovenberg, in "Microbial Iron Metabolism," J.B. Neilands, Ed., Academic Press, New York, Chap. 8 (1974).
10. R. Wickramasinghe, *Enzyme,***17**, 227 (1974).
11. R. Wickramasinghe and E.N. McIntosh, *Enzyme,***17**, 210 (1974).

12. D.O. Hall, R. Cammack and K.K. Rao, in "Iron in Biochemistry and Medicine", A. Jacobs and M. Worwood, Eds., Academic Press, New York, Chap. 8 (1974).
13. B.B. Buchanan and D.I. Arnon, *Advan. Enzymology*, **33**, 119 (1970).
14. D.C. Yoch and R.C. Valentine, *Ann. Rev. Microbiol.*, **26**, 139 (1972).
15. M. Llinás, *Struct. Bonding (Berlin)*, **17**, 135 (1973).
16. J.C.M. Tsibris and R.W. Woody, *Coord. Chem. Rev.*, **5**, 417 (1970).
17. R. Mason and J.A. Zubieta, *Angew. Chem. Int. Ed.*, **12**, 390 (1973).
18. L.H. Jensen, *Ann. Rev. Biochem.*, **43**, 471 (1974).
19. R.H. Sands and W.R. Dunham, *Quart. Rev. Biophys.*, **7**, 443 (1975).
20. D.O. Hall, R. Cammack and K.K. Rao, in "Theory and Experiment in Exobiology", A.W. Schwartz, Ed., Wolters-Noordhoff Publishing, Gröningen, The Netherlands, Vol. 2, pp. 67-85 (1972).
21. K.T. Yasunobu and M. Tanaka, *Syst. Zool.*, **22**, 570 (1973).
22. R.H. Wickramasinghe, *Space Life Sci.*, **4**, 341 (1973).
23. D.O. Hall, R. Cammack and K.K. Rao, *Nature*, **233**, 136 (1971); *Space Life Sci.*, **4**, 455 (1973); *Orig. Life*, **5**, 363 (1974).
24. D.O. Hall, R. Cammack and K.K. Rao, *Sci. Progr. Oxford*, **62**, 285 (1975).
25. W. Lovenberg, Ed., "Iron-Sulfur Proteins", Vols. 1 and 2, Academic Press, New York (1973); Vol. 3 (1976).
26. R.H. Holm, *Endeavour*, **34**, 38 (1975).
27. R.H. Holm and J.A. Ibers, in "Iron-Sulfur Proteins", W. Lovenberg, Ed., Vol. 3, Academic Press, New York, Chap. 7, (1976).
28. L.H. Jensen, in "Iron-Sulfur Proteins", W. Lovenberg, Ed., Vol. 2, Academic Press, New York, Chap. 4 (1973); K.D. Watenpaugh, L.C. Sieker, J.R. Herriott and L.H. Jensen, *Acta Crystallogr.*, B**29**, 943 (1973).
29. C.W. Carter, Jr., J. Kraut, S.T. Freer, Ng. Xuong, R.A. Alden and R.G. Bartsch, *J. Biol. Chem.*, **249**, 4212 (1974); C.W. Carter, Jr., J. Kraut, S.T. Freer and R.A. Alden, *ibid.*, **249**, 6339 (1974).
30. S.T. Freer, R.A. Alden, C.W. Carter, Jr. and J. Kraut, *J. Biol. Chem.*, **250**, 46 (1975).
31. E.T. Adman, L.C. Sieker and L.H. Jensen, *J. Biol. Chem.*, **248**, 3987 (1973); E. Adman, K.D. Watenpaugh and L.H. Jensen, *Proc. Nat. Acad. Sci. U.S.A.*, **72**, 4854 (1975); E.T. Adman, L.C. Sieker and L.H. Jensen, *J. Biol. Chem.*, **251**, 3801 (1976).

32.   K.T. Yasunobu and M. Tanaka, in "Iron-Sulfur Proteins",
       W. Lovenberg, Ed., Vol. 2, Academic Press, New York, Chap.
       2 (1973).
33.   R.W. Lane, J.A. Ibers, R.B. Frankel and R.H. Holm, *Proc. Nat.
       Acad. Sci. U.S.A.*,**72**, 2868 (1975).
34.   R.W. Lane, J.A. Ibers, R.B. Frankel, G.C. Papaefthymiou and
       R.H. Holm, *J. Am. Chem. Soc.*, in press.
35.   J.J. Mayerle, R.B. Frankel, R.H. Holm, J.A. Ibers, W.D.
       Phillips and J.F. Weiher, *Proc. Nat. Acad. Sci. U.S.A.*,**70**,
       2429 (1973).
36.   J.J. Mayerle, S.E. Denmark, B.V. DePamphilis, J.A. Ibers and
       R.H. Holm, *J. Am. Chem. Soc.*,**97**, 1032 (1975).
37.   W.O. Gillum, R.B. Frankel, S. Foner and R.H. Holm, *Inorg.
       Chem.*,**15**, 1095 (1976).
38.   T. Herskovitz, B.A. Averill, R.H. Holm, J.A. Ibers, W.D.
       Phillips and J.F. Weiher, *Proc. Nat. Acad. Sci. U.S.A.*,**69**,
       2437 (1972).
39.   B.A. Averill, T. Herskovitz, R.H. Holm and J.A. Ibers, *J. Am.
       Chem. Soc.*,**95**, 3523 (1973).
40.   M.A. Bobrik, L. Que, Jr. and R.H. Holm, *J. Am. Chem. Soc.*,
       96, 285 (1974).
41.   R.H. Holm, W.D. Phillips, B.A. Averill, J.J. Mayerle and T.
       Herskovitz, *J. Am. Chem. Soc.*,**96**, 2109 (1974).
42.   B.V. DePamphilis, B.A. Averill, T. Herskovitz, L. Que, Jr.
       and R.H. Holm, *J. Am. Chem. Soc.*,**96**, 4159 (1974).
43.   R.H. Holm, B.A. Averill, T. Herskovitz, R.B. Frankel, H.B.
       Gray, O. Siiman and F.J. Grunthaner, *J. Am. Chem. Soc.*,**96**,
       2644 (1974).
44.   L. Que, Jr., M.A. Bobrik, J.A. Ibers and R.H. Holm, *J. Am.
       Chem. Soc.*,**96**, 4168 (1974).
45.   L. Que, Jr., J.R. Anglin, M.A. Bobrik, A. Davison and R.H.
       Holm, *J. Am. Chem. Soc.*,**96**, 6042 (1974).
46.   R.B. Frankel, B.A. Averill and R.H. Holm, *J. Phys. (Paris)*,
       35, C6-107 (1974).
47.   S-P. W. Tang, T.G. Spiro, C. Antanaitis, T.H. Moss, R.H.
       Holm, T. Herskovitz and L.E. Mortenson, *Biochem. Biophys.
       Res. Commun.*,**62**, 1 (1975).
48.   G.R. Dukes and R.H. Holm, *J. Am. Chem. Soc.*,**97**, 528 (1975).
49.   C.L. Hill, J. Renaud, R.H. Holm and L.E. Mortenson, results
       to be published.
50.   R.B. Frankel, T. Herskovitz, B.A. Averill, R.H. Holm, P.J.
       Krusic and W.D. Phillips, *Biochem. Biophys. Res. Commun.*,**58**,
       974 (1974).
51.   R.W. Lane, A.G. Wedd, W.O. Gillum, E. Laskowski, R.H. Holm,
       R.B. Frankel and G.C. Papaefthymiou, results to be published.

52. D.G. Holah and D. Coucouvanis, *J. Am. Chem. Soc.*,**97**, 6917 (1975); A. Kostikas, V. Petrouleas, A. Simopoulos, D. Coucouvanis and D.G. Holah, *Chem. Phys. Lett.*,**38**, 582 (1976).

53. J.R. Anglin and A. Davison, *Inorg. Chem.*,**14**, 234 (1975).

54. T.C. Bruice, R. Maskiewicz and R. Job, *Proc. Nat. Acad. Sci. U.S.A.*,**72**, 231 (1975).

55. R.C. Job and T.C. Bruice, *Proc. Nat. Acad. Sci. U.S.A.*,**72**, 2478 (1975).

56. R.G. Shulman, P. Eisenberger, W.E. Blumberg and N.A. Stombaugh, *Proc. Nat. Acad. Sci. U.S.A.*,**72**, 4003 (1975); D.E. Sayers, E.A. Stern and J.R. Herriott, *J. Chem. Phys.*,**64**, 427 (1976).

57. R. Cammack, *Biochem. Biophys. Res. Commun.*,**54**, 548 (1973).

58. W.V. Sweeney, A.J. Bearden and J.C. Rabinowitz, *Biochem. Biophys. Res. Commun.*,**59**, 188 (1974).

59. W.V. Sweeney, J.C. Rabinowitz and D.C. Yoch, *J. Biol. Chem.*, **250**, 7842 (1975).

60. W. Lovenberg and B.E. Sobel, *Proc. Nat. Acad. Sci. U.S.A.*, **54**, 193 (1965).

61. W.A. Eaton and W. Lovenberg, in "Iron-Sulfur Proteins", W. Lovenberg, Ed., Vol. 2, Academic Press, New York, Chap. 3 (1973).

62. K. Tagawa and D.I. Arnon, *Biochim. Biophys. Acta*,**153**, 602 (1968).

63. J. Rawlings, O. Siiman and H.B. Gray, *Proc. Nat. Acad. Sci. U.S.A.*,**71**, 125 (1974).

64. W.A. Eaton, G. Palmer, J.A. Fee, T. Kimura and W. Lovenberg, *Proc. Nat. Acad. Sci. U.S.A.*,**68**, 3015 (1971).

65. S.G. Mayhew, D. Petering, G. Palmer and G.P. Foust, *J. Biol. Chem.*,**244**, 2830 (1969).

66. W.D. Phillips, M. Poe, J.F. Weiher, C.C. McDonald and W. Lovenberg, *Nature*, **227**, 574 (1970).

67. G. Palmer, W. R. Dunham, J.A. Fee, R.H. Sands, T. Iizuka and T. Yonetani, *Biochim. Biophys. Acta*,**245**, 201 (1971).

68. I. Salmeen and G. Palmer, *Arch. Biochem. Biophys.*,**150**, 767 (1972); W.R. Dunham, G. Palmer, R.H. Sands and A.J. Bearden, *Biochim. Biophys. Acta*,**253**, 373 (1971).

69. R.E. Anderson, W.R. Dunham, R.H. Sands, A.J. Bearden and H.L. Crespi, *Biochim. Biophys. Acta*,**408**, 306 (1975).

70. K.K. Rao, M.C.W. Evans, R. Cammack, D.O. Hall, C.L. Thompson, P.J. Jackson and C.E. Johnson, *Biochem. J.*,**129**, 1063 (1972).

71. C.L. Thompson, C.E. Johnson, D.P.E. Dickson, R. Cammack, D.O. Hall, U. Weser and K.K. Rao, *Biochem. J.*,**139**, 97 (1974).

72. K.K. Rao, R. Cammack, D.O. Hall and C.E. Johnson, *Biochem. J.*, **122**, 257 (1971); R. Cammack, K.K. Rao, D.O. Hall and C.E. Johnson, *ibid.*,**125**, 849 (1971).

73. E. Münck, P.G. Debrunner, J.C.M. Tsibris and I.C. Gunsalus, *Biochemistry*, **11**, 855 (1972).

74. J. Peisach, W.E. Blumberg, E.T. Lode and M.J. Coon, *J. Biol. Chem.*, **246**, 5877 (1971).

75. W.E. Blumberg and J. Peisach, *Arch. Biochem. Biophys.*, **162**, 502 (1974).

76. J.A. Peterson and M.J. Coon, *J. Biol. Chem.*, **243**, 329 (1968).

77. K. Dus, H. DeKlerk, D. Sletten and R.G. Bartsch, *Biochim. Biophys. Acta*, **140**, 291 (1967).

78. J-S. Hong and J.C. Rabinowitz, *J. Biol. Chem.*, **245**, 4982 (1970).

79. M. Cerdonio, R-H. Wang, J. Rawlings and H.B. Gray, *J. Am. Chem. Soc.*, **96**, 6534 (1974).

80. B.C. Antanaitis and T.H. Moss, *Biochim. Biophys. Acta*, **405**, 262 (1975).

81. W.D. Phillips and M. Poe, in "Iron-Sulfur Proteins", W. Lovenberg, Ed., Vol. 2, Academic Press, New York, Chap. 7 (1973); W.D. Phillips, C.C. McDonald, N.A. Stombaugh and W.H. Orme-Johnson, *Proc. Nat. Acad. Sci. U.S.A.*, **71**, 140 (1974).

82. D.P.E. Dickson, C.E. Johnson, R. Cammack, M.C.W. Evans, D.O. Hall and K.K. Rao, *Biochem. J.*, **139**, 105 (1974).

83. D.P.E. Dickson and R. Cammack, *Biochem. J.*, **139**, 105 (1974).

84. N.A. Stombaugh, R.H. Burris and W.H. Orme-Johnson, *J. Biol. Chem.*, **248**, 7951 (1973).

85. R. Mathews, S. Charlton, R.H. Sands and G. Palmer, *J. Biol. Chem.*, **249**, 4326 (1974).

86. R.E. Anderson, G. Anger, L. Petersson, A. Ehrenberg, R. Cammack, D.O. Hall, R. Mullinger and K.K. Rao, *Biochim. Biophys. Acta*, **376**, 63 (1975).

87. G. Palmer, in "Iron-Sulfur Proteins", W. Lovenberg, Ed., Vol. 2, Academic Press, New York, Chap. 8 (1973).

88. R. Cammack, *Biochem. Soc. Trans.*, **3**, 482 (1975).

89. E.T. Lode, C.L. Murray and J.C. Rabinowitz, *J. Biol. Chem.*, **251**, 1683 (1976).

90. J-S. Hong and J.C. Rabinowitz, *J. Biol. Chem.*, **245**, 6582 (1970).

91. M.A. Bobrik, G. Wong, K.O. Hodgson and R.H. Holm, results to be published.

92. R. Maskiewicz, T.C. Bruice and R.G. Bartsch, *Biochem. Biophys. Res. Commun.*, **65**, 407 (1975).

93. L. Que, Jr., R.H. Holm and L.E. Mortenson, *J. Am. Chem. Soc.*, **97**, 463 (1975).

94. D.O. Hall, K.K. Rao and R. Cammack, *Biochem. Biophys. Res. Commun.*, **47**, 798 (1972).

95. W.O. Gillum, L.E. Mortenson, J-S. Chen and R.H. Holm, *J. Am. Chem. Soc.*, in press.
96. R.N. Mullinger, R. Cammack, K.K. Rao, D.O. Hall, D.P.E. Dickson, C.E. Johnson, J.D. Rush and A. Simopoulos, *Biochem. J.*,**151**, 75 (1975).
97. W.H. Orme-Johnson and L.C. Davis, in "Iron-Sulfur Proteins", W. Lovenberg, Ed., Vol. 3, Academic Press, New York, Chap. 2 (1976).
98. D.L. Erbes, R.H. Burris and W.H. Orme-Johnson, *Proc. Nat. Acad. Sci. U.S.A.*,**72**, 4795 (1975).
99. R.W. Estabrook, K. Suzuki, J.I. Mason, J. Baron, W.E. Taylor, E.R. Simpson, J. Purvis and J. McCarthy, in "Iron-Sulfur Proteins", W. Lovenberg, Ed., Vol. 1, Academic Press, New York, Chap. 8 (1973).
100. W.H. Orme-Johnson, private communications, 1974-76.
101. C.C. McDonald, W.D. Phillips, W. Lovenberg and R.H. Holm, *Ann. N.Y. Acad. Sci.*,**222**, 789 (1973).
102. T. Fujinaga, K. Izutsu and S. Sakura, *Pure Appl. Chem.*,**44**, 117 (1975).
103. J. Cambray, R.W. Lane and R.H. Holm, unpublished results.
104. *Cf.*, *e.g.*, C.M. Bell, E.D. McKenzie and J. Orton, *Inorg. Chim. Acta*,**5**, 109 (1971); L.G. Stadtherr and R.B. Martin, *Inorg. Chem.*,**11**, 92 (1972); A. Tomita, H. Hirai and S. Makishima, *ibid.*,**4**, 760 (1968).
105. M.F. Perutz and H. Raidt, *Nature*,**255**, 256 (1975).
106. E. Adman, K.D. Watenpaugh and L.H. Jensen, *Proc. Nat. Acad. Sci. U.S.A.*,**72**, 4854 (1975).
107. L.E. Mortenson and J-S. Chen, in "Microbial Iron Metabolism", J.B. Neilands, Ed., Academic Press, New York, Chap. 11 (1974).
108. J-S. Chen and L.E. Mortenson, *Biochim. Biophys. Acta*,**371**, 283 (1974).
109. G. Nakos and L.E. Mortenson, *Biochemistry*,**10**, 2442 (1971); *Biochim. Biophys. Acta*,**227**, 576 (1971).
110. J-S. Chen, L.E. Mortenson and G. Palmer, in "Iron and Copper Proteins", K.T. Yasunobu, H.F. Mower and O. Hayaishi, Eds., Plenum Press, New York, pp. 68-82 (1976).
111. P.H. Gitlitz and A.I. Krasna, *Biochemistry*,**14**, 2561 (1975).
112. R.H. Haschke and L.L. Campbell, *J. Bacteriol.*,**105**, 249 (1971).
113. J. Legall, D.V. Dervartanian, E. Spilker, J.P. Lee and H.D. Peck, *Biochim. Biophys. Acta*,**234**, 525 (1971).
114. T. Yagi, *J. Biochem. (Tokyo)*,**68**, 649 (1970).
115. T. Yagi, K. Kimura, H. Daidoji, F. Sakai, S. Tamura and H. Inokuchi, *J. Biochem. (Tokyo)*,**79**, 661 (1976).

116. T. Yagi, M. Tsuda and H. Inokuchi, *J. Biochem. (Tokyo)*, **73**, 1069 (1973).
117. J.F. Harrod, S. Ciccone and J. Halpern, *Can. J. Chem.*, **39**, 1372 (1961).
118. J. Halpern and B.R. James, *Can. J. Chem.*, **44**, 671 (1966).
119. R. Malkin, P.J. Aparicio and D.I. Arnon, *Proc. Nat. Acad. Sci. U.S.A.*, **71**, 2362 (1974).
120. E. Münck, H. Rhodes, W.H. Orme-Johnson, L.C. Davis, W.J. Brill and V.K. Shah, *Biochim. Biophys. Acta*, **400**, 32 (1975).
121. D.J. Steenkamp and J. Mallinson, *Biochim. Biophys. Acta*, **429**, 705 (1976); D.J. Steenkamp, private communication.
122. K. Uyeda and J.C. Rabinowitz, *J. Biol. Chem.*, **246**, 3111 (1971).
123. V. Massey, in "Iron-Sulfur Proteins", W. Lovenberg, Ed., Vol. 1, Academic Press, New York, Chap. 10 (1973).
124. L.M. Siegel, M.J. Murphy and H. Kamin, *J. Biol. Chem.*, **248**, 251 (1973).
125. M.J. Murphy, L.M. Siegel, H. Kamin and D. Rosenthal, *J. Biol. Chem.*, **248**, 2801 (1973).
126. M.J. Murphy, L.M. Siegel, S.R. Tove and H. Kamin, *Proc. Nat. Acad. Sci. U.S.A.*, **71**, 612 (1974).
127. K.A. Davis and Y. Hatefi, *Biochemistry*, **10**, 2509 (1971).
128. T. Ohnishi, J.C. Salerno, D.B. Winter, J. Lim, C.A. Yu, L. Yu and T.E. King, *J. Biol. Chem.*, **251**, 2094 (1976); T. Ohnishi, J. Lim, D.B. Winter and T.E. King, *ibid.*, **251**, 2105 (1976).
129. T.P. Singer, M. Gutman and V. Massey, in "Iron-Sulfur Proteins", W. Lovenberg, Ed., Vol. 1, Academic Press, New York, Chap. 9 (1973).
130. J.S. Olson, D.P. Ballou, G. Palmer and V. Massey, *J. Biol. Chem.*, **249**, 4363 (1974).
131. K.V. Rajagopalan, I. Fridovich and P. Handler, *J. Biol. Chem.*, **237**, 922 (1962); K.V. Rajagopalan, P. Handler, G. Palmer and H. Beinert, *J. Biol. Chem.*, **243**, 3784 (1968).
132. R.E. Miller and E.R. Stadtman, *J. Biol. Chem.*, **247**, 7407 (1972).
133. P.J. Aparicio, D.B. Knaff and R. Malkin, *Arch. Biochem. Biophys.*, **169**, 102 (1975).
134. W.G. Zumft, *Biochim. Biophys. Acta*, **276**, 363 (1972).
135. P. Forget, *Eur. J. Biochem.*, **42**, 325 (1974).
136. J. Van't Riet and R.J. Planta, *Biochim. Biophys. Acta*, **379**, 81 (1975); J. Van't Riet, J.H. Van Ee, R. Wever, B.F. Van Gelder and R.J. Planta, *ibid.*, **405**, 306 (1975).
137. R.A. Clegg, *Biochem. J.*, **153**, 533 (1976).

138. D.O. Hall, K.K. Rao and R.N. Mullinger, *Biochem. Soc. Trans.*, **3**, 472 (1975).
139. J.B. Howard, T. Lorsbach and L. Que, *Biochem. Biophys. Res. Commun.*, **70**, 582 (1976).

structural studies of ionophores and their ion-complexes

J.D. DUNITZ AND MAX DOBLER

Organic Chemistry Laboratory, Swiss Federal Institute of Technology (ETHZ), CH-8092 Zürich, Switzerland

1. INTRODUCTION

Once the metabolic activity of certain antibiotics had been
found to depend on the presence or absence of specific alkali
cations in the culture medium *(1,2)* it did not take long to establish that the antibiotics in question had the ability to combine
selectively with such cations *(3)* and transport them, as lipid-
soluble complexes, across natural and synthetic lipid membranes
(4,5,6). Broadly speaking, there are two groups of such antibiotics, distinguished both by structural differences and by
differences in their effects on mitochondrial respiration. The
valinomycin group (valinomycin, enniatins, nonactin and homo-
logues - Fig. 1) are macrocyclic molecules, built from regular
sequences of aminoacid or hydroxyacid sub-units, with a characteristic alternation of chirality of these sub-units. They are
neutral molecules containing no ionizable groups and have little
or no ability to catalyze proton transport across natural or
synthetic membranes. The nigericin group (nigericin, monensin,
grisorixin, dianemycin, X-206, X-537A, A204, etc. - a few new
ones are discovered every year nowadays) are polycyclic polyethers with a terminal carboxyl group at one end of the molecule
and one or two hydroxyl groups at the other (Fig. 2).

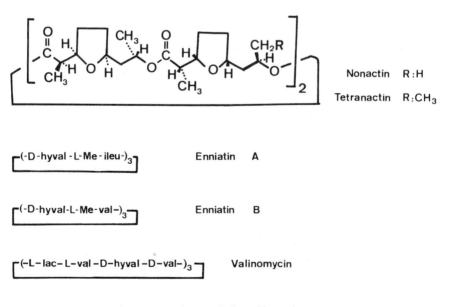

Nonactin   R:H

Tetranactin   R:CH₃

⌐(-D-hyval-L-Me-ileu-)₃⌐   Enniatin   A

⌐(-D-hyval-L-Me-val-)₃⌐   Enniatin   B

⌐(-L-lac-L-val-D-hyval-D-val-)₃⌐   Valinomycin

Fig. 1. Ionophores of the valinomycin group.

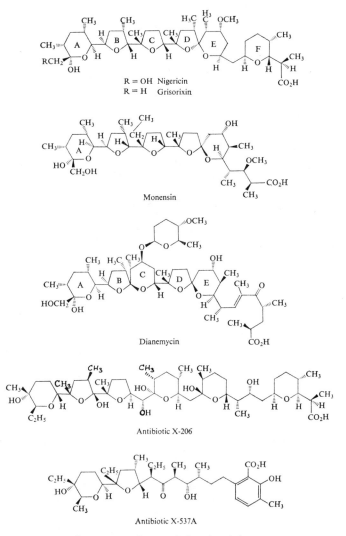

R = OH Nigericin
R = H   Grisorixin

Monensin

Dianemycin

Antibiotic X-206

Antibiotic X-537A

Fig. 2. Ionophores of the nigericin group.

At physiological pH the carboxyl group is ionized, giving the molecule a negative charge. The molecule is maintained in a cyclic conformation by hydrogen bonding between hydroxyl groups and carboxylate anion. This group of compounds is able to couple H^+/K^+ transport across membranes. The important feature of both groups of antibiotics is their ability to make natural membranes leaky towards cations, thus disturbing the balance required for correct functioning of the various parts of the cell. Other antibiotics, notably gramicidins and amphotericins, may make the cell membranes leaky by formation of channels or pores.

In this paper we shall try to review some aspects of the structures of these ion-carriers or ionophores, as they have been called, from a rather personal point of view. Our review will be selective, dealing with points that happen to interest us, rather than encyclopaedic, for in the available space we cannot possibly attempt to cover all aspects of the subject, nor are we competent to do so. Moreover, we hope we shall be excused for drawing examples mainly from results from our own laboratory. Naturally, it is with these results that we are most familiar. In trying to relate the structures of these compounds to their ion-specific properties, it is essential to have detailed information about the conformations of the uncomplexed molecules as well as of their metal complexes. The most extensive available information of this kind refers to the actin group, so we might begin here. What are the structures of these compounds and what can be learnt from them?

2.   VALINOMYCIN GROUP

2.1 ACTINS – When we began to study the crystalline potassium thiocyanate complex of nonactin in 1967, the cation specificity and the membrane permeability of nonactin complexes had already been established. What was lacking was a structural basis for understanding these phenomena. We had already encountered non-actin itself several years earlier when a preliminary X-ray study *(7)* (unit cell and space group) helped to establish that the molecule is a cyclic tetramer of nonactic acid. Nonactin has a 32-membered ring, built from 20 carbon and 12 oxygen atoms, but no one at the time guessed that this could provide suitable accommodation for a metal ion. In retrospect we were not very clever. It was only ten years later, after the clues had been provided by the biochemists, that our colleague Willy Simon suc-ceeded in crystallizing the potassium thiocyanate complex, and within a few months the main features of its structure were clearly established *(8)*.

The potassium ion sits at the centre of a nearly perfect cube (Fig. 3) formed by four ether-type oxygen atoms from tetrahydro-furane rings and four carbonyl oxygen atoms from the ester groups, all at distances of 2.75 - 2.85 Å, only slightly larger than the sum of the ionic radius of K^+ and the van der Waals radius of O (2.73 Å). This symmetrical arrangement of oxygen atoms is at-tained by twisting the 32-membered ring of the organic ligand

into a shape that resembles the seam of a tennis ball. The K^+
ion sits at the centre of the ball, while the methyl substituents
and the methylene carbons of the tetrahydrofurane rings form the
outside surface, which thereby consists mainly of hydrophobic
groups. The molecule as a whole has a crystallographic dyad axis
but it retains approximately the higher point group symmetry S_4
in the crystal, the four monomeric units being alternatingly of
opposite sense of chirality. The net result is that the K^+ ion
is enclosed in a polar cavity within a little ball of fat. This
method invented by Nature for transporting K^+ ions across lipid
membranes seems so simple and obvious that one wonders why nobody
had thought of it before!

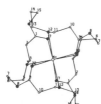

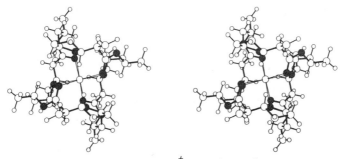

Fig. 3.   Stereoview of K^+-nonactin complex.

The Na^+ complex of nonactin (9) has a very similar conformation to the K^+ complex (Fig. 4). Indeed, if one does not look too
closely, they appear to be almost identical. There is an important difference, however. Whereas the eight metal-oxygen distances
in the K^+ complex are nearly equal, the distances in the Na^+ complex fall into two clearly distinct sets; the four tetrahydrofurane oxygens remain roughly at the same distance from the centre of
the ball ($Na^+...O$, 2.77 Å) but the four carbonyl oxygens are
drawn appreciably closer to the centre ($Na^+...O$, 2.42 Å). It is
clear that a uniform contraction of the cube to bring all eight

oxygens to a distance of about 2.35 - 2.40 Å from the centre
would lead to intolerable overcrowding of non-bonded atoms. The
carbonyl oxygens with their less crowded surroundings seem to be
able to approach the cation better than the ether oxygens.

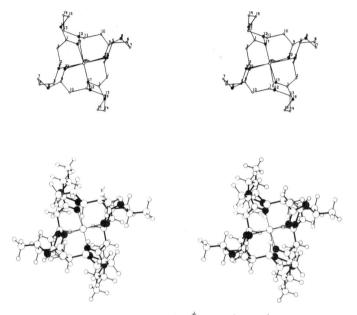

Fig. 4. *Stereoview of* Na^+*-nonactin complex.*

In the NH_4^+ complex *(10)*, the cube of oxygen atoms is again
(Fig. 5) irregular but in just the opposite sense - the carbonyl
oxygens are now appreciably further from the centre (3.01 - 3.16
Å) while the tetrahydrofurane oxygens remain at approximately the
same distance (2.83 - 2.89 Å) from the centre as in the K^+ and
Na^+ complexes. In the course of the X-ray analysis of the NH_4NCS
complex of nonactin *(10)*, the hydrogen atoms of the ammonium ion
could be located fairly clearly in electron-density difference
maps. The N-H bonds point approximately towards the four surrounding ether-type oxygens (N-H...O, 161-171°), the nearer set,
so there is little doubt that the ammonium ion is effectively
hydrogen-bonded to these atoms.
    A parallel series of X-ray studies on crystalline alkali
cation and ammonium complexes of tetranactin, the higher homologue
of nonactin with four methyl groups replaced by ethyl groups, has
led to very similar results *(11,12,13)*. In all cases, the molecule retains approximate S_4 symmetry with the cation at the

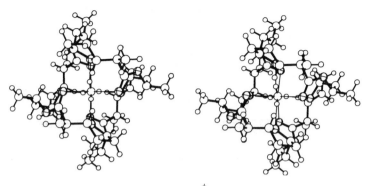

Fig. 5. Stereoview of NH_4^+-nonactin complex.

centre, surrounded by a cube of oxygens that is either regular
(as in the Rb^+ and K^+ complexes) or deformed in just the same way
as the corresponding Na^+ and NH_4^+ complexes of nonactin. The four
ethyl substituents, on the outside surface of the tetranactin
molecule, provide better shielding of the ester oxygen atoms and
thus serve to enhance the lipid solubility of the complexes. A
summary of the $M^+...O$ distances observed in nonactin and tetranactin complexes is given in Table 1. It is seen that the distances to ether-type oxygens do not vary by more than about 0.1 Å
from 2.85 Å, the mean distance in the K^+ complex, whereas the
distances to carbonyl oxygen show a much greater variability.
Before we discuss the bearing of these results on the relative
stabilities of the complexes, a few words have to be said about
the conformations of the uncomplexed organic ligands, which are
quite different from those of the metal complexes.

TABLE 1: *CATION-OXYGEN DISTANCES (IN Å) OBSERVED IN VARIOUS
NONACTIN AND TETRANACTIN COMPLEXES. THE TWO ENTRIES FOR K^+-
TETRANACTIN COMPLEXES REFER TO DIFFERENT CRYSTAL MODIFICATIONS*

| | Nonactin | | Tetranactin | |
	$M^+...O<$	$M^+...O=$	$M^+...O<$	$M^+...O=$
Rb^+	---	---	2.90 - 2.98	2.88 - 2.93
K^+	2.81 - 2.88	2.73 - 2.81	2.85 - 2.93	2.77 - 2.80
			2.83 - 2.92	2.75 - 2.81
Na^+	2.74 - 2.79	2.40 - 2.44	2.70 - 2.94	2.43 - 2.45
NH_4^+	2.83 - 2.89	3.01 - 3.16	2.86 - 2.93	2.97 - 3.05

The elucidation of the detailed structure of uncomplexed non-actin presented certain difficulties because the crystal structure is of the OD (order-disorder) type. However, even from the pre-liminary X-ray study *(7)*, the presence of a short 5.7 Å periodi-city was enough to show that the uncomplexed nonactin molecule must be a rather flat, plate-like object, quite unlike the roughly spherical molecule that occurs in the metal complexes. When the crystal structure was finally established *(14)*, the differences in overall shape turned out indeed to be quite con-siderable, as may be seen by comparing Fig. 6 with Figs. 3, 4, 5. Nevertheless, there are also some striking similarities; the free molecule, like its metal complexes, shows approximate S_4 symmetry and detailed examination reveals that the conformational change involved in passing from the free to the complexed molecule is achieved mainly by making major changes in only 8 of the 32 tor-sion angles in the macrocyclic ring, i.e. two per monomeric unit.

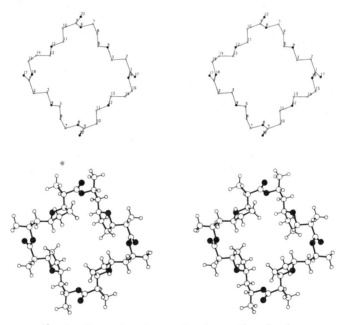

Fig. 6. Stereoview of uncomplexed nonactin molecule.

The bonds involved are C(1)-C(2) and C(2)-C(3). As a result of these changes the hole at the centre of the uncomplexed molecule becomes appreciably larger, the central cavity being now walled by the four tetrahydrofurane rings, with their oxygen atoms in an approximately square arrangement, about 4.5 Å from the molecular centre. The ester oxygens are about 1 Å further from the centre.

The tetranactin molecule *(15)* has a different shape, which is illustrated in Fig. 7. Two long, antiparallel chains are linked at both ends by two of the four ester groups, the other two having quite different environments in the middles of the chains; the S_4 symmetry has been lost. In this arrangement, two opposite ethyl groups and two opposite carbon atoms of tetrahydrofurane rings approach each other (to about 3.95 Å) transannularly, which presumably gives rise to favourable intramolecular van der Waals contacts that stabilize the observed conformation.

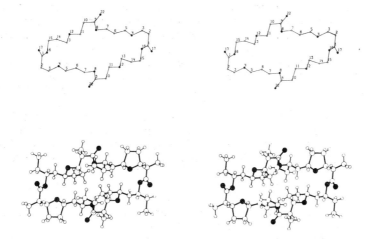

Fig. 7. Stereoview of uncomplexed tetranactin molecule.

According to force-field calculations reported by Sakamaki, Iitaka and Nawata *(13)*, the elongated conformation of Fig. 7 is about 6-7 kcal mol^{-1} more stable than the S_4 conformation both for nonactin and tetranactin, but the calculations refer, of course, to unsolvated molecules in the gas phase. The solid-state infrared spectra of nonactin and tetranactin show distinct differences, whereas the solution spectra (CHCl$_3$ and CCl$_4$) are similar, both showing rather broad absorption bands. The Raman carbonyl band of uncomplexed nonactin in CHCl$_3$ solution is also markedly broadened *(16)*, indicating either that the molecule adopts the elongated conformation or that more than one conformation is present. On the other hand, nmr spectra of nonactin *(17, 18)* and of tetranactin *(19)* in solution show that the four sub-units are magnetically equivalent, even at -110°C, indicating that these conformations are rapidly interconverted to give a time-averaged S_4 symmetry. On balance, it can be concluded that in solution both molecules are rather flexible and that the S_4 conformation, if not actually preferred, is at least present in appreciable concentration.

For a molecule to act as an ion-carrier, there must be a delicate balance between the binding energy of the ion in its cage and its hydration energy in solution, which amounts to 60-120 kcal mol^{-1} for alkali cations. A large difference between these quantities would mean either that the ion could never escape from its hydration shell or that, once enclosed in its cage, it would have no tendency to go back into solution. There must also be a low enough activation energy to allow the cation to enter or leave its cage at a reasonably fast rate. Indeed, the specific complexation rates of actins (measured for Na$^+$ complexation in methanol) are about 10^8M^{-1}sec^{-1} (20), approaching the rate to be expected for a diffusion-controlled process. The decomplexation rates are several orders of magnitude slower, corresponding to stability constants in the range 10^3-10^5M^{-1} for the different cations (NH$_4$$^+$ > K$^+$ > Rb$^+$ > Na$^+$). Clearly, the complexation mechanism must be concerted in the sense that loss of solvation at each step is compensated by formation of new bonds to the carrier molecule. This requires that the carrier molecule should be flexible enough to undergo rapid conformational change, as indeed seems to be the case.

The conformational change from uncomplexed nonactin to its complexes is quite drastic. It involves essentially a change from outward to inward pointing carbonyl and tetrahydrofurane oxygen atoms, which is brought about by large changes in two of the eight torsion angles in the monomeric unit (C(1)-C(2) and C(2)-C(3)), plus minor changes in the others. On the other hand, the changes that occur within the series of Na$^+$, K$^+$ and NH$_4^+$ complexes are quite smooth. As mentioned earlier, the carbonyl oxygens are closer to the centre of the cube than the ether oxygens in the Na$^+$ complex, almost equidistant in the K$^+$ complex, and further from the centre in the NH$_4^+$ complex. The overall expansion of the cube is associated mainly with an outward displacement of the carbonyl oxygens which is brought about by a gradual rotation of the plane of the ester groups, involving changes of 20° - 30° in the torsion angles around C(1)-C(2), C(7)-O(8), and symmetry-equivalent bonds. The changes in the remaining 24 torsion angles of the ring amount only to a few degrees. The overall changes in torsion angles with respect to the uncomplexed molecule (Fig. 8) are smallest for the NH$_4^+$ complex, greatest for the Na$^+$ complex, thus providing an interesting structural analogy to the ion-selectivity order (NH$_4^+$ > K$^+$ > Na$^+$).

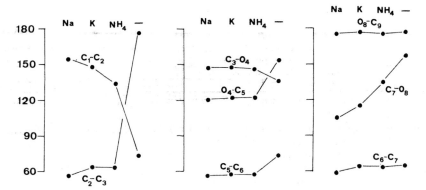

Fig. 8. *Torsion angles in 32-membered ring of free nonactin and its Na^+, K^+ and NH_4^+ complexes, averaged over S_4 symmetry in each case.*

From these structural correlations, a tentative mechanism for the complexation process can be proposed. Inspection of space-filling models shows that a hydrated K^+ cation can be inserted into the central cavity of a nonactin molecule with formation of hydrogen bonds between its water molecules and the various oxygen atoms of the prospective ligand. The actual complexation can then occur by a concerted process in which removal of the water molecules is coupled to conformational changes in the ligand, leading to gradual envelopment of the cation in the central cavity. The four ester-group oxygens are particularly well disposed to form hydrogen bonds with four equatorial water molecules of an octahedrally coordinated cation. By rotating the planes of the ester groups to bring the carbonyl oxygens inwards, the ester oxygens are tilted towards the outside of the molecule. The removal of water molecules attached to these atoms from the cation can thus occur simultaneously with the building up of strongly attractive ion-dipole interactions between the cation and the remaining eight oxygen atoms of the nonactin molecule. When this scenario is played out with space-filling models, the process is reminiscent of the envelopment of an insect or bird by one of those tropical carnivorous plants!

2.2 VALINOMYCIN - The cyclododecadepsipeptide valinomycin is perhaps the best known member of the macrocyclic group of ion-carriers. Because of its extraordinary K^+/Na^+ specificity it has been much used in electrode systems of high selectivity for K^+ in the presence of Na^+, discriminating against the latter by a factor of about 5000 or more *(21,22)*. Like nonactin, the uncomplexed molecule of valinomycin appears to enjoy considerable conformational flexibility in solution. Various kinds of spectroscopic evidence (infrared, Raman, nmr, optical rotatory dispersion, circular dichroism, ultrasonic absorption) have been interpreted in terms of an equilibrium mixture of at least three and possibly more conformations, in amounts that appear to depend strongly on the polarity of the solvent *(24, 26-30)*.

The crystal structure of a $KAuCl_4$ complex of valinomycin *(23)* showed that the K^+ ion is coordinated octahedrally to the six carbonyl oxygens of the ester groups, with the other six carbonyl oxygens (of the amide groups) engaged in a belt of hydrogen bonds with the amide hydrogen atoms of the valine residues as partners. A structure of this type, involving polypeptide β-structure elements, has also been derived on the basis of force-field calculations *(24)*. More detailed structural information has recently become available through the crystal structure analysis of a KI_3/KI_5 complex *(25)*, the result of which is depicted in Fig. 9.

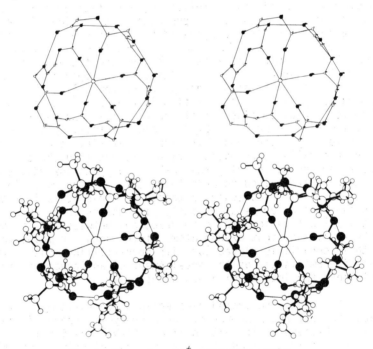

Fig. 9.   Stereoview of K^+-valinomycin complex.

If the differences between the side-chains of the L-lactic and D-hydroxyisovaleric acid residues are neglected, the complex is seen to have S_6 symmetry in good approximation. The six $K^+...O$

distances lie in the range 2.67 - 2.83 Å, roughly the same as in
the cubically coordinated nonactin and tetranactin complexes, and
the N-H...O hydrogen bond distances are 2.88 - 2.98 Å, with
N-H...O angles of 130-140°.

Three crystal structure analyses of free valinomycin have
been reported, two of them *(31,32)* dealing apparently with the
same triclinic crystal modification, which can be obtained from
polar solvents (methanol/water or acetone) as well as nonpolar
ones (n-octane). The unit cell contains two symmetry-independent
molecules with virtually the same conformation. This conformation is essentially the same as the one that occurs in the
previously studied crystal modification *(33)* crystallized from
n-octane, which was described as monoclinic and as containing
disordered solvent molecules. Karle *(31)* believes that this
modification is merely a twinned form of the triclinic modification. However, this uncertainty does not affect the general form
of the uncomplexed valinomycin molecule, as found in these studies, which is illustrated in Fig. 10. It is clear that the
conformation does not show, even approximately, a three-fold
axis, in contrast to the K$^+$-complex and all models proposed on
the basis of spectral data in which this symmetry element is retained, although it does show an approximate centre of symmetry,

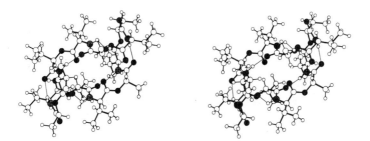

Fig. 10. Stereoview of uncomplexed valinomycin molecule.

again neglecting the difference between the L-lac and D-hyval side chains. The six amide hydrogens are still engaged in hydrogen bonding to carbonyl oxygens, but now two of the carbonyl acceptors belong to ester groups rather than amide groups. It is these 1-5 hydrogen bonds (the others are 1-4) that distort the cylindrical conformation of Fig. 9 into one with an elliptical cross-section in which none of the free carbonyl oxygens point directly towards the central cavity (Fig. 10). Another crystal modification (orthorhombic) of uncomplexed valinomycin has been obtained by crystallization from dimethyl sulphoxide *(31)* and yet another (monoclinic) from aqueous dioxane *(32)*, but their structures have not yet been established. A similar tendency to polymorphism was encountered in studies of K^+ valinomycin complexes crystallized under slightly different conditions *(25)*.

2.3 ENNIATINS - The K^+-complex of the cyclohexadepsipeptide enniatin B *(34)* consists of a rather flat disc with the central cation coordinated to all six carbonyl oxygen atoms in a flattened octahedral arrangement (Fig. 11). The six isopropyl side chains all point outwards to the lipophilic exterior of the disc, which is possible only by the alternating chirality of the subunits, D-α-hydroxyisovaleric acid and L-N-methylvaline. The role of the N-methyl substituents is presumably to prevent transannular hydrogen bonding in the free ligand. The hexagonal crystals of the KI-enniatin B complex are of poor quality and they have some unusual properties that made the crystal structure analysis rather difficult. The large, disc-shaped cation-complexes form stacks parallel to the *c* axis, and the I^- ions, introduced to facilitate the structure analysis, are statistically disordered in the channels between these stacks; because of the resultant smearing out of their electron-density peaks, the I^- ions hardly contribute to the scattering at all! On account of these difficulties, a satisfactory interpretation of the three-dimensional Patterson function

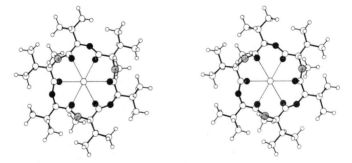

Fig. 11.   Stereoview of K^+-enniatin B complex.

could not be achieved, and Fig. 11 is actually based on an interpretation of the electron-density projection down the c axis.
While it is probably correct in its essential features, some of
the details may have to be amended when better data from more
suitable crystalline complexes become available. In particular,
it is not certain that the K^+-ions sit at the centres of the disc-
like molecules; they might possibly sit between successive molecules in a stack, being coordinated to the three upward pointing
carbonyls of the molecule below and the three downward pointing
carbonyls of the molecule above. In solution, sandwich-type
(enniatin)$_2$K$^+$ complexes appear to exist in equilibrium with the
1:1 complexes and to play a role in the ion-transport across
membranes (35).

Molecules of uncomplexed enniatin B show a threefold rotation
axis, the highest symmetry compatible with the structural formula,
in the crystalline state (36). The observed conformation (Fig.
12) is similar to that in the K$^+$ complex, the differences involving only slight changes in the orientation of the ester and amide
groups and of the isopropyl substituents. While this conformation seems to be preferred in polar solvents, another conformation of lower symmetry, with non-equivalent monomeric units, is
reported to occur in non-polar solvents (37).

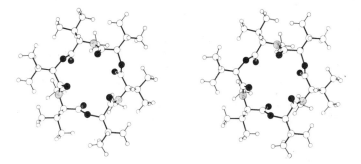

Fig. 12. Stereoview of uncomplexed enniatin B molecule.

To close our discussion of the structural features of this
group of ionophores (actins, valinomycin, enniatins) we remark
that, whereas the uncomplexed molecules appear to be rather
flexible, the corresponding complexes show highly regular coordination of the enclosed cations - cubic coordination for the
actins, octahedral for valinomycin and the enniatins. The ring
skeletons also show correspondingly high symmetries, at least
approximately - S_4 for the 32-membered ring in nonactin and
tetranactin complexes, S_6 for the 36-membered ring in the K^+-
valinomycin complex and for the 18-membered ring in the K^+-

enniatin B complex (neglecting the distinction between amide and ester groups). These high symmetries clearly provide optimal environments for complexation of spherical cations. Moreover, for the side-chains to be equivalent (*e.g.* all pointing outwards, as in the enniatin B complex), the rotatory-reflection symmetry of the ring skeletons imposes regular alternation in the chirality of the sub-units. This is then the structural basis for this characteristic feature of the valinomycin group of ionophores.

3.   NIGERICIN GROUP

In contrast to the structural features of the valinomycin group, the coordination polyhedra found in the ion-pair complexes of the nigericin group are typically formed by rather irregular arrangements of five to seven ether-type oxygen atoms. The carboxylate anion, which is always present, is sometimes involved in the coordination, as in nigericin itself *(38)*, sometimes not. Because of the additional ion-pair contribution to the binding energy, the details of the coordination environment do not seem to be so critical as in the complexes of the valinomycin group, where the ligand molecule is electrically uncharged.

3.1   MONENSIN – Characteristic features of the nigericin group are well illustrated by a comparison between the structures of the monensin Ag^+ salt *(39)*, which is isomorphous with the K^+ salt, and the free acid *(40)*. In the Ag^+ salt, the anion forms a macrocycle secured by a pair of hydrogen bonds between the oxygen atoms of the negatively charged carboxylate group and the two hydroxyl groups at the other end of the molecule, as shown in Fig. 13. The Ag^+ cation, enclosed in the resulting cavity, is coordinated to 6 oxygen atoms (4 ether, 2 hydroxyl) at distances of 2.4 - 2.7 Å in an irregular arrangement.

Irregular coordination polyhedra are much more difficult to describe in words than regular ones. There are actually seven topologically distinct polyhedra with six vertices, of which only two, the regular octahedron $(4)_6$ and the trigonal prism $(3)_6$, have attracted the attention of structural chemists (see Reference *(41)* for description of nomenclature). In keeping with the custom of describing low-symmetry coordination polyhedra as distorted versions of more symmetric figures, there is a tendency to refer all irregular arrangements of six ligand atoms to one or other of these two. However, the arrangement that occurs in the monensin-Ag^+ salt does not resemble either of these two figures very closely, as can be seen by inspection of Fig. 13. The actual arrangement corresponds more closely to a low-symmetry version of a second kind of octahedron $(5)_2(4)_2(3)_2$ (also known as a bi-capped tetrahedron) with maximum symmetry C_{2v}.

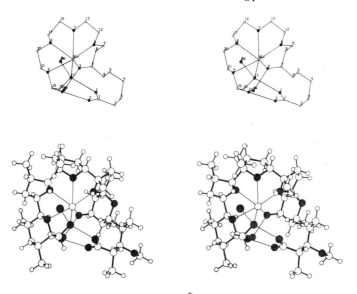

Fig. 13. Stereoview of Ag^+-monensin complex.

The spiro junction and the numerous ring substituents severely restrict the conformational freedom of the monensin molecule. It is not surprising, therefore, that the conformation of uncomplexed monensin *(40)* is rather similar to that of the complexed anion. There are, however, some noticeable differences, mainly in the number and arrangement of hydrogen bonds and hence in the relative positions of the oxygen atoms surrounding the cavity. One side of the cavity (the lower side in the stereographic representation of Fig. 14) is drawn together by two intramolecular hydrogen bonds O(10)H...O(4)H...O(6), and the opposite side is dilated by the presence of a water molecule hydrogen-bonded to O(1), O(7) and O(11). Some of the distances between oxygen atoms surrounding the cavity change by more than 1 Å on passing from Ag^+ salt to free acid, yet changes in individual torsion angles do not exceed 17° and are mostly much smaller. This shows how shape and function of a biologically important molecule can undergo significant alteration by the combined effect of a large number of small, cooperative structural changes associated with only minimal energy cost.

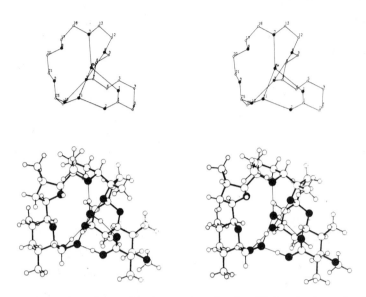

Fig. 14. Stereoview of uncomplexed monensin molecule with bound water molecule.

Monensin crystallizes from absolute ethanol as a monohydrate, and the bound water molecule may play a part in the complexation mechanism. It is situated on the more open, polar side of the molecule and could well function as one of the water molecules in the hydration sphere of a cation during the initial complexation step. It is easy to imagine that the next step would be displacement of the bound water molecule, possibly as H_3O^+, by the partially hydrated metal ion, which would then be coordinated on one side by three oxygens of the anion and on the other side by water molecules. The subsequent steps, further dehydration of the cation with simultaneous penetration into the cavity, can take place with a minimum of atomic reorganization and hence presumably with low activation energy.

3.2 MISCELLANEOUS - Other antibiotics of the nigericin group, whose structures have been found by X-ray analysis of heavy-metal salts include grisorixin *(42)*, alborixin *(43)*, dianemycin *(44)*, X-204A *(45)*, X-206 *(46)*, X-537A *(47)*, and probably a few others. In the case of antibiotic X-206, a subsequent X-ray analysis of the free molecule *(48)* revealed that the previously proposed structure of the Ag$^+$ salt *(46)* was incorrect, lacking three methyl groups and with one hydroxyl group incorrectly identified as methyl. This ought to dispel any idea that structures derived by X-ray analysis are above suspicion! On the whole, the structural features described for monensin are also present in the other members of the group.

4. BOROMYCIN AND DERIVATIVES

We turn now to another compound which does not, strictly speaking, belong to the nigericin group, but which shows many of the characteristic properties of an anionic ion-carrier. It is the compound obtained from boromycin by mild hydrolytic cleavage of D-valine. Boromycin itself is an interesting compound, an antibiotic obtained from certain strains of *streptomyces anti-bioticus*; it appears to be the first well defined organic compound found in Nature that contains boron, and its chemical formula, as determined by X-ray analysis of Rb^+ and Cs^+ salts of its des-valine hydrolysis product *(49)*, is shown in Fig. 15. The molecule is a Böeseken complex of boric acid with a macrodiolide consisting of two almost identical halves. The overall shape of the anion is roughly spherical with a lipophilic surface and a cleft lined with oxygen atoms. The cations are housed in this cleft, coordinated by eight oxygen atoms in irregular coordination at distances of 2.80 - 3.17 Å (Rb^+ salt). The coordination polyhedron does not resemble any of the standard types for 8-coordination - cube, square antiprism or rhombic dodecahedron - and corresponds most closely to No. 47 in the list of the 257 non-isomorphous polyhedra with eight vertices *(41)*.

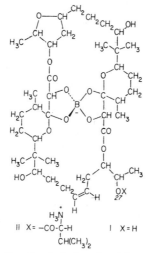

Fig. 15. Chemical formula of boromycin (II) and des-valine hydrolysis product (I).

One might have expected that it is the spiro boron junction
that is mainly responsible for maintaining the secondary structure of these molecules. However, the overall conformation of
des-boron-des-valine-boromycin *(50)* is remarkably similar to that
of the boron-containing molecule (Fig. 16), the changes in the
torsion angles of the 28-membered ring being all less than 13°.
In particular, the oxygen-lined cleft that houses the counter-ion
to the negatively charged boron atom in the Böeseken complex is
also present in the des-boron derivative, where it accommodates
a water molecule, surrounded by eight oxygen atoms in much the
same arrangement as in the Rb^+ salt.

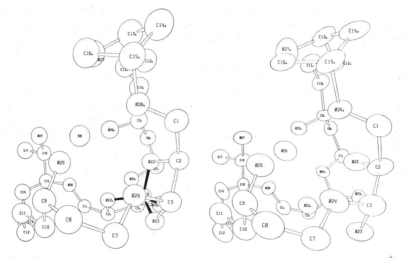

Fig. 16. Perspective view of main 28-membered ring in des-valine-boromycin Rb^+
salt (left) and des-valine-des-boron-boromycin (right) (with bound water mole-
cule). Oxygen substituents are shown but others have been omitted for the sake
of clarity.

In boromycin itself the negative charge of the spiro boron
atom is neutralized by the $-NH_3^+$ group of a D-valine residue. Of
the three hydroxyl groups available as possible esterification
sites, only O(27) would permit the $-NH_3^+$ group to be accommodated
in the oxygen-lined cleft, which would then be plugged by the
lipophilic isopropyl side-chain of the amino-acid. The biological
significance of this structural feature is unknown but it may become clearer when the biochemical function of boromycin is better
understood.

5. CROWN ETHERS AND CRYPTATES

So far, we have discussed exclusively naturally occurring ionophores, but almost simultaneously with the discovery of the ionophoric properties of these molecules, synthetic compounds with very similar properties were becoming available - macrocyclic polyethers, usually known as "crown" ethers *(51,52)* and macro-bicyclic polyethers containing tertiary amino groups at the bridgeheads, usually known as "cryptates" *(53)*. Excellent reviews of the factors influencing the design of these synthetic complex-ing agents and of the structures of such complexes are available *(54,55)*. Apart from their intrinsic interest as complexing agents, the synthetic compounds are useful models for studying the various factors involved in ionophoric activity, since the number, nature and relative arrangement of the coordination sites can be systematically varied. Here we discuss only the results of some structural studies on the unsubstituted 18-crown-6 (1,4, 7,10,13,16-hexaoxacyclooctadecane) and its alkali cation com-plexes. The objective was to show how this simple, potentially highly symmetrical but flexible ligand might adapt its conforma-tion to the different steric requirements of alkali cations.

The conformations of the hexaether found in the complexes with K^+, Rb^+ and Cs^+ thiocyanates *(56,57)* have approximate D_{3d} symmetry and are virtually identical, the torsion angles about C-C bonds being close to $65°$ in all three cases, those about the C-O bonds being close to $180°$. The six oxygen atoms are alter-nately about 0.20 Å above and below their mean plane to form a nearly planar hexagon of side approximately 2.80 Å. The K^+ ion sits at the centre of such a hexagon ($K^+...O$, 2.77 - 2.83 Å, almost the same as the $K^+...O$ (ether) distances in nonactin and tetranactin (Table 1)). The coordination is completed by two terminal atoms of thiocyanate anions at 3.19 Å; these anions occupy crystallographic centres of symmetry and thus appear to be completely disordered. A view of the crystal structure is shown in Fig. 17. Although the macrocycle has the same D_{3d} conformation in the RbNCS and CsNCS complexes, the crystal struc-tures are different. The cations are displaced from the mean plane of the hexagon, by 1.19 Å (Rb^+) and 1.44 Å (Cs^+) so that the $M^+...O$ distances fall into two sets in each case (Rb^+, 2.93 - 3.00 Å and 3.01 - 3.15 Å; Cs^+, 3.04 - 3.10 Å and 3.16 - 3.27 Å). Each cation is additionally coordinated to two bridging thio-cyanate anions ($Rb^+...N$, 3.02 Å; $Cs^+...N$, 3.31 Å) to produce dis-crete dimeric units, as seen in Fig. 18. In the crystalline Rb^+ complex, the anions still show some degree of disorder, with a 4:1 preference for the nitrogen atoms to occupy the additional coordination sites; in the Cs^+ complex the additional coordination

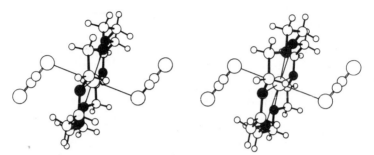

Fig. 17. Stereoview of KNCS complex of 18-crown-6. The K⁺ and NCS ions occupy crystallographic centres of inversion (NCS disordered).

sites are exclusively occupied by nitrogen. Since the radius of the cavity (1.40 Å) is only slightly smaller than the standard ionic radius of Rb^+ (1.48 Å), one might have expected this cation to occupy the centre of a perhaps slightly dilated hexagon of oxygens. However, in the (Rb^+, Na^+)NCS complex of a closely similar dibenzo-18-crown-6 the Rb^+ atom lies 0.94 Å out of the mean plane and is coordinated to the nitrogen atom of a thiocyanate anion at 2.94 Å in addition to the six oxygens at 2.86 - 2.94 Å *(58)*.

Whereas Rb^+ and Cs^+ are too large to fit into the central cavity, Na^+, with ionic radius of 0.95 Å, is too small to fill it. Several possibilities can be imagined:

(a) the hexagon of oxygen atoms is uniformly contracted, which would lead to serious O...O repulsions;

(b) the cation sits at the centre of the hexagon with non-optimal Na^+...O distances of 2.8 Å *(58)*;

(c) the cation is displaced from the centre of the hexagon, making shorter contacts with some oxygens, longer ones with others *(59)*;

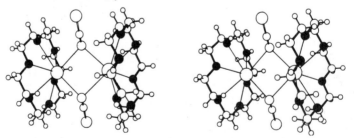

Fig. 18. Stereoview of dimeric structural unit of CsNCS complex of 18-crown-6 (the RbNCS complex is very similar).

(d) the cation "rattles" in the cavity, a possibility that would be expressed in unusually large vibrational amplitudes of the central atom;

(e) the hexaether adopts a different conformation that is more compatible with the coordination preferences of the cation *(60)*.

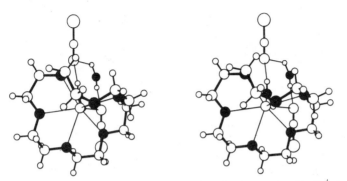

Fig. 19. Stereoview of hydrated NaNCS complex of 18-crown-6. The Na$^+$ ion has pentagonal bipyramidal coordination.

In the hydrated NaNCS complex of the unsubstituted hexaether, possibility (e) is fulfilled, the change in conformation (Fig. 19) being quite drastic *(61)*. One of the oxygen atoms is drawn out of the plane of the other five by nearly 2 Å to give a somewhat irregular pentagonal pyramidal coordination of the cation (Na$^+$... O distances, 2.45 Å (apical O) to 2.62 Å). The water molecule, at 2.32 Å, is situated at the opposite apex, to complete a pentagonal bipyramid. The irregular conformation of the hexaether required to produce this 5 + 1 + 1 coordination looks quite different from the D_{3d} conformation, yet the relative positions of 13 of the 18 ring atoms are left essentially unaltered. Who would have predicted this structure - from molecular mechanics calculations, for example - or derived it from spectroscopic evidence? In the Ca(NCS)$_2$ complex, the Ca^{+2} ion, with approximately the same radius at Na$^+$, occurs in an undistorted or only slightly distorted 6 + 2 coordination *(62)*.

The unsubstituted hexaether does not show the symmetrical D_{3d} conformation, neither in solution *(52)* nor in the solid state *(63)*. Instead, it adopts a centrosymmetric conformation (shown in Fig. 20) containing three different types of sub-unit. Here the cavity at the centre of the molecule has shrunk to nothing - the opposite sides being in contact with one another. This is in contrast to the behaviour of 1,10-diaza-4,7,13,16-dioxacyclooctadecane, where a pair of opposite oxygens have been replaced by

nitrogen atoms. Here the uncomplexed molecule *(64)* has virtually
the same conformation as in the KNCS complex *(65)*, with a hole at
its centre.

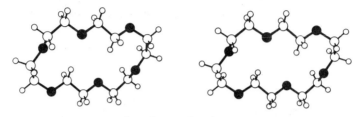

Fig. 20. Stereoview of uncomplexed 18-crown-6 molecule.

The stability of a given cation-ionophore complex depends on
the difference between the "binding energy" of the cation to the
organic ligand and the solvation energy of the cation in water
(or other solvent). While hydration energies are available for
the common cations *(66)* and are moderately well understood *(67)*,
binding energies to complex organic ligands are not. They would
seem to depend on a large number of factors - the number and
nature of the binding sites (charge, polarity, polarisability,
van der Waals radius) and of the bonds between these sites, which
may permit greater or lesser flexibility of the molecule as a
whole. Some of these factors have been discussed by Krasne and
Eisenman *(68)*, others by Simon and Morf *(69,70)* and by Lehn *(54)*.
The structures observed for the alkali cation complexes of
18-crown-6 suggest that some of the simple models that have been
introduced should be regarded with considerable reservation. For
example, the concept of the equilibrium cavity (Gleichgewichts-
hohlraum), applied to these complexes, could be taken to imply
that the central cavity of the D_{3d} conformation should be modified more or less, depending on the size of the cation it accommodates. It is true that K^+, which has the optimal fit, forms
the most stable complex with log K_S = 6.1 in methanol at 25° C
(71), and that the cations that do not fit so well form less
stable complexes (log K_S = 4.3 and 4.6 for Na^+ and Cs^+, respectively *(71)*. However, for cations larger than K^+ the size and
shape of the cavity remains essentially unaltered; the larger
ions penetrate as far as they can and make up for the less efficient coordination with the ligand by interacting with anions.
For cations smaller than K^+, a completely different arrangement
of coordinating sites may be adopted, as in the Na^+ complex of
18-crown-6, or it may not, as in the Na^+ complex of the dibenzo
derivative *(58)*, which has almost the same stability constant in
methanol *(71)*.

6. CONCLUDING REMARKS

This lecture has covered only a small fraction of the structural information on ionophores and their cation complexes, and has hardly touched the extensive information available on complexation rates, stability constants and membrane permeabilities, not to mention the isolation and biochemical properties of these compounds. The field is very broad and it is one where several traditional areas of research - natural product chemistry; synthetic organic chemistry, inorganic chemistry, physical chemistry - all merge into one another. It has certainly provided us with a number of fascinating problems over the last 9 years, and we hope we have shown something of how a preoccupation with the structural aspects may contribute to the problem as a whole.

The computer-drawn plots shown in this article have all been prepared with the ORTEP programme *(72)*.

ACKNOWLEDGEMENT. Our work on this topic has been generously supported over the years by the Swiss National Fund for the Advancement of Scientific Research.

REFERENCES

1. C. Moore and B.C. Pressman, *Biochem. Biophys. Res. Comm.*,**15**, 562 (1964).
2. S.N. Graven, H.A. Lardy, D. Johnson and A. Rutter, *Biochemistry*,**5**, 1726 (1966).
3. Z. Štefanac and W. Simon, *Chimia*,**20**, 436 (1966).
4. P. Mueller and D.O. Rudin, *Biochem. Biophys. Res. Comm.*,**26**, 398 (1967).
5. T.E. Andrioli, M. Tieffenberg and D.C. Tosteson, *J. Gen. Physiol.*,**50**, 2527 (1967).
6. A.A. Lev and E.P. Buzhinsky, *Cytology (U.S.S.R.)*,**9**, 102 (1967).
7. J. Dominguez, J.D. Dunitz, H. Gerlach and V. Prelog, *Helv. Chim. Acta*,**45**, 129 (1962).
8. B.T. Kilbourn, J.D. Dunitz, L.A.R. Pioda and W. Simon, *J. Mol. Biol.*,**30**, 559 (1967); M. Dobler, J.D. Dunitz and B.T. Kilbourn, *Helv. Chim. Acta*,**52**, 2573 (1969).
9. M. Dobler and R.P. Phizackerley, *Helv. Chim. Acta*,**57**, 664 (1974).
10. K. Neupert-Laves and M. Dobler, *Helv. Chim. Acta*,**59**, 614 (1976).
11. Y. Iitaka, T. Sakamaki and Y. Nawata, *Chem. Letters (Japan)*, 1225 (1972).

12. Y. Nawata, T. Sakamaki and Y. Iitaka, *Chem. Letters (Japan)*, 151 (1975).
13. T. Sakamaki, Y. Iitaka and Y. Nawata, *Acta Crystallogr.*,**B32**, 768 (1976).
14. M. Dobler, *Helv. Chim. Acta*,**55**, 1371 (1972).
15. Y. Nawata, T. Sakamaki and Y. Iitaka, *Acta Crystallogr.*,**B30**, 1047 (1974).
16. I.M. Asher, G.D.J. Phillies and H.E. Stanley, *Biochem. Biophys. Res. Comm.*,**61**, 1356 (1974).
17. J.H. Prestegard and S.I. Chan, *Biochemistry*,**8**,3921 (1969).
18. E. Pretsch, M. Vasák and W. Simon, *Helv. Chim. Acta*,**55**, 1098 (1972).
19. Y. Kyogoku, U. Masaharu, H. Akutsu and Y. Nawata, *Biopolymers*, **14**, 1049 (1975).
20. R. Winkler, *Structure and Bonding*,**10**, 1 (1972).
21. L.A.R. Pioda and W. Simon, *Chimia*,**23**, 72 (1969).
22. M.S. Frant and J.W. Ross, Jr., *Science*,**167**, 987 (1970).
23. M. Pinkerton, L.K. Steinrauf and P. Dawkin, *Biochem. Biophys. Res. Comm.*,**53**, 512 (1969).
24. D.F. Mayers and D.W. Urry, *J. Am. Chem. Soc.*,**94**, 77 (1972).
25. K. Neupert-Laves and M. Dobler, *Helv. Chim. Acta*,**58**, 432 (1975).
26. V.T. Ivanov, I.A. Laine, N.D. Abdulaev, L.B. Senyavina, E.M. Popov, Yu A. Ovchinnikov and M.M. Shemyakin, *Biochem. Biophys. Res. Comm.*,**34**, 803 (1969).
27. M. Ohnishi and D.W. Urry, *Biochem. Biophys. Res. Comm.*,**36**, 194 (1969); *Science*,**168**, 1091 (1970).
28. D.H. Hayes, A. Kovalsky and B.C. Pressman, *J. Biol. Chemistry*, **244**, 502 (1969).
29. D.J. Patel and A.E. Tonelli, *Biochemistry*,**12**, 486 (1973).
30. T. Funk, F. Eggers and E. Grell, *Chimia*,**26**, 637 (1972).
31. I.L. Karle, *J. Am. Chem. Soc.*,**97**, 4379 (1975).
32. G.D. Smith, W.L. Duax, D.A. Langs, G.T. de Titta, J.W. Edmonds, D.C. Rohrer and C.M. Weeks, *J. Am. Chem. Soc.*,**97**, 7242 (1975).
33. W.L. Duax, H. Hauptman, C.M. Weeks and D.A. Norton, *Science*, **176**, 911 (1972); W.L. Duax and H. Hauptman, *Acta Crystallogr.*, **B28**, 2912 (1972).
34. M. Dobler, J.D. Dunitz and J. Krajewski, *J. Mol. Biol.*,**42**, 603 (1969).
35. V.T. Ivanov, A.V. Evstratov, L.V. Sumskaya, E.I. Melnik, T.S. Chumburidze, S.L. Portnova, T.A. Balashova and Yu A. Ovchinnikov, *F.E.B.S. Letters*,**36**, 65 (1973).
36. C. Kratky and M. Dobler, unpublished work.
37. Yu A. Ovchinnikov, V.T. Ivanov, A.V. Evstratov, V.F. Bystrov, N.D. Abdullaev, E.M. Popov, G.M. Lipkind, S.F. Arkhipova,

E.S. Efremov and M.M. Shemyakin, *Biochem. Biophys. Res. Comm.*, **37**, 668 (1969).

38. L.K. Steinrauf, M. Pinkerton and J.W. Chamberlin, *Biochem. Biophys. Res. Comm.*,**33**, 29 (1968); T. Kubota, S. Matsutani, M. Shiro and H. Koyama, *Chem. Commun.*, 1541 (1968); M. Shiro and H. Koyama, *J. Chem. Soc. B*, 243 (1970).

39. A. Agtarap, J.W. Chamberlin, M. Pinkerton and L.K. Steinrauf, *J. Am. Chem. Soc.*,**89**, 5737 (1967); M. Pinkerton and L.K. Steinrauf, *J. Mol. Biol.*,**49**, 533 (1970).

40. W.K. Lutz, F.K. Winkler and J.D. Dunitz, *Helv. Chim. Acta*,**54**, 1103 (1971).

41. D. Britton and J.D. Dunitz, *Acta Crystallogr.*,**A29**, 362 (1973).

42. M. Alleaume and D. Michel, *Chem. Commun.*, 1422 (1970); 175 (1972).

43. M. Alleaume, B. Busetta, C. Farges, P. Gachon, A. Kergomard and T. Staron, *Chem. Commun.*, 411 (1975).

44. E.W. Czerwinski and L.K. Steinrauf, *Biochem. Biophys. Res. Comm.*,**45**, 1284 (1971); L.K. Steinrauf, E.W. Czerwinski and M. Pinkerton, *ibid.*, 1279.

45. N.D. Jones, M.O. Chaney, J.W. Chamberlin, R.L. Hamill and S. Chen, *J. Am. Chem. Soc.*,**95**, 3399 (1973).

46. J.F. Blount and J.W. Westley, *Chem. Commun.*, 927 (1971).

47. S.M. Johnson, J. Herrin, S.J. Lin and I.C. Paul, *J. Am. Chem. Soc.*,**92**, 4428 (1970); C.A. Maier and I.C. Paul, *Chem. Commun.*, 181 (1971); E.C. Bissell and I.C. Paul, *ibid.*, 967 (1972).

48. J.F. Blount and J.W. Westley, *Chem. Commun.*, 533 (1975).

49. J.D. Dunitz, D.M. Hawley, D. Miklos, D.N.J. White, Y. Berlin, R. Marusic and V. Prelog, *Helv. Chim. Acta*,**54**, 1709 (1971).

50. W. Marsh, J.D. Dunitz and D.N.J. White, *Helv. Chim. Acta*,**57**, 10 (1974).

51. C.J. Pederson, *J. Am. Chem. Soc.*,**89**, 7107 (1967);**92**, 386 (1970).

52. J. Dale and P.O. Kristiansen, *Acta Chem. Scand.*,**26**, 1471 (1972).

53. J.M. Lehn and J.P. Sauvage, *Chem. Commun.*, 40 (1971).

54. J.M. Lehn, *Structure and Bonding*,**16**, 1 (1973).

55. M.R. Truter, *Structure and Bonding*,**16**, 71 (1973).

56. P. Seiler, M. Dobler and J.D. Dunitz, *Acta Crystallogr.*,**B30**, 2744 (1974).

57. M. Dobler and R.P. Phizackerley, *Acta Crystallogr.*,**B30**, 2746 (1974).

58. D. Bright and M.R. Truter, *J. Chem. Soc. B.*, 1544 (1970).

59. M.A. Bush and M.R. Truter, *J. Chem. Soc. B.*, 1440 (1971).

60. M. Mercer and M.R. Truter, *J. Chem. Soc. Dalton*, 2215 (1973).

61. M. Dobler, J.D. Dunitz and P. Seiler, *Acta Crystallogr.*,**B30**, 2741 (1974).

62. J.D. Dunitz and P. Seiler, *Acta Crystallogr.*, **B30**, 2750 (1974).
63. J.D. Dunitz and P. Seiler, *Acta Crystallogr.*, **B30**, 2739 (1974).
64. M. Herceg and R. Weiss, *Bull. Soc. Chim. Fr.*, 549 (1972).
65. D. Moras, B. Metz, M. Herceg and R. Weiss, *Bull. Soc. Chim. Fr.*, 551 (1972).
66. R.M. Noyes, *J. Am. Chem. Soc.*, **84**, 513 (1962).
67. C.S.G. Phillips and R.J.P. Williams, "Inorganic Chemistry", Vol. I, Oxford Univ. Press, New York, 1965, p. 160.
68. S. Krasne and G. Eisenman in "Membranes", Vol. 2, G. Eisenman, Ed., Decker, New York, 1973, p. 277.
69. W. Simon and W.E. Morf in "Membranes", Vol. 2, G. Eisenman, Ed., Decker, New York, 1973, p. 329.
70. W.E. Morf and W. Simon, *Helv. Chim. Acta*, **54**, 794 (1971); W. Simon, W.E. Morf and P.C. Meier, *Structure and Bonding*, **16**, 113 (1973).
71. H.K. Frensdorff, *J. Am. Chem. Soc.*, **93**, 600 (1971).
72. C.K. Johnson, ORTEP Report ORNL-3794, Oak Ridge National Laboratory, Oak Ridge, Tennessee (1965).

metal-OH and its ability to hydrolyse (or hydrate) substrates of biological interest

DAVID A. BUCKINGHAM

Research School of Chemistry, The Australian National University, Canberra, Australia

1. INTRODUCTION

In recent years there has been a rapid development in the understanding of how enzymes function. Specifically, the position of the reactive site in the large protein molecule usually involved, the sequence of reactions which go on there, and some clues as to the mechanism, have been forthcoming *(1)*. Important in these processes is the hydrolysis (or hydration) of carbonyl and phosphoryl type substrates such as CO_2, carbon esters, phosphate esters and anhydrides, and peptides *(1)*. Many of these processes are part of a more complex transfer property of the enzyme but some are complete in themselves. It is perhaps not so surprising now that a significant number of these systems require a divalent metal ion *(2-4)*. More surprising is the finding that this usually turns out to be Zn^{2+}. Thus the high concentration of Zn^{2+} in mammals (2.4 g per 70 Kg in humans) is exceeded only by that of Fe^{2+}/Fe^{3+} (5.4 g per 70 Kg), and most of it is earmarked for enzyme function. Indeed, without zinc we would not survive for long and its deficiency leads to severe metabolic disorders which can be reversed most rapidly by the simple application of Zn^{2+} salts. The greater becomes the detailed understanding of these proteolytic processes, the more significant appears the role of the metal ion. It is on work we have done in our laboratory in this area that I wish to speak of today.

2. BACKGROUND

A metal ion, such as Zn^{2+}, can facilitate hydrolysis or nucleophilic displacement by some other species in essentially two ways: direct polarization of the substrate and external attack, or by the generation (by ionization) of a particularly reactive basic reagent. The scheme below gives these possibilities:

SCHEME 1 - Metal Ion Catalysis of Displacement Reactions on RCOY
         Species

(1) *"Lewis" Acidity*

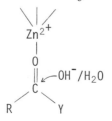
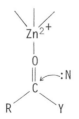

This mechanism can be modified slightly to include polarization
by coordination to the leaving group, but this property is probably useful only with phosphate ester type substrates (incipient
metaphosphate production)

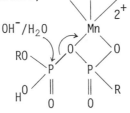

(2) *Ionization to Generate Coordinated Nucleophile*

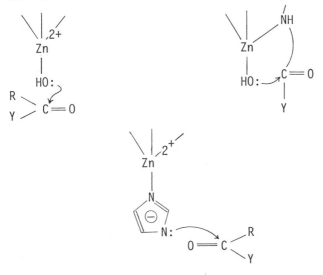

The direct polarization of the carbonyl function generates a more
electrophilic carbon centre. Different metal ions have different
abilities in this respect, depending largely on their overall
charge, size, coordination number, and ease of displacement of
(usually) a coordinated water molecule.

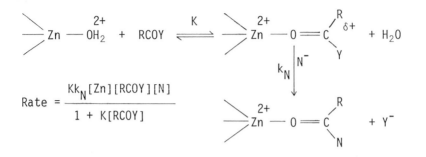

$$\text{Rate} = \frac{Kk_N[\text{Zn}][\text{RCOY}][\text{N}]}{1 + K[\text{RCOY}]}$$

For Co(III) complexes, this leads to an acceleration in the nucleophilic part (k_N) of 10^4 - 10^6 for carbonyl-bound esters and
amides, and we have reason to believe that aquated divalent metal
ions such as Cu^{2+}_{aq}, Co^{2+}_{aq} and Zn^{2+}_{aq} will not be much less able,
although an unfavorable binding constant (K) is usually involved.
The beauty of the Co(III) systems is that it allows us to separate the two factors involved. Water exchange, and substitution
in the coordination sphere is in general a very slow process
($t_{\frac{1}{2}} \approx$ mins to hrs) which means, of course, that Co(III) is useless
in the enzyme situation.
    Rate data given in Table 1 show that Co^{3+} and H^+ have a simi-
lar polarizing power (rate increase of 50-100) when two atoms
removed from the carbonyl function, but that H^+ is much more
effective than Co^{3+} (by $\sim 10^4$) when attached directly to the carbonyl oxygen (rate increases of $\sim 10^{11}$ and 10^7 respectively).
The difficulty here is that, whereas the Co(III) situation is
100% effective at pH 7, impossibly strong acid conditions (10^2 -
10^6 M!!) are required to realise the H^+ property. There also
appears some effect of ring size, with hydrolysis of the six-
membered β-alanine ester chelate *(5)* being accelerated by $\sim 10^5$
and the comparable five-membered chelate *(6)* involving a factor
of $\sim 10^7$; an increase of 10^4 is involved in the hydrolysis of
the O-bound dimethyl formamide complex *(7)*.

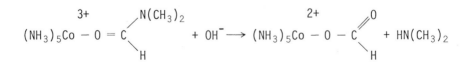

TABLE 1: VARIATION IN THE RATES[a] OF GLYCINE ESTER AND DIPEPTIDE HYDROLYSIS DEPENDING ON LOCATION OF "LEWIS ACID"

Esters

Substrate	k_{H_2O} (M^{-1}sec^{-1})	k_{OH} (M^{-1}sec^{-1})
$NH_2CH_2CO_2C_2H_5$		0.6
$NH_3^+CH_2CO_2C_2H_5$	$\sim 1 \times 10^{-10}$	24
$(NH_3)_5Co^{3+}-NH_2CH_2CO_2C_2H_5$		50
$(en)_2Co^{3+}$ chelate ester	1.3×10^{-4}	1×10^7
$NH_3^+CH_2C(O^+H)OC_2H_5$	$\sim 6^{b}$	
$Cu^{2+}aq\ -NH_2CH_2CO_2C_2H_5 \rightleftharpoons K$ (en) chelate		
$Cu^{2+}aq$ chelate ester	0.8×10^{-6c}	0.8×10^{5c}
$NH_2CH_2CH_2CO_2C_3H_7$		2×10^{-2}
$(en)_2Co^{3+}$ chelate ester OC_3H_7		5×10^3

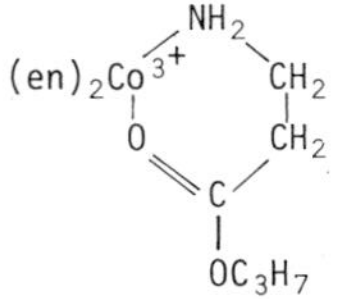

Amides

Substrate	k_{H_2O} (M^{-1}sec^{-1})	k_{OH} (M^{-1}sec^{-1})
$NH_2CH_2CONH_2$		2.2×10^{-3}
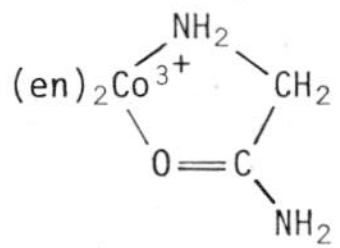 $(en)_2Co^{3+}$ chelate amide (NH_2)	10^{-9}(est.)	25
$NH_3^+CH_2C(O^-H)-NH_2$	$\sim 10^{-5}$	
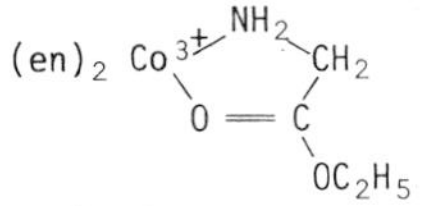 $NH_2CH_2CONHCH_2CO_2^-$		6×10^{-6}
$(en)_2Co^{3+}$ chelate $NHCH_2CO_2^-$		3
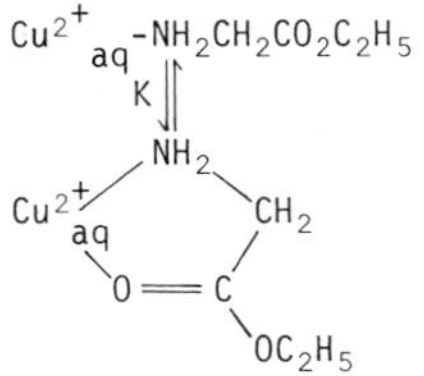 $NH_3^+CH_2CONHCH_2CO_2H$	$\sim 2 \times 10^{-4}$	

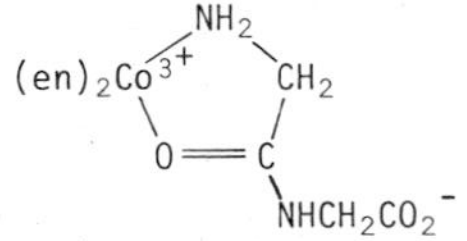

a At 25°; μ(usually) = 1.0 but sometimes not.

b Assuming pKa = -6.5 (ester); -1.78 (glyNH₂); -3.1 (glygly).

c $k_{obs} = kK$ (i.e., the second-order rate constant has built into it an (unfavorable) equilibrium constant, K ≈ 0.1, estimate).

$$k_{OH} = 1.1 \ M^{-1}sec^{-1} \ (25°\mu = 1.0)$$

$$\Delta H^{\neq} = 13.8 \pm 0.5 \ kcal \ mol^{-1}$$

$$\Delta S^{\neq} = -12 \pm 1 \ cal \ deg^{-1}mol^{-1}$$

$$O = C \begin{array}{l} \diagup N(CH_3)_2 \\ \diagdown H \end{array} + OH^- \longrightarrow HCO_2^- + HN(CH_3)_2$$

$$k_{OH} = 10^{-4} \ M^{-1} \ sec^{-1}$$

$$\Delta H^{\neq} = 14.2 \ kcal \ mol^{-1}$$

$$\Delta S^{\neq} = -28 \ cal \ deg^{-1}mol^{-1}$$

All these differences arise from large increases in $\Delta S^{\neq}$ ($\Delta H^{\neq}$ remains essentially unchanged), but an understanding of the effects responsible remains obscure. Furthermore, if an equilibrium constant of 0.1 is assumed for chelation of the monodentate glycine ethyl ester in the Cu^{2+}_{aq} reaction *(8)*(Table 1), then $k_{OH} \approx$ 1 x $10^6 M^{-1}sec^{-1}$ (hydrolysis by OH^-) which is within a factor of 10 of that found for Co^{3+}; a similar result holds for the H_2O reaction ($k_{H_2O} \approx$ 4 x $10^{-4}sec^{-1}$). These combined facts support the view that *divalent metal ions increase solvolytic hydrolysis rates by ~10^4 when a mechanism involving direct polarization of the carbonyl function is responsible but these effects can be modified by changes in the solvation properties of the species involved.* However, more work is needed in this area. From the Co(III) promoted reactions we have excellent evidence that the rate determining step in ester (and amide) hydrolysis is elimination of the alcohol (or amine) function from the tetrahedral intermediate *(5)*. This complicates any analysis of the $\Delta H^{\neq}$ and $\Delta S^{\neq}$ factors. However, a similar catalysis is

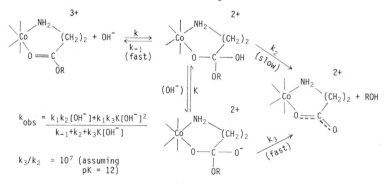

$$k_{obs} = \frac{k_1k_2[OH^-]+k_1k_3K[OH^-]^2}{k_{-1}+k_2+k_3K[OH^-]}$$

$$k_3/k_2 \approx 10^7 \ (assuming \ pK = 12)$$

found in nitrile hydration where no leaving group is involved, and the effect still resides largely in $\Delta S^{\neq}$ *(9)*.

$$\langle O \rangle\text{—CN} + \text{OH}^- \longrightarrow \langle O \rangle\text{—}\overset{\displaystyle \|}{\underset{\displaystyle O}{C}}\text{—NH}_2$$

$$k_{OH} = 8\times10^{-6}M^{-1}sec^{-1}(25^\circ)$$
$$\Delta H^{\neq} = 19.9 \text{ kcal mol}^{-1}$$
$$\Delta S^{\neq} = -15 \text{ cal deg}^{-1}mol^{-1}$$

$$\langle O \rangle\text{—CNCo}^{3+}(NH_3)_5 + \text{OH}^- \longrightarrow \langle O \rangle\text{—}\overset{\displaystyle \|}{\underset{\displaystyle O}{C}}\text{—NHCo}^{2+}(NH_3)_5$$

$$k_{OH} = 19 \text{ M}^{-1}sec^{-1}$$
$$\Delta H^{\neq} = 16.5 \text{ kcal mol}^{-1}$$
$$\Delta S^{\neq} = +3 \text{ cal deg}^{-1}mol^{-1}$$

This provides a short summary of some results found for systems *directly activated* by the metal ion but, in this lecture, I wish to concentrate on the alternative mode of activation depicted in Scheme 1, *viz.* the ability of a coordinated species (OH, OH_2, imidazole) to act as the nucleophilic agent. It has often been said that such a process would not be very efficient since, whereas coordination to the metal ion will undoubtedly increase the acidity of a water molecule (for example) and thereby produce a higher concentration of "effective" hydroxide in neutral aqueous solutions (pH 6-8), it might also be expected to decrease the basicity towards the carbonyl function *by a similar margin.* Such a result would mean a net zero, or marginal, change in reactivity compared with the solvent species. It is the purpose of this lecture to demonstrate that this is not entirely so.

I will deal with our results in two parts by considering the simple bimolecular processes first:

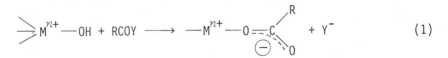

$$\eqno(1)$$

followed by a discussion of some of the more facile intramolecular processes:

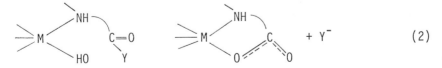

$$\eqno(2)$$

To facilitate a close examination of these reactions we have used, in the main, Co(III)-hydroxo complexes. There are several reasons for this. Firstly and most importantly, they are usually slow to exchange their ligands in aqueous solution, including the aquo group, ("inert" in the Taube classification), and this allows us to use ^{18}O tracers to follow the path of the coordinated aquo or hydroxo group and thereby distinguish between the direct nucleophilic and general base paths for hydrolysis. Secondly, this inertness to exchange means that we know what coordination environment we are dealing with even in complicated situations; we can use mixed ligand complexes without fear of disproportionation, and with a good understanding of the isomerization and asymmetric properties about the metal center. Sometimes such properties can be used to advantage. It allows us to vary the overall charge on the complex (anionic or neutral ligands) while retaining a close-to-octahedral geometry about the metal center; it makes possible crystal structure verification of reactants, products, and intermediates in many cases. Indeed, the tremendous wealth of information concerning Co(III) chemistry is at our disposal, and we have had some experience in this area.

These advantages bring with them obvious problems when one considers their parallel (or lack of it) in the enzymic processes. Thus the Co(III) promoted reactions are stoichiometric rather than catalytic with the product of hydrolysis or hydration remaining firmly bound to the metal center.

In this respect, it is perhaps not too surprising that oxidation of active Co(II) carbonic anhydrase *(10)* and Co(II) alkaline phosphatase *(11)* to their corresponding Co(III) enzymes results in complete inhibition of the hydrolysis process (p-nitrophenylacetate, or p-nitrophenylphosphate) although some evidence exists that substrate binding to the enzyme still occurs. Also, oxidation of Co(II) carboxypeptidase to the Co(III) enzyme *(12)* inhibits peptidase activity entirely, although some esterase activity seems to remain. However, we believe that the Co(III) systems imitate the labile metals in their ability to polarize adjacent substrate molecules and to activate coordinated nucleophilic groups. The following essentially unpublished results support this claim.

3. BIMOLECULAR REACTIONS INVOLVING M-OH SPECIES

$$\underset{}{\overset{}{\text{>}}}M-OH \ + \ RCOY \ \xrightarrow{\ k_{MOH}\ } \ \underset{}{\overset{}{\text{>}}}M-O-\underset{\overset{\|}{O}}{C}R \ + \ YH$$

Questions we asked of the reactions were:
(1) Do the MOH species react directly with the carbonyl substrates (esters, CO_2, anhydrides, amides, nitriles, etc.) as suggested by the above equation?
(2) Is there a general base component k_{gb} (indirect path)?
(3) How effective is M-OH *vs.* M-OH$_2$?
(4) Is the decrease in basicity of the M-OH species (pKa, 6-10 of conjugate acid) compared with OH$^-$ (pKa, 15.7) paralleled by a similar decrease in nucleophilicity towards the carbonyl center, *i.e.*, does a simple Brønsted correlation hold for these reactions, $\Delta \log k_{MOH} = \beta \, \Delta$ pKa + const.?
(5) Is the charge on M^{n+} important, or is the charge on the M-OH complex as a whole important?
(6) Do variations in coordination number, stereochemistry, ligand type and charge, transition metal or non-transition metal, have effects on k_{MOH} which are not apparent in the pKa?
(7) How important is the solvent, do variations in k_{MOH} reside in $\Delta H^{\neq}$ or $\Delta S^{\neq}$ effects, and are they paralleled by similar changes in the pKa?

We already knew that MOH species react rapidly with certain electron deficient species. Thus the $(NH_3)_5Co-OH_2{}^{3+}$ or $(NH_3)_5$ $CoOH^{2+}$ ions retain the Co-O bond in reactions with N_2O_3 *(13)*, NCO^- *(14)*, $SeO_3{}^{2-}$ *(15)*, CrO_4 *(16)*, $MoO_4{}^-$, $WO_4{}^-$ *(17)*, $HReO_4$ *(18)*, $H_2AsO_4{}^-/HAsO_4{}^{2-}$ *(19)*, acacH *(20)*, CO_2 *(21)*, and such reactions are fast ($t_{\frac{1}{2}} \leqslant$ sec or mins) compared with normal rates of substitution at a Co(III) center. An interesting comparison would be between these rates and those of solvent catalysed O-exchange.

3.1 PROPIONIC ANHYDRIDE REACTION (22), ACTIVATED ESTERS, CH_3CHO AND CO_2 -

$$M-OH^{n+} \ + \ C_2H_5COOCOC_2H_5 \ \xrightarrow{\ k_{MOH}\ } \ M-\underset{\overset{\|}{O}}{O}C-C_2H_5 \ + \ C_2H_5CO_2H$$

The following MOH^{n+} complexes were used: $(NH_3)_5CoOH^{2+}$, $(NH_3)_5$ $CrOH^{2+}$, $(NH_3)_5RuOH^{2+}$, $(NH_3)_5IrOH^{2+}$, $(H_2O)_5CrOH^{2+}$, *trans*-$(NH_3)_4$ $Co(NO_2)(OH)^{1+}$ and $(CN)_5CoOH^{3-}$. Most reactions were followed spectrophotometrically under pseudo first-order conditions of metal complex and the data were collected over a pH range which included the pKa of the $M-OH_2^{n+}$ species, $cf.$ Fig. 1. The observed rate law takes the form

$$k_{obs} = \frac{k_{MOH} \; Ka \, [MOH]_T}{Ka \, + \, [H^+]} \; + \; k_{H_2O} \; + \; k_{OH} [OH^-]$$

where $[MOH]_T$ is the total metal complex concentration, and Ka is the independently measured acidity constant of the aquo ligand; k_{H_2O} $(9.9 \times 10^{-4} sec^{-1})$ and k_{OH} $(500 \; M^{-1}sec^{-1})$ correspond to the lyate solvolytic terms and the values obtained here were in agreement with those measured by pH-stat titration in the absence of metal complex.

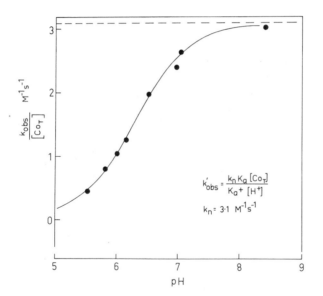

Fig. 1. *Comparison of observed data (•) with calculated curve for the reaction of $(NH_3)_5CoOH^{2+}$ with propionic anhydride at $25°$ and $\mu = 1.0$ $(NaClO_4)$. Values of k_n and pK_a used in the fit are $3.1 \; M^{-1}sec^{-1}$ and $6.35 \; M$ respectively.*

In many cases the product $M-OCOC_2H_5^{n+}$ was separated quantitatively by ion-exchange methods and, in some cases, it was isolated as an ionic salt and completely characterized. For $(NH_3)_5CoOH^{2+}$

[18]O-tracer studies showed the *complete retention of the hydroxo oxygen atom* in the propionate complex, and product analysis on several systems showed that k_{MOH} corresponds to the direct nucleophilic path with *general-base catalysis by M-OH being absent.* This result also holds for the other reactions mentioned below. It is clear from the nature of the rate law (and in particular from data obtained at low pH compared with the pKa) that $M\text{-}OH_2{}^{n+}$ *species are unreactive* towards the anhydride; this result was also found in the other systems.

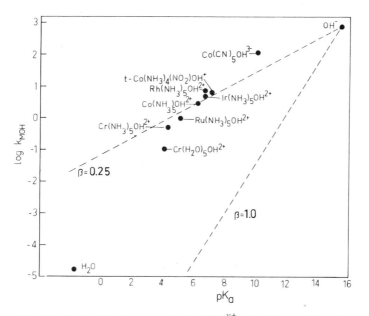

Fig. 2. *Plot of log k_{MOH} vs. pK_a (of $MOH_2{}^{n+}$) for the reaction of various metal hydroxide complexes with propionic anhydride at 25° and $\mu = 1.0$ ($NaClO_4$). The observed slope of about 0.25 differs appreciably from that expected of a parallel correspondence ($\beta = 1.0$).*

A plot of log k_{MOH} *vs.* pKa of the aquo complex (Fig. 2) shows that all M-OH^{n+} species fall on a fairly smooth curve which includes OH$^-$ and H$_2$O. If H$_2$O is not included, a good linear relationship holds with a Brønsted coefficient $\beta \simeq 0.25$. Thus the second order rate constant k_{MOH} is simply related to the pKa of M-OH$_2{}^{(n+1)+}$ and is not especially sensitive to it; certainly a one-to-one correspondence ($\beta = 1.0$) does not obtain. No anomalous effects on k_{MOH} occur even though the overall charge on the M-OH species varies from +2 to -3 and a variety of non-participating ligands (H_2O, NH_3, CN^-, NO_2^-), and first, second and third row transition metals, were used.

Similar Brønsted plots were obtained for 4-nitrophenylacetate (4-NPA) and 2,4-dinitrophenylacetate (2,4-DNPA), Fig. 3, and from literature k_{MOH} values obtained by others for CO_2 *(21,23,24)* and CH_3CHO *(25)* (Fig. 4).

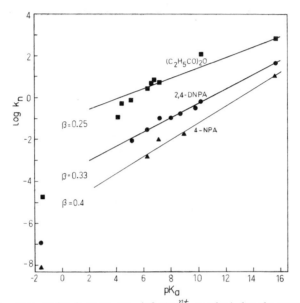

Fig. 3. *Plot of log k_{MOH} vs. pK_a (of MOH_2^{n+} species) for the reactions of metal hydroxide species with 4-nitrophenylacetate (4-NPA), 2,4-dinitrophenylacetate (2,4-DPA), and propionic anhydride ($25°$, $\mu = 1.0$ ($NaClO_4$)). The experimental points for 2,4-DPA and 4-NPA are (in order of increasing pK_a): H_2O, $(NH_3)_5CrOH^{2+}$, $(NH_3)_5CoOH^{2+}$, trans-$(NH_3)_4Co(NO_2)OH^{1+}$, $(NH_3)_5Pt$ $(NH_2)^{3+}$, $(NTA)CuOH^{2-}$, $(NTA)ZnOH^{2-}$, $(CN)_5CoOH^{3-}$, OH^- and H_2O, $(NH_3)_5CoOH^{2+}$, trans-$(NH_3)_4Co(NO_2)OH^{1+}$, $(NTA)CuOH^{2-}$, OH^- respectively. For propionic anhydride the experimental points are as given in Fig. 2.*

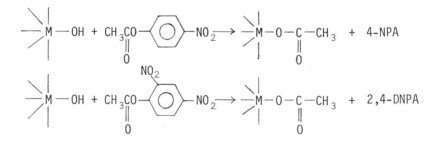

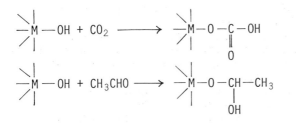

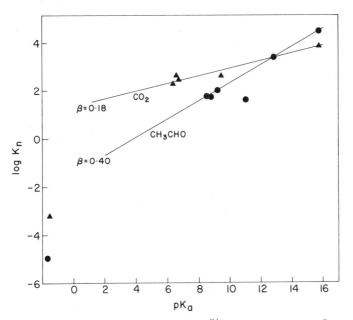

Fig. 4.   Plots of log k_{MOH} vs. pK_a (of $MOH_2{}^{n+}$ species) for Co_2 ($25°$) and acetaldehyde ($0°$) hydration.   For CO_2 the experimental points correspond to (in order of increasing pK_a): H_2O, $(NH_3)_5CoOH^{2+}$, $(NH_3)_5IrOH^{2+}$, $(NH_3)_5$ $RhOH^{2+}$, $(glygly)CuOH^{1-}$ (Ref. 59) and OH^- (Refs. 21 & 23).   For CH_3CHO the corresponding experimental points represent H_2O, $Zn(des-diMeCR)OH^+$, $Zn(N-MeCR)OH^+$, $Zn(CR)OH^+$, $(H_2O)_5ZnOH^+$ (questionable data), $Cu(CR)(OH)^+$ and OH^- (Ref. 25 and page 176 of text).

In these reactions, the range of M-OH^{n+} species was extended to include several exchange labile complexes of Cu^{2+} and Zn^{2+}, although care was necessary in such cases to choose complexes of certain pKa and well-established composition. It is clear from the Figs. 2-4 that rather smooth relationships between k_{MOH} and pKa hold for all these reactions. In a number of cases where "inert" MOH species were involved, the reaction product was isolated (4-NPA, 2,4-DNPA) and careful experiments were done to check for any general base component.

The following conclusions came out of these studies:

(1) M-OH^{n+} species are effective catalysts in hydration reactions (CO_2, aldehydes) and in hydrolyses reactions where good leaving groups are involved.

(2) M-OH$_2$$^{(n+1)+}$ species are ineffective compared with M-OH^{n+} (by at least a factor of 10^3, and probably more).

(3) The direct nucleophilic reaction (k_{MOH}) is preferred over any indirect general base process (by at least a factor of 10^2); no general base path (k_{gb}) has been observed.

(4) Exchange labile M-OH^{n+} species have similar k_{MOH} values to their non-labile counterparts.

(5) The k_{MOH} value is simply related to the pKa of the aqua ligand

$$d(\Delta \log k_{MOH}) = \beta \, d(\Delta pKa)$$

with β values varying from 0.18 (CO_2), 0.25 (propionic anhydride), 0.33 (2,4-DNPA), 0.4 (4-NPA). This holds irrespective of overall charge, metal ion, non-participating ligands, and coordination geometry.

The k_{MOH} values are increased substantially in non-aqueous solution. Table 2 gives k_{MOH} and activation data for the hydrolysis of propionic anhydride and 2,4-DNPA in H_2O, DMSO, and DMF.

TABLE 2: ACTIVATION PARAMETERS AND THE EFFECT OF SOLVENT ON THE HYDROLYSIS OF PROPIONIC ANHYDRIDE AND 2,4-DINITROPHENYLACETATE

Hydrolytic Species	pKa	k_N ($M^{-1}sec^{-1}$)	$\Delta H^{\neq}$ (kcal/mole^{-1})	$\Delta S^{\neq}$ (cal deg^{-1}mole^{-1})
Propionic Anhydride				
H_2O	-1.7	1.8×10^{-5}	9.3	-49
$(NH_3)_5Co-OH$ (in H_2O)	6.31	3.1	7.0	-33
$(NH_3)_5Co-OH$ (in DMSO)	~10	375	0.6	-46
OH^-	15.7	500	7.4	-20
2,4-DNPA				
H_2O	-1.7	1.2×10^{-7}		
$(NH_3)_5Co-OH$ (in H_2O)	6.31	3.1×10^{-2}		
$(NH_3)_5Co-OH$ (in DMSO)	~10	40	3.9	-38
$(NH_3)_5Co-OH$ (in DMF)	~10	165		
OH^-	15.7	37		

The basicity of M-OH towards H^+ (pK_b) and the carbonyl center (k_{MOH}) are both substantially increased with probably the former being more pronounced. Also, whereas in water the rate difference between OH^- and M-OH^{n+} resides entirely in a more negative $\Delta S^{\neq}$, in DMSO a decrease in $\Delta H^{\neq}$ is responsible. Solvation of the different species undoubtedly plays a major role in the activation process, but it is impossible at this time to unravel the effects since both ground state and transition state properties need to be considered. However, the results are in line with those found for the more usual nucleophilic organic displacement reactions *(26)*.

It is interesting to compare the reactivities of the M-OH^{n+} species with the more general reactions of organic and inorganic amines and oxygen bases *(27)*, and Fig. 5 shows this for 2,4-DNPA. The apparent difference in slopes ($\beta \simeq 0.8$ oxyanions; 0.33) means that the k_{MOH} values are less dependent on changes in pKa and this may be of some value in a biological situation. For example, the MOH species appear to have similar reactivities to other oxyanions of pKa $\simeq 5$-6 (*i.e.*, carboxylic acid groups) but are less effective (by ~ 10-10^2) than the comparable amine base.

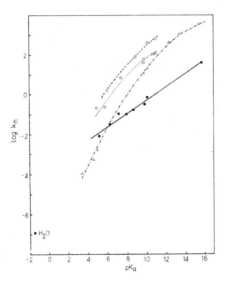

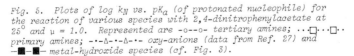

Fig. 5. Plots of log k_N vs. pK_a (of protonated nucleophile) for the reaction of various species with 2,4-dinitrophenylacetate at 25° and $\mu = 1.0$. Represented are -o--o- tertiary amines; $\cdots\square\cdots\square\cdots$ primary amines; $-\cdot-\triangle-\cdot-\triangle-\cdot-$ oxy-anions (data from Ref. 27) and $-\blacksquare-\blacksquare-$ metal-hydroxide species (cf. Fig. 3).

These results have general implications for the reactions of
metal coordinated nucleophiles and may be important considerations for metalloenzymes. Firstly, the relative insensitivity of
k_{M-N} to pKa suggests that a metal-bound nucleophile (M-N) is
superior in a rate sense than the uncoordinated nucleophile (N)
at pH's where the latter is largely protonated. Although this is
demonstrated above for OH⁻ only, there is becoming available evidence to show that M-NHR⁻ and M-SR⁻ species are also very efficient nucleophiles in the neutral pH region (particularly in
intramolecular processes).

The metal catalysed hydrolytic reactions may have relevance
to biological hydration (CO_2) and hydrolysis (carboxypeptidase,
alkaline phosphatase, leucine aminopeptidase) reactions and the
results given above provide experimental support for the often
quoted hypothesis that a prime function of a metal ion in a biological system is to provide a useful concentration of a potent
nucleophile at a biologically acceptable pH. With these results
in mind, let us now consider some intramolecular reactions involving a coordinated lyate species.

4.   INTRAMOLECULAR REACTIONS INVOLVING COORDINATED OH OR H_2O

The intramolecular attack of a coordinated water or hydroxo
group is not unknown in coordination chemistry. The most clear-
cut examples arise in Co(III) complexes where, for the reasons I
have already outlined, a tag can be kept on the oxygen atoms.
Notable among these reactions are the following:

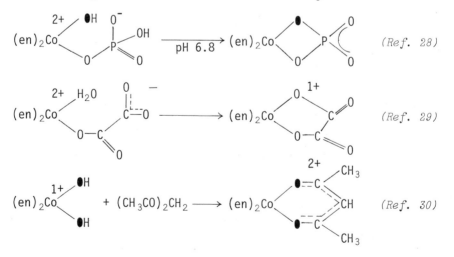

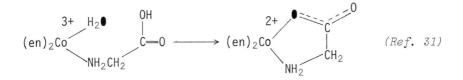

(Ref. 31)

Particularly interesting from the point of view of this lecture is the last reaction. The pH-rate profile is given in Fig. 6 and three inflections can be seen; these have been assigned to the processes:

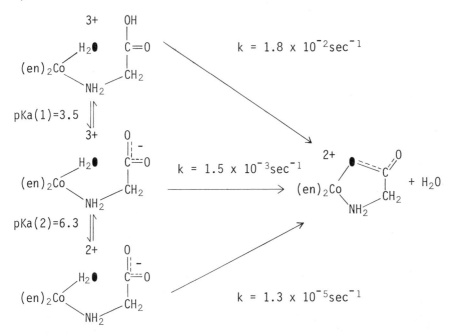

The rate constants given above correspond to 25° and $\mu = 1.0$ (NaClO$_4$). Almost certainly a coordinated hydroxide is a better nucleophile than an aqua ligand, and the more rapid reaction for the aqua complexes suggest that protonation of the leaving group (OH$_2$ in this case) must be an important facet of the overall process. We will see this property developed in the examples following but, compared with the bimolecular reactions described in the first part of this lecture, the properties of the leaving group gain in importance here - often they become overall rate controlling. In the absence of the metal 0-exchange in the glycinate

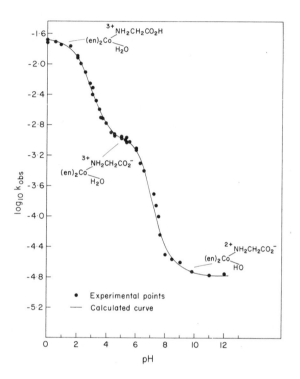

Fig. 6. Plot of log k_{obs} vs. pH for the intramolecular condensation of monodentate glycinate in the cis-$[Co(en)_2(glyOH)$
$(OH/H_2O)]^{3+,2+,+}$ ions at 25°, $\mu = 1.0$ (NaClO_4).

moiety is very slow indeed, $k \simeq 10^{-14}$ sec^{-1} *(32)* so that the intramolecular process affords an enormous acceleration (of 10^{11}) in the neutral pH range.

Another example of intramolecular attack of a lyate species is found in the hydrolysis of glycine esters *(33)*. The required adjacent *cis*-hydroxo group is most easily generated by base hydrolysis of the corresponding *cis*-chloro, or bromo, complex (see diagram, following page) but it is not seen directly at the intermediate stage since subsequent cyclization to the hydrolysed glycinate chelate is very rapid under the reaction conditions. However, ^{18}O-tracer studies have shown that about half the overall reaction occurs *via* this route, with the remainder forming the chelated ester, which also rapidly hydrolyses. The two paths give a different positioning of the ^{18}O-label and this can be distinguished by the (subsequent) exchange properties in aqueous acidic solution; the *exo*-O exchanges in 0.1 M HClO_4 (25°) with a half-life of ∿8 days, whereas the *endo*-O is practically inert. This again displays the versatility of the Co(III) chemistry in these more complex situations.

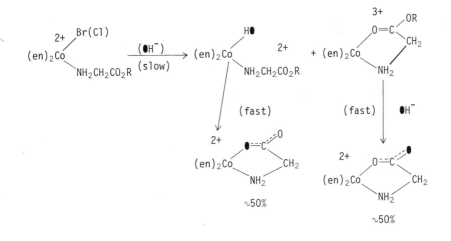

The corresponding monodentate amides provide even better examples *(34)*. Here the *cis*-hydroxo and chelated amide complexes are slower to hydrolyse and their subsequent reactions may be isolated. The following examples give some idea of the division of products formed in the base hydrolysis reactions ($\mu = 1.0$ (NaClO$_4$), 25°), $k_{obs} = k_{OH}[OH^-]$.

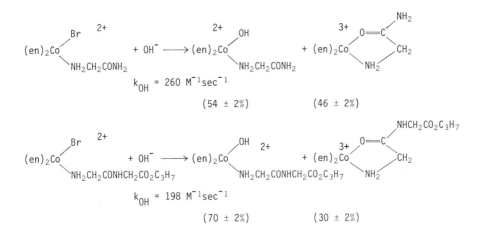

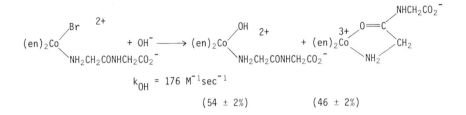

$$k_{OH} = 176 \ M^{-1}sec^{-1}$$

(54 ± 2%)   (46 ± 2%)

The same species can be formed by HOCl induced oxidation of co-ordinated bromide and the advantage here is that they retain their optical configurations about the metal center; in the base hydrolysis process, some 50% racemization occurs in both products.

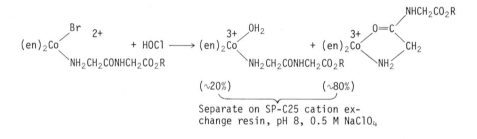

(∼20%)   (∼80%)

Separate on SP-C25 cation ex-change resin, pH 8, 0.5 M NaClO₄

Subsequent hydrolysis of the *cis*-aquo or hydroxo ions turns out to be most intriguing and I want to spend some time describing it. The pH-rate profiles for the monodentate glycine amide and glycyl-glycine isopropyl ester are given in Figs. 7 and 8.

SCHEME 2

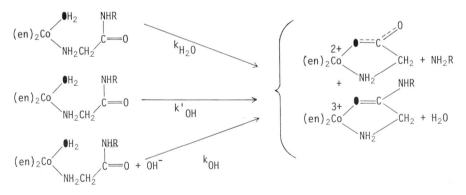

The peculiar bell-shaped curves fit the rate law

$$k_{obs} = \left[\frac{1}{K_a + [H^+]} \ k_{H_2O}[H^+] + k'_{OH} + k_{OH}K_a[OH^-] \right]$$

with K_a being the acidity constant of the aqua ligand (pKa = 6.0, 6.1 respectively), and k_{H_2O}, k'_{OH}, and k_{OH} being the rate constants for the processes in Scheme 2 (preceding page). Values for the rate constants are given in Table 3 along with the reaction products.

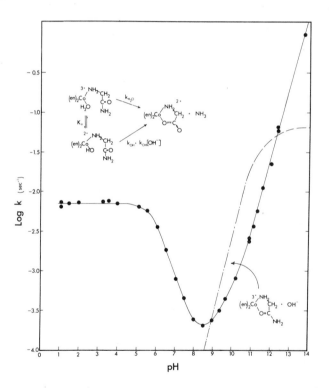

Fig. 7. *pH-rate profile (−•−•−) for the intramolecular hydrolysis of monodentate glycine amide in the cis-*[Co(en)$_2$(glyNH$_2$)(OH/H$_2$O)]$^{3+,\,2+}$ *ions at 25° and μ = 1.0 (NaClO$_4$). For comparison is given the experimental curve (−•−•−) for hydrolysis of chelated glycine amide in the* [Co(en)$_2$(glyNH$_2$)]$^{3+}$ *ion as a function of pH. It will be noted that between pH 9 and 12.5 the later reaction is some 10 times faster than the intramolecular process.*

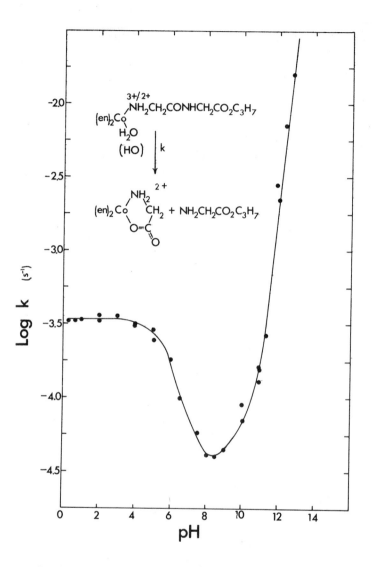

Fig. 8. *pH-rate profile for the intramolecular condensation of coordinated H_2O and OH^- with monodentate glycylglycine isopropyl-ester in the cis-$[Co(en)_2(glyglyOC_3H_7)(OH/H_2O)]^{3+,2+}$ ions at $25°$, $\mu = 1.0$ ($NaClO_4$). Contrary to that indicated by the scheme (inset) both $[Co(en)_2(glyO)]^{2+}$ and $[Co(en)_2(glyglyOC_3H_7)]^{3+}$ are formed at pHs $\leqslant 8$. Above pH 9 the latter product is hydrolysed at a faster rate than the intramolecular reaction and only $[Co(en)_2(glyO)]^{2+}$ is observed.*

TABLE 3: RATE CONSTANTS AND PRODUCTS FOR THE INTRAMOLECULAR REACTIONS OF COORDINATED GLYCINAMIDE, GLYCYLGLYCINE ISOPROPYL ESTER, AND GLYCYLGLYCINATE $(25°, \mu = 1.0 \ (NaClO_4))$

	R = -H	R = $-CH_2CO_2C_3H_7$		R = $-CH_2CO_2^-$	
k_{H_2O} (sec^{-1})	7.1×10^{-3}	3.3×10^{-4}		3.3×10^{-4}	
k'_{OH} (sec^{-1})	1.6×10^{-4}	3.8×10^{-5}		8.2×10^{-5}	
k_{OH} (M^{-1}sec^{-1})	1.7	0.17		0.1	
Products:	R = 0; R =$-NH_2$	R = 0;	R = $-NHCH_2CO_2C_3H_7$	R = 0;	R = $-NHCH_2CO_2^-$
pH 1.0	100 0	46	54	-	-
pH 4.0	100 0	44	56	77	23
pH 8.0	100 0	45	55	76	24

An examination of the above shows that we are now dealing with a more complex situation. Not only does hydrolysis occur (it is complete in the glycinamide case, R = H) but some chelated amide is also produced *via* a process involving oxygen exchange. This suggests the formation of a common intermediate which divides to eliminate amine or water; I will return to this aspect again later.

Optical changes which accompany the reaction are given in Fig. 9, and it can be seen that formation of the chelate rings involve an increase in rotation at all wavelengths. Indeed, close to full retention of configuration about the metal obtains in both products. It is clear that we are *not* dealing with in-organic chemistry at the metal center - the metal just provides a convenient means of holding the reacting groups in juxtaposition.

As we found for the glycine ester, ^{18}O-tracers provide an exact means of following the history of the coordinated H_2O and OH groups. For the glycine amide system ≥90% incorporation of the bound label occurs under all pH conditions (*cf*. Scheme 2). For the dipeptide ester, a similar result obtains for both products in alkaline solution (pH 8.77), but in acid (pH 3.77) the aqua group is itself displaced, possibly by a S_N2 process, in forming the chelated dipeptide. Hydrolysis *via* the coordinated species remains a competitive process, however, with full retention of the coordinated label in this part.

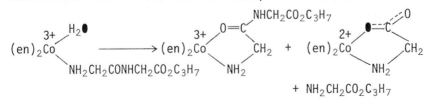

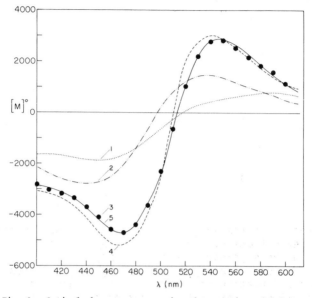

*Fig. 9. Optical changes accompanying the reaction of (+)589-
[Co(en)2(glyglyOC3H7)Br](NO3)2 ([M]589 + 816°, pH 4) curve (1)
....., with HOCl to give optically pure (+)589-[Co(en)2(glygly-
OC3H7)](OH)]2+ ([M]589 + 706°, pH ~8.5) curve (2)—·—··—, and
(+)589-[Co(en)2(glyglyOC3H7)]3+ ([M]5 + 1615°, pH ~6, 1.0M NaClO4)
curve (3) —•—•— .*

The combined results give a good deal of information concerning the mechanism. In acid solution, attack of the aqua group is
almost certainly rate determining, with elimination of protonated
amine occurring as a subsequent fast step. Further support for
this comes from the absence of catalysis by general acids or bases,

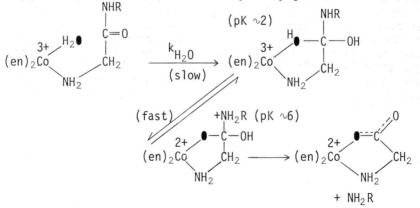

a result which is very different from that obtained in alkaline solution (below). It is significant that loss of the amine is preferred over water under acid conditions.

In alkaline solution, where the coordinated hydroxo group is the attacking species, a very different pattern of events is found. Now considerable buffer catalysis occurs and we have concluded, after an extensive study of these reactions, that general base catalysed deprotonation of the alcohol function, formed following the rapid attack of the hydroxo ligand, is rate determining. This deprotonation may occur either in a stepwise manner (as indicated) or by a concerted process with the alcoholate ion appearing as a transition state complex *(35)*. Consistent with this proposal is the product analysis results which show a steady increase in the amount of hydrolysed product with increasing buffer concentration (dipeptide).

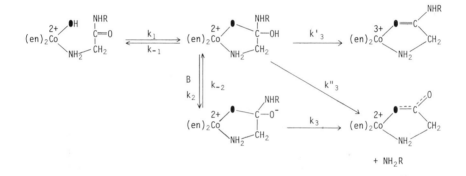

Representative kinetic plots for the reactions of *cis*-[Co(en)$_2$ (OH)(glyNH$_2$)]$^{2+}$ are given in Fig. 10 and demonstrate that buffers act as general bases; similar plots were obtained for the dipeptide ester complexes. Fig. 11 gives a plot of log k_{gb} *vs*. pKa for different buffer types and the trend $B^{3-} > B^{2-} > B^-$ obtains for the oxygen bases. It will be noted that OH$^-$ appears to be somewhat less effective than predicted (possibly due to diffusion controlled (stepwise) removal of the proton), but H$_2$O is more reactive if the base independent term k'_{OH} (1.6 x 10^{-4}sec^{-1}) is represented in the same fashion; the latter may, however, be a measure of the more direct path without deprotonation (k'_3, k''_3).

The most dramatic effect of buffers, however, is found with the *bifunctional* catalysts HPO$_4$$^{2-}$, HAsO$_4$$^{2-}$, and HCO$_3$$^-$ (Fig. 11), and we feel that these processes are of considerable significance. A perhaps clearer demonstration of their effect on the rate can be seen from the half-lives given in Table 4 for the reaction:

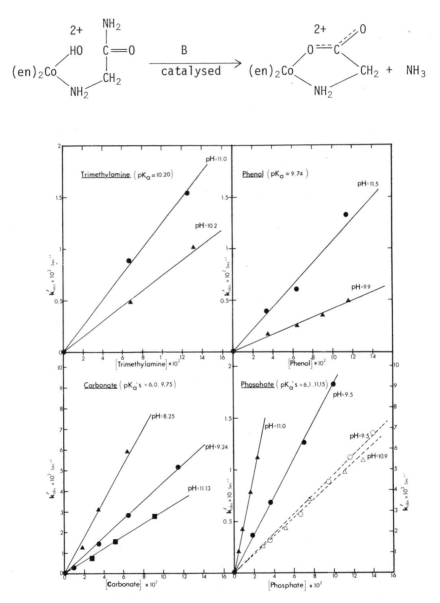

Fig. 10. *Representative plots of k_{obs} vs. buffer concentration for the intramolecular reactions of cis-[Co(en)$_2$(glyNH$_2$)OH]$^{2+}$ (———), at 25°, μ = 1.0 (NaClO$_4$), and cis-[Co(en)$_2$(glyglyOC$_3$H$_7$)OH]$^{2+}$ (— — — —) at 25°, μ = 2.0 (NaClO$_4$).*

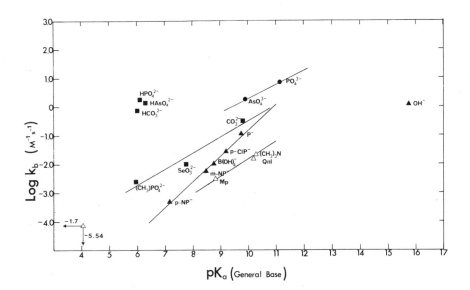

Fig. 11. *Plot of log k (buffer) vs. pK_a (of conjugate acid) for the buffer catalysed intramolecular hydrolysis of cis-[Co(en)₂(glyNH₂)OH]²⁺ at 25°, μ = 1.0 (NaClO₄).*

Obviously such buffers have a remarkable effect and, unlike the monofunctional ones, they catalyse *both* the formation of the hydrolysed product *and* the oxygen-exchanged chelated amide. The former process is favored, however, as can be seen from Fig. 12 which gives products for the reaction:

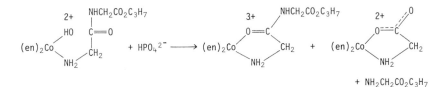

About equal amounts are formed in the absence of the buffer, Table 3, but on increasing the concentration, the amount of hydro-lysed product also increases. Such catalysis probably occurs by a process in which HPO_4^{2-} binds to both the hydroxyl and amine functions of the tetrahedral intermediate, helping eliminate a developing protonated leaving group by concerted deprotonation of the other.

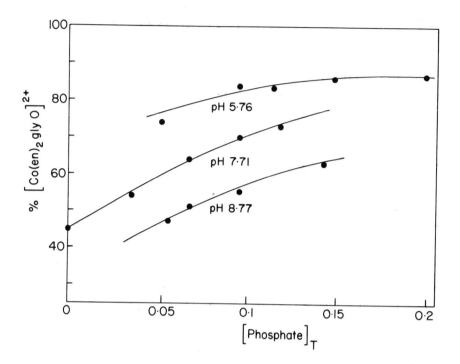

Fig. 12. *Increase in the formation of* $[Co(en)_2(glyO)]^{2+}$ *with increasing phosphate concentration in the intramolecular reaction of cis-*$[Co(en)_2(glyglyOC_3H_7)OH/H_2O]^{2+, 3+}$ *at* 25°, $\mu = 2.0$ (NaClO$_4$).

TABLE 4:   INTRAMOLECULAR GENERAL BASE CATALYSED
HYDROLYSIS OF $[Co(en)_2OH(glyNH_2)]^{2+}$ AT pH 7.5,
25°, $\mu = 1.0$ (NaClO$_4$)

Catalyst (1.0 M)	$t_{\frac{1}{2}}$ sec
OH$^-$ ($\sim 2 \times 10^{-6}$ M)	2000
Oxygen Bases, 3–	7
2–	60
1–	700
Nitrogen Bases	700
Bifunctional Bases	0.5
(HPO_4^{2-}, $HAsO_4^{2-}$, HCO_3^-)	

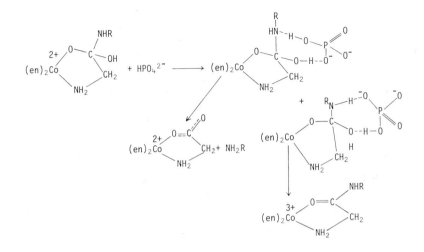

Compared with $CH_3PO_4^{2-}$, which does not have this capacity (and behaves quite normally, Fig. 11), such bifunctional ability gives an additional increase in rate of $\sim 10^3$.

The Tables 5 and 6 summarize the acceleration provided by the direct polarization mechanism

$$\text{>}C = 0 - M + OH^-$$

and the intramolecular attack of bound hydroxide

$$M - 0 \overset{H}{\cdots} \text{>}C = 0 \quad .$$

The former, requiring solvent hydroxide as the nucleophilic species, provides a rate increase of 10^4-10^5 over that found in the absence of the metal. The latter in the absence of buffers is somewhat less effective at pH 7 (10^3-10^4) but buffers result in a tremendous rate enhancement, 10^{10}-10^{11} at pH 7. Under slightly more acidic conditions, pH 4-5, where the bound aqua group is involved, the full efficiency of the intramolecular process is revealed in a similar rate increase; the peptide is not effectively hydrolysed by water at all at these pH's in the absence of the metal, and possibly stabilization of the chelate ring and/or solvation properties of the transition state are responsible.

Such rates match, or exceed, the turn over numbers found in carboxypeptidase and trypsin proteases under identical conditions of pH and temperature. The mechanistic analogy with chymotrypsin is close, with the acylation step proposed for the enzyme *(36)*

TABLE 5:   COMPARISON OF RELATIVE RATES OF HYDROLYSIS OF GLYCINE AMIDE AT 25°

$NH_2CH_2CONH_2 + OH^-$ $k = 2.2 \times 10^{-3}\ M^{-1}\ s^{-1}$	Difference	$NH_2CH_2CONH_2 + OH^-$ $k = 2.2 \times 10^{-3}\ M^{-1}\ s^{-1}$	Difference
$\begin{array}{c} Co\diagdown\overset{NH_2}{\underset{O=C}{\diagup}}CH_2 + OH^- \\ \diagdown NH_2 \end{array}$		$\begin{array}{c} Co\diagup NH_2CH_2CONH_2 \\ \diagdown OH \quad + HPO_4{}^{2-} \end{array}$	
$k = 25\ M^{-1}\ s^{-1}$	10^4	$k = 1.8\ M^{-1}\ s^{-1}$	10^{10} (at pH 7)
$\begin{array}{c} Co\diagup NH_2CH_2CONH_2 \\ \diagdown OH \quad + OH^- \end{array}$		$\begin{array}{c} Co\diagup NH_2CH_2CONH_2 \\ \diagdown OH_2 \end{array}$	
$k = 1.24\ M^{-1}\ s^{-1}$	10^3	$k = 3.5 \times 10^{-3}\ s^{-1}$	10^9 (at pH 5)

TABLE 6:   COMPARISON OF RELATIVE RATES OF HYDROLYSIS OF GLYCYLGLYCINE AT 25°

$NH_2CH_2CONHCH_2CO_2{}^- + OH^-$ $k = 8 \times 10^{-6}\ M^{-1}\ sec^{-1}$	Difference	$NH_2CH_2CONHCH_2CO_2{}^- + OH^-$ $k = 8 \times 10^{-6}\ M^{-1}\ sec^{-1}$	Difference
$\begin{array}{c} Co\diagdown\overset{NH_2}{\underset{O=C}{\diagup}}CH_2 + OH^- \\ \diagdown NHCH_2CO_2C_3H_7 \end{array}$		$\begin{array}{c} Co\diagup NH_2CH_2CONHCH_2CO_2C_3H_7 \\ \diagdown OH \quad + HPO_4{}^2 \end{array}$	
$k = 2\ M^{-1}\ sec^{-1}$	3×10^5	$k = 5 \times 10^{-2}\ M^{-1}\ sec^{-1}$	10^{11} (at pH 7)
$\begin{array}{c} Co\diagup NH_2CH_2CONHCH_2CO_2C_3H_7 \\ \diagdown OH \quad + OH^- \end{array}$		$\begin{array}{c} Co\diagup NH_2CH_2CONHCH_2CO_2C_3H_7 \\ \diagdown OH_2 \end{array}$	
$k = 0.1\ M^{-1}\ sec^{-1}$	10^4	$k = 3.3 \times 10^{-4}\ sec^{-1}$	10^{11} (at pH 5)

being very similar to that found here; Fig. 13(a), (b) schematically depict the two situations. Thus we have not only found some interesting chemistry but have modelled the enzyme rather closely. Unfortunately, our Co(III) systems are stoichiometric rather than catalytic, but the moral is there.

One property which has not been mentioned is that, whereas the intramolecular reactions are general base catalysed, hydrolysis *via* direct polarization of the carbonyl function is not,

a) **Acylation step**

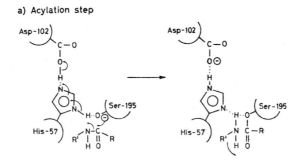

b) <u>Acylation Step</u> — General Base Catalysis

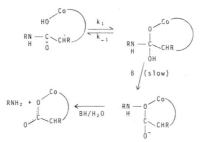

Fig. 13. Comparison of the acylation step for the chymotrypsin catalysed hydrolysis of an amide substrate (D.M. Blow, J.J. Birktoft and B.S. Hartley, Nature, 221, 337 (1969)) and the general base catalysed hydrolysis of an amide substrate by an adjacent hydroxyl group in the Co(III) model.

the rate law taking the form $k_{obs} = k_{OH}K_b[OH^-]/(K_b + [OH^-])$ under all conditions *(37)*.

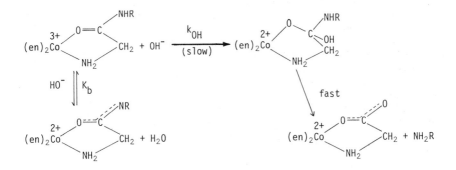

This implies that the rate determining step involves the attack of OH⁻ at the bound carbonyl function, which is a different result from that found for the similar chelated esters. Reasons for this change from rate determining elimination of the alcohol to rate determining addition of hydroxide for the amide are not clear at the present time.

For the remainder of the lecture, I wish to look in some detail at one enzyme which involves a hydration process. Carbonic anhydrase provides a good example since current discussion revolves around either a directly activated bound substrate mechanism and one in which a coordinated hydroxide is involved.

5. CARBONIC ANHYDRASE (CA) AND ITS HYDRATION OF CO_2 AND HYDROLYSIS OF p-NITROPHENYL ESTERS

These enzymes are widely distributed in nature (plants, bacteria, mammals) and they carry out a very important function; in animals, human CA(C) and bovine CA (A and B) are involved in the fast hydration of CO_2 (in the tissues) and the dehydration of HCO_3^- (in the lungs) and the observed turnover numbers of $\sim 10^6$ sec^{-1} are among the fastest known for any enzyme:

$$CO_2 \ + \ H_2O \ \underset{EH}{\overset{E}{\rightleftharpoons}} \ HCO_3^- \ + \ H^+$$

Briefly *(38)*, these enzymes consist of a single strand polypeptide chain of approximately 260 amino acid residues and one essential Zn^{2+} atom (MW $\simeq$ 29,000). The high resolution crystal structures of the Human C (high activity) and Human B (low activity) forms show remarkable similarities in both secondary and tertiary structure, with the Zn^{2+} atom in the active site cavity and coordinated to three histidine residues. The most significant differences in the primary structure occur in the immediate vicinity of the zinc and this probably has a direct bearing on their activities ($K_m(CO_2)$ differ by ~ 2; k_{cat} differ by ~ 10). From this point of view, it is perhaps worth mentioning that the crystals for both structures were grown under conditions (pH ~ 8.5) in which the enzymes are in their active forms. Fig. 14(a) shows an idealized model of the C enzyme with the active site clearly delineated, and Fig. 14(b) shows a closer view of the residues surrounding the Zn^{2+} (His 117, 93, 95 coordinated) as well as further out residues (His 63, 128; Thr 197; Leu 198). Between the Zn^{2+} atom and His-128, there are 8 peaks of residual electron density which are thought to be H_2O molecules; one is adjacent to Zn^{2+}

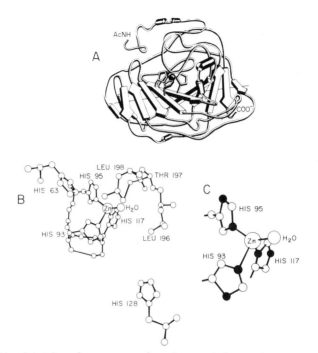

Fig. 14. (a) Schematic representation of some of the tertiary structure and peptide chain folding in human CA (C), with the active site depicted by a dark ball coordinated by three imidazole residues. (b) An ORTEP drawing of the active site region showing the coordinated His 117, 93, and 95 residues and those further out (His 63, 128; Thr 197; Leu 198). (c) Model of coordination environment of Zn^{2+} atom showing N_1 coordination of His 117 and N_3 coordination of His 93, 95. (From Ref. 38).

(200 pm) and probably coordinated to it to form a rather distorted tetrahedral aqua-complex (deviations of $\sim 20°$ from tetrahedral). Fig. 14(c) shows a model of this coordination environment and is given here to emphasize that two of the imidazole residues are coordinated *via* N_3 (93, 95) while His-117 is bound *via* N_1; the latter is the more usual coordinating site for metal ions in simple histidine complexes, presumably because 5-membered rather than 6-membered rings are involved, but coordination through N_3 is nearly always found in metalloproteins.

Although the reversible hydration of CO_2 is the only known physiological function of these enzymes, they also catalyse all the reactions considered in the previous section - particularly the hydration of aldehydes and the hydrolysis of activated phenyl esters. A comparison is given in Table 7 between the turnover numbers for CO_2, CH_3CHO and 4-nitrophenylacetate (4-NPA) hydrolysis and the rates of normal solvent hydrolysis at pH 7.6 (pH of maximum enzyme efficiency) and it is obvious that significant rate enhancements are involved; 10^8 (CO_2), 2×10^5 (4-NPA), 3×10^4 (CH_3CHO).

TABLE 7:   RATES FOR THE BOVINE CA CATALYSED, AND THE UNCATALYSED (SOLVENT), HYDROLYSIS OF THREE CARBONYL SUBSTRATES COMPARED WITH THE $Zn\text{-}OH^+$ (pKa 7.6) REACTION

	CA (pH 7.6, 25°)		H_2O/OH^- (pH 7.6, 25°)	$Zn\text{-}OH^+$ (pKz 7.6, μ =1.0)
	k_{enz} ($M^{-1}sec^{-1}$)	k_{cat} (sec^{-1})	$k_{OH}[OH^-]+k_{H_2O}$ (sec^{-1})	$k_{ZnOH}(M^{-1}sec^{-1})^a$
$CO_2 \longrightarrow HCO_3^-$	4.7×10^7	7.1×10^5	9.5×10^{-3}	4×10^2
$CH_3CHO \longrightarrow CH_3CH(OH)_2^{\,b}$	1.5×10^3	8.8×10^2	2.8×10^{-2}	30
4-NPA $\longrightarrow p\text{-NP}+CH_3CO_2^-$	9.6×10^2	1.3	6.4×10^{-6}	3.6×10^{-3}

^{a}*Taken from Brønsted plot (Figs. 2-4).* ^{b}*Data at 0°.*

The kinetic data has been interpreted in terms of the Michaelis-Menten parameters K_m and k_{enz}

$$E + sub \;\underset{1/K_m}{\rightleftharpoons}\; E.\,sub \;\xrightarrow{k_{cat}}\; E + prod$$

$$\frac{Rate}{[E]_T} = k_{obs} = \frac{k_{cat}[sub]}{K_m+[sub]} \;;\; k_{enz} = \frac{k_{cat}}{K_m}$$

$k_{obs} = k_{cat}$ when $[sub]\gg K_m$ and this type of fitting of kinetic data in enzymology is done with little regard for other possible schemes. A more careful examination of this aspect is, I feel, warranted. However, comparisons between the second-order rate for the enzyme reaction (k_{enz}) and k_{ZnOH} for a model $Zn\text{-}OH^+$ species of pKa 7.6 shows a somewhat closer agreement although the differences are still substantial; $\sim10^5$ (CO_2, 4-NPA), ~50 (CH_3CHO). Clearly, the protein part of the molecule plays an important function in addition to the metal species if it is agreed that the nucleophilicity of the latter is similar to that of the model system. We will return to discuss this aspect in more detail later.
    There is a substantial body of evidence which implicates the immediate environment about the Zn atom in both the hydration

(CO$_2$, aldehydes) and hydrolysis (4-NPA) reactions and it appears that very similar, if not identical, mechanisms are involved. No attempt will be made to review the evidence here, as many excellent summaries are available *(3,4,38)*. What I wish to do is to look critically at two mechanistic proposals that are currently attractive and to make some comments based on my own observations on the inorganic model systems. Also I will consider a third, not-too-likely alternative, which has not been put forward previously.

5.1 Zn-OH$^+$ MECHANISM - This is the most favored proposal and it has the advantage of being direct and simple; the *coordinated OH* moiety is the active hydrating or hydrolysing agent as well as the group undergoing displacement from the coordination sphere by monovalent anion inhibitors. All the experimental data that I am aware of can be accommodated by this mechanism. Two possible paths can be considered: the *direct attack* or the coordinated OH at the carbonyl function (k$_{MOH}$) and the indirect *general base* catalysis and solvolysis by an intervening solvent molecule (k$_{gb}$) - the latter process has the advantage of not disturbing the coordination sphere of the metal.

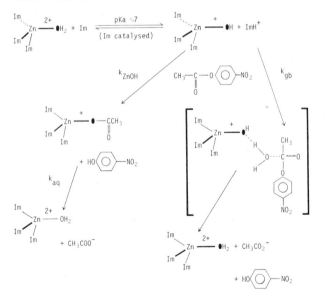

SCHEME 3 - ZnOH$^+$ Mechanism for CA

NOTES ON SCHEME 3

1. General base mechanisms have not been observed.
2. pKa = 9.1 (octahedral ion); pKa (enzyme) = 7.6.
3. $k_{ZnOH} = 3.6 \times 10^{-3}$ M^{-1} sec^{-1} for isolated $ZnOH^+$ species of pKa = 7.6
   k_{enz} (enzyme) = 9.6×10^2 M^{-1} sec^{-1}.
4. $k_{aq} = 10^6$ sec^{-1} for $ZnOAc^+$; k_{cat} (enzyme) = 1.3 sec^{-1}. (For CO_2 hydration, this step likely to be rate determining.)

Scheme 3 on the preceding page outlines the alternatives and summarizes the main points made below.

Both the active Zn^{2+} and Co^{2+} enzymes show a pH-rate profile requiring the titration of an acidic group of pKa 6.8 - 7.6 (depending on substrate, ionic strength, etc.) to generate the active conjugate base form of the enzyme. A wide range of spectroscopic techniques (visible on Co^{2+} enzyme, ^{31}Cl nmr, epr, CD/ORD) have been used to infer that this titratable group is bound to the metal *(38)*, and the magnitude of the effects observed suggest that it is unlikely to be in the second coordination sphere. The binding of anions results in inactivation, and the pKa for the titratable group obtained from these studies agrees with that obtained kinetically for the working enzyme. This is excellent evidence that the reactive group and that replaced by anions are one and the same, and suggests either a coordinated OH^- or a coordinated imidazolate anion. The former would agree with the observed strong binding of monovalent anions *including OH^-* (smaller dissociation constants K_i than normally found in simple Zn^{2+} complexes), whereas the latter would require that all other anions *except OH^-* bind strongly - a somewhat unlikely result.

Most simple $Zn-OH_2$ complexes have pKa's in the range 9 - 10, Table 8; many other values are unreliable due to precipitation of Zn-hydroxide species at these pH's. However, distorted 5-coordinate complexes are known with substantially lower pKa's, and that below provides a good example *(39)*.

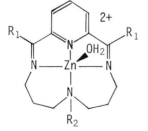

Abbrev.	R_1	R_2	pKa (25°, $\mu \simeq 0.01$)
CR	CH_3	H	8.69
N-MeCR	CH_3	CH_3	8.12
des-diMeCR	H	H	8.13

TABLE 8: pKa VALUES OF Zn-OH$_2$ AND M(II)-IMIDAZOLE COMPLEXES

Complex	pKa	Conditions	Comments References
Aquo Complexes			
Zn^{2+}_{aq}	9.12	2 M NaCl	40
	9.26	3 M NaCl	40
	8.96	$\mu = 0$, 25°	41
$Zn(NTA)(H_2O)^-$	9.81	$\mu = 1.0$, 25°	42
	9.99	$\mu = ?$, 25°	pmr 43
$Zn(tren)(H_2O)_2{}^{2+}$	11.1	$\mu = ?$, 25°	pmr 43
$Zn(EIDA)(H_2O)^+$	8.7	30°	44
Imidazole-Type Complexes			
$Co(\ell\text{-His})_2$	12.5		45
$Co(Histamine)_2{}^{2+}$	13		45
$Co(Im)_4{}^{2+}$	12.5		45
$Cu(\ell\text{-His})_2$	11.7		45
$Pd(en)(\ell\text{-His})^+$	10.8		46
$(NH_3)_5CoIm^{3+}$	10.0	25°, $\mu = 1.0$ (NaClO$_4$)	47

It is therefore not difficult to accommodate a distorted tetra-hedral active site geometry with a pKa of 6.8-7.6. The main points here are that lowering the coordination number will in-crease the acidity of a bound group, and that this may also be influenced by a distorted geometry. Certainly the high overall turnover numbers of the enzyme ($\sim10^6$ sec^{-1} for CO$_2$) require both the protonation/deprotonation rates to be rapid at pH ~7.0,

$$ZnAH^{2+} + B \rightleftharpoons ZnA^+ + BH$$
$$(pKa \sim 7)$$

and this, in turn, requires some special general buffer mediated proton transfer process, probably by an adjacent uncoordinated histidine residue, rather than direct solvent OH$^-$ or H$_2$O partici-pation. This is to some extent supported by specific modifica-tion studies (by bromopyruvate) of the His-63 residue of the Human C enzyme (within ~6 Å of the Zn but not coordinated) which reduces considerably - but does not eliminate entirely - esterase and CO$_2$ activity *(48)*.

The next question to consider is the exact role of the Zn-OH species. Several points argue against a general base process,

k_{gb}. Firstly, we have shown in the first part of this lecture that the identical reactions (propionic anhydride, 2,4-DPA, 4-NPA) in the absence of the protein involve only direct nucleophilic attack. This was proven for the "non-labile" MOH species, and the absence of anomalously low k_{MOH} values for the labile systems (as shown by the linear Brønsted correlations) implies a similar result. That is, Zn-OH$^+$ and Co-OH$^+$ complexes are effective nucleophiles for the biological substrates in question and it seems unlikely that such efficient enzymes would use the less effective general base path. Secondly, a general base mechanism is invariably a poor substitute in a rate sense for the corresponding direct process. They are usually found *only* when the reactant is itself not a sufficiently good base to displace the leaving group from the tetrahedral intermediate or transition state *(49)*.

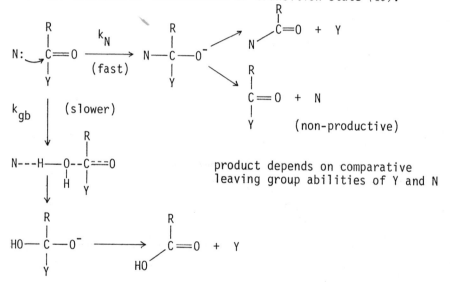

product depends on comparative leaving group abilities of Y and N

For CO_2 and aldehyde hydrations, no leaving group is involved and a direct nucleophilic reaction must be favored; for reactive esters such as 4-NPA the phenolate anion (OR$^-$) is certainly a better leaving group than Zn^{2+}— O$^-$, especially since the latter has to be protonated to act in this capacity (microscopic reversibility). This is an unfavorable equilibrium process in neutral or slightly alkaline reaction conditions (pKa $\sim$3-4 (est)). Also, ^{13}C nmr studies on the hydration (CO_2 substrate) and dehydration (HCO$_3^-$ substrate) reactions of the active Co(II) enzymes *(50,51)* have shown a weakly bound substrate site within 3.8 ± 0.2 Å and

4.0 ± 0.2 Å, respectively of the Co atom (Human B enzyme).

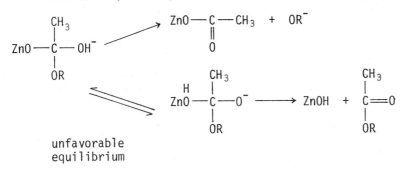

unfavorable
equilibrium

If this is a measure of the reactive ground state geometry then
the substrate is almost certainly too close to the metal to allow
an intervening H_2O molecule to be accommodated in the second coordination sphere. Henkens, in the poster session of this meeting, puts forward a substrate bound mechanism (I have modified
Henkens model *(51)* in the scheme below) and this fits the recent
elegant work of Silverman and Tu *(52)* who show that the oxygen

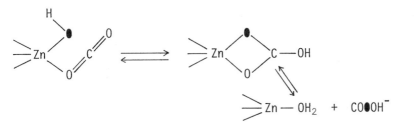

atom of the hydrating water molecule undergoes several transfers
between CO_2 and HCO_3^- before it becomes fully equilibrated with
the bulk solvent. A general base mechanism is not nearly as
satisfying in this respect, particularly if solvent in the active
site region is in rapid equilibrium with that outside. Finally,
it might be argued that, if monovalent anion inhibition results
from direct coordination to the metal (as it seems to) rather
than binding at a position further removed, then such species
might also be expected to act as a general base albeit with a
modified reactivity, which is not in accord with experiment.

Combined, these arguments add up to a strong case for the direct nucleophilic role for Zn-OH.

The rate constants for the bimolecular processes in water using the model systems, k_{ZnOH}, do not approach those found in the enzyme; from the Brønsted plots of Figs. 3-4, we estimate that a Zn-OH$^+$ species of pKa 7.6 will have second order rate constants (M^{-1} sec^{-1}) of 400 (CO_2), 30 (CH_3CHO) and 3.6 x 10^{-3} (4-NPA) which are 10^5-10^2 smaller than those for the enzyme, Table 7. However, these rates are considerably faster than those for unperturbed solvent species (H_2O, OH^-). Thus some additional activation by the protein portion of the enzyme is necessary, and the second part of this lecture was devoted to demonstrating how this can be achieved by an intramolecular process in which the substrate and M-OH species are held in juxtaposition.

The immediate product of the hydrolysis reaction, the E-Zn-OCOCH$_3$$^+$ (or E-Zn-OCO$_2$H$^+$) species has to undergo rapid aquation to regenerate E-Zn-OH$_2$$^{2+}$. For Zn_{aq}-OCOCH$_3$$^+$ and Zn_{aq}-Cl$^+$, these reactions have rates of the order of 10^6 sec^{-1}, and a similar rate might be anticipated for the similarly charged acylated enzyme; it is difficult to see how the enzyme could drastically affect this process, although enforced protonation of the leaving group by an adjacent protonated imidazole residue (His-63) may be structurally possible (the basicity of the coordinated leaving group is poor by comparison with Im (pKa ∿7.0)).

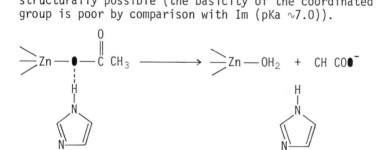

We are therefore stuck with rates of ∿10^6 sec^{-1} or so, and it may be significant that this is close to the maximum turnover number for CO_2 hydration (k_{cat} = 7 x 10^5 sec^{-1}); the release of HCO_3^- from the metal could well be rate limiting. Similarly, k_{cat} for dehydration (8 x 10^4 sec^{-1} (C); 4 x 10^5 sec^{-1} (Bovine)) can be equated to a HCO_3^- concentration of ∿10^{-3} M (C); ∿10^{-2} M (Bovine), if the (ligand independent) substration rate of ≈3 x 10^7 M^{-1} sec^{-1} holds in the enzyme situation; these are not unreasonable HCO_3^- concentrations in the tissue. For the non-physiological ester and aldehyde reactions, with much lower turnover numbers (k_{cat} (4-NPA) 1.3 sec^{-1}, Table 7) aquation of the acylated enzyme is almost certainly not rate controlling; k_{cat} probably measures the acylation process, $k_{cat} = k_{ZnOH}$.

5.2 Zn-IMIDAZOLE MECHANISM - This mechanism has also taken two forms; that proposed independently by Sakar *(53)* and Pesando *(54, 55)* whereby a coordinated imidazolate anion is involved in a general base catalysed reaction, and the direct nucleophilic path suggested for ester hydrolysis *(47)*. Scheme 4 represents these proposals and summarizes the main points made in the following section.

SCHEME 4 - Zn-Im Mechanism for CA Catalysed
Hydrolysis of 4-NPA

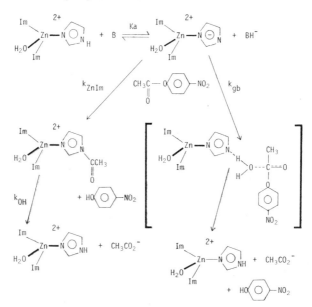

NOTES:

1. *General base mechanisms have not been observed.*

2. *pKa = 11.5 (octahedral ion); pKa (uncoordinated His) = 14.2; pKa (enzyme) = 7.6.*

3. $k_{ZnIm} = 4 \times 10^{-2}~M^{-1}~sec^{-1}$ *for* $ZnIm^+$ *of pKa = 7.6 (≈10× faster than* $ZnOH^+$*);* k_{enz} *(enzyme) = 2 × 10² M⁻¹ sec⁻¹.*

4. $k_{OH} = 1 \times 10^{-3}~sec^{-1}$ *for Zn-ImAc⁺* $\}$ k_{cat} *(enzyme) = 1.3 sec⁻¹*
    $k_{OH} = 2.2 \times 10^{-4}~sec^{-1}$ *for ImAc* $\}$ *at pH 7.6*

    Expect to see acylated enzyme intermediate.

For CO_2 the direct nucleophilic role by imidazole is most unlikely; the resulting carbamate intermediate would undoubtedly cleave heterolytically to reform the initial reactants rather than hydrolyse to generate HCO_3^-. This is the behavior of all known alkyl and aryl carbamates in aqueous solution *(56)*, and is especially likely in the present case where the protonated $Zn\text{-}Im^{2+}$ moiety provides an excellent leaving group from the developing stable CO_2 species. Indeed, ^{18}O studies show just such a behavior for coordinated carbamates, with the $(NH_3)_5M\text{-}NHCO_2H^{2+}$ ions

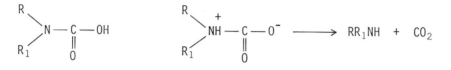

(M = Co(III), *(57)*, Rh(III) *(58)*) cleaving to the hexammine and CO_2 rather than adding water and generating HCO_3^-. The rate law for this reaction requires a simple process such as that given by

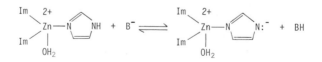

the above equation, and the reactions are relatively rapid, $k_{obs} = Kk = 0.38\ sec^{-1}$ (Co^{3+}) *(57)*. There is again good justification here for considering the metal species as replacing H^+. Thus, whether coordinated or not, the direct attack of imidazole on CO_2 is unlikely to produce HCO_3^-. Ester hydrolysis could well be another story, since acetylimidazole would be the intermediate involved. This system readily undergoes rapid hydrolysis under appropriate conditions: I will look at this situation a little closer below.

The major difficulty with the coordinated imidazole mechanism concerns its pKa, since it is necessary to ionize the proton in

order to generate the nucleophile - the other nitrogen is tied up
by the metal.

The pKa of an isolated imidazole or histidine residue is
14.2 - 14.4, and in order that this correspond to the titratable
group in the enzyme, it needs to be lowered to ~7.5. There is
good experimental evidence now to show that this is out of
the question in simple metal complexes, and the magnitude of the
additional effects required of the enzyme (some 5 pKa units) in
my opinion rules out this proposal. However, some authors continue to support it and one such mechanism put forward recently
(54,55) is given in Fig. 15.

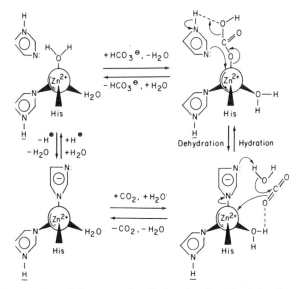

*Fig. 15. Pesando's proposed mechanism for the CA catalysed
hydration of CO_2 based on pmr titration data (Ref. 55). Note
that a deprotonated imidazole ligand acts as a general base
in the addition of a solvent molecule with the concomitant
displacement of itself by the HCO_3^- moiety: no role is ascribed
to the coordinated water molecule.*

Our arguments, like those of Martin's *(59)*, are based largely on
the pKa's of well-established complexes. Table 9 lists pKa's of
some Co(III) systems and clearly shows that the polarizability
afforded by the metal decreases rapidly with distance from the
acidic site. Of particular significance in the present context
is the pKa of 10.0 for $(NH_3)_5CoImH^{3+}$ - a reduction of only 4
units over the uncoordinated molecule. A similar reduction is

found in $(NH_3)_5CoOCO_2H^{2+}$ where the acidic group is also 3 atoms removed from the metal. Further removed, the metal has little effect on the pKa; when closer, the acidity increases rapidly; a similar effect is found for both saturated and conjugated ligand systems.

TABLE 9: pKa's OF COORDINATED LIGANDS IN $(NH_3)_5Co^{3+}$—L

Ligand	Free Acid pKa	Ligand pKa	Ligand	Free Acid pKa	Ligand pKa
$-OH_2$	15.7	6.3	$-NH_2CH_2C\ OH$ ($\parallel$ O)	2.3^a	2.3
$-OHCH_3$	15.9	5.6			
$-OCOH$ ($\parallel$ O)	9.8	6.7	$-NH_2CH_2CH_2C\ OH$ ($\parallel$ O)	3.6^a	3.9
$-OC-COH$ ($\parallel$ $\parallel$ $O\ O$)	3.85	2.1	$-NH_3$	~35	~16
$-OC-CH_2-COH$ ($\parallel$ O $\parallel$ O)	5.3	4.8	$-NH_2CH_2CH_2NH_3^+$	9.9	7.3
$-O-P-OH$ (OH$\mid$ $\parallel$ O)	6.5	3.6	$-N\diagdown\diagup NH$	14.2	10.0
$-NH-C-OH$ ($\parallel$ O)	-	0.4	$-N\diagdown\diagup \overset{+}{NH}$	1.90	~1.0

^{a}pKa for the $NH_3^+(CH_2)_{2,1}CO_2H$ species.

We support the view that such effects *are quite general in their applicability to metal complexes, and determine the pKa in those cases where interactions involving π-orbitals on the metal and the ligand are absent, or are unimportant.* Such is the case for all first-row transition metal complexes in their normal (+2, +3) oxidation states and certainly Co^{2+}, Ni^{2+}, Cu^{2+} and Zn^{2+} do not enter into significant π-bonding with the common coordinating groups found in enzymes. For the second or third row transition elements, and for first row complexes containing metals in unusual oxidation states, π-interactions can play an important part in both the bonding and the pKa (*e.g.*, Ru(II)). Such is not the case in biological systems where filled electron shells and coordinatively saturated ligand systems determine the chemistry. Here the basicity of a coordinated ligand is modified largely by the polarizing power of the metal and, as we have seen above, Co^{2+}, Ni^{2+}, Cu^{2+} and Zn^{2+} are somewhat less effective in this respect than Co^{3+}. Such arguments based on the Co^{3+} situation are supported by experiment as can be seen by the pKa's listed in the second part of Table 8. Our estimate of 11.5 for the pKa of a $Zn-Im^{2+}$ complex (based on a linear correlation between the pKa's of $(NH_3)_5CoOH_2{}^{3+}$, $Zn_{aq}-OH_2{}^{2+}$, H_2O and $(NH_3)_5CoImH^{3+}$, ImH) is in fair agreement (ionic strength differences) with that observed for the various Co^{2+}-imidazole complexes; no reliable data on a $Zn-Im^{n+}$ complex exists. Also, and as a separate argument, if the immediate surroundings of the active site are such as to provide a more basic environment, it would seem unreasonable to expect this to affect a coordinated imidazole group without having a similar, or at least a measurable effect, on the adjacent, potentially more acidic, coordinated water molecule. Only one pKa is observed in the enzyme up to pH 11. Other arguments, concerning the observed pKa in the enzyme and anion competition studies, have already been discussed.

Other results show that the nucleophilicity of a coordinated imidazole anion is not substantially different from that of free imidazole. By comparing the reactions below, it can be seen that H^+ is slightly more effective (a factor of ~10 is involved) than $(NH_3)_5Co^{3+}$ in the reaction with 4-NPA, and by comparing these with the second order rate constant for the imidazolate anion it is clear that such "acids" bound to N_1 reduce the nucleophilicity

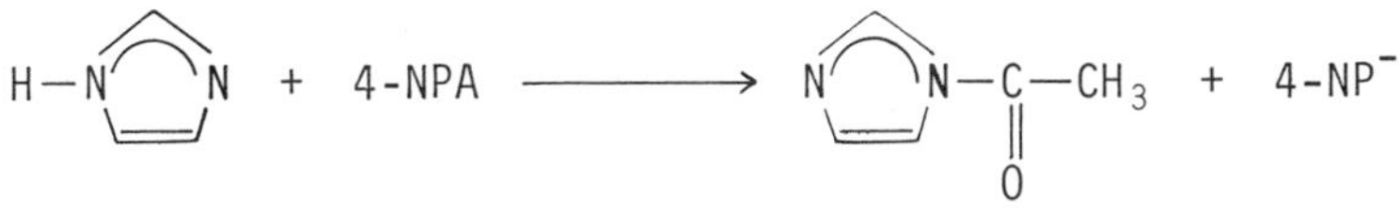

$$k = 0.6 \text{ M}^{-1} \text{ sec}^{-1} \ (25^\circ) \ (pKa = 7.2)$$

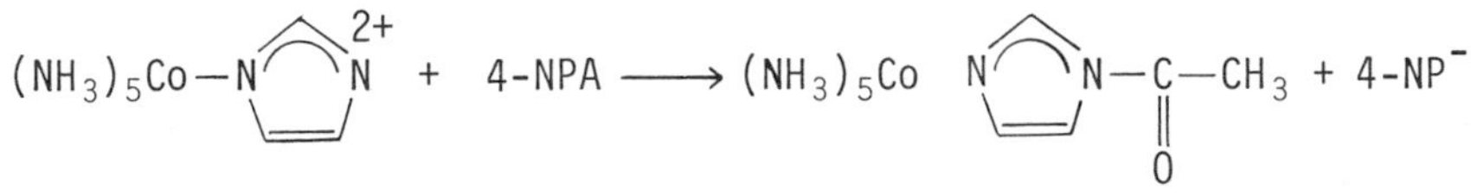

$$k = 9 \text{ M}^{-1} \text{ sec}^{-1} \ (25^\circ, \ \mu = 1.0) \ (pKa = 10.0)$$

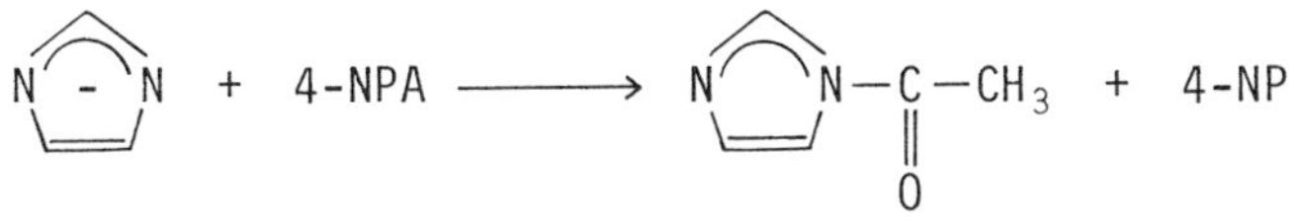

$$k \approx 3160 \text{ M}^{-1} \text{ sec}^{-1} \ (\text{estimated, Ref. 60})(pKa = 14.2)$$

of the anion at N_3 by 10^3-10^2, not an overly large effect when compared with some of the large rate differences we have been talking about in this lecture. Thus, bearing in mind the pKa differences (7.2 *vs*. 10.0 *vs*. 14.2) it is apparent that the un-coordinated imidazole group is somewhat more effective molecule for molecule (by $\sim10^2$) than $(NH_3)_5CoIm^{2+}$ in the pH range for carbonic anhydrase and 4-NPA ($\sim7-8$). It therefore seems clear that the nucleophilic capabilities depend, as does the pKa, on distance from the metal center, and that at 3 atoms removed such a relationship seems to hold.

This reduction in reactivity towards 4-NPA parallels the variation in pKa and a good Brønsted line can be drawn through the points ImH, $(NH_3)_5CoIm^{2+}$ and Im^- (Fig. 16); more generally, the imidazole nucleophiles fit reasonably well into a Brønsted relationship governing a wide range of different types of amines, Fig. 16. For such relationships, with slopes of less than unity, it is obvious that the more effective nucleophiles at neutral pH's will be those with a pKa in the same region - for a species of higher pKa, the larger k_N value is more than offset by a decrease in its effective concentration at the lower pH, $k_{obs} = kK_a[N]_T/(K_a + [H^+])$.

From Fig. 16 we predict that a $Zn-Im^+$ species of pKa 7.6 (CA pKa towards 4-NPA), 11.5 (estimate, see above), and 12.5 (observed in various Co-Im species), will have k_N values of $\sim4 \times 10^{-2}$, 50 and 300 M^{-1} sec^{-1} respectively and efficiencies (k_{obs}) of 4×10^{-2}, 5×10^{-3} and 3×10^{-3} sec^{-1} at pH 7.5. These values fall far short of the observed rate for the enzyme of 1.3 sec^{-1} (25°) and some additional means of acceleration would need to be forthcoming. A similar result was found for the M-OH species (above) but there the rate difference between the enzyme and that predicted from the Brønsted curve was a little larger. This is also suggested in the more general case by Fig. 16. Amine nucleophiles (both coordinated and non-coordinated) are some $10-10^2$ times more effective than M-OH species of similar pKa, at least in the pH range 7-10 ($\beta_N = 0.8$, 0.5 respectively). This result is chemically reasonable and suggests an alternative mechanism - which will be outlined in the following section.

We have mentioned above the possibility of a direct nucleophilic role in CO_2 hydration and have dismissed it on the grounds of heterolytic cleavage of the resulting carbamate. However, for 4-NPA direct attack of a coordinated imidazole moiety would lead to an acetylimidazole intermediate and the hydrolysis of this needs to be considered. Firstly for the enzyme, attempts to detect an initial burst of 4-NP$^-$ which would herald the formation of acetylimidazole, have been unsuccessful; this is a particularly

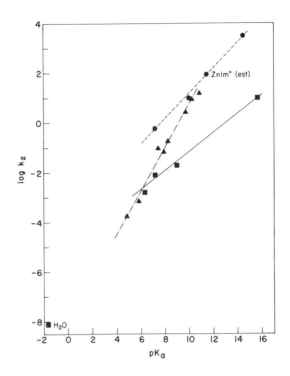

Fig. 16. Rate-basicity correlations for the hydrolysis of 4-nitrophenylacetate at 25° and μ = 1.0 M. The experimental points for imidazolate species (—o—o—) are (in order of increasing pK$_a$): ImH, (NH$_3$)$_5$CoIm^{2+}, ZnIm$_{aq}^+$ and Im$^-$. For metal-hydroxide species (— — —) the experimental points are as given in Fig. 3. The remaining data (—•Δ—•Δ—•Δ) represent experimental points for reaction with a variety of primary amines (Ref. 27).

sensitive test since 4-NP$^-$ can be easily observed spectrophotometrically at 400 nm. Such a species would therefore need to be rapidly hydrolysed - at least 10 times faster than the observed turnover number; $k_{obs} \geq 13$ sec^{-1} at pH 7.6 (25°). In the absence of the enzyme, the hydrolysis of acetylimidazole is not particularly rapid at neutral pH's; $k_{obs} = 3 \times 10^{-4}$ sec^{-1} at pH 7.6. Hydrolysis follows a bell-shaped curve, rate = $k_H[AcImH^+]$ + $k[AcIm] + k_{OH}[AcIm][OH^-]$, Fig. 17, ($k_H = 2.22 \times 10^{-2}$ sec^{-1}, k = 4.6×10^{-5} sec^{-1}, $k_{OH} = 2.95 \times 10^2$ M^{-1} sec^{-1}, 25°, μ = 1.0) and obviously a catalytic effect is required irrespective of the pH. Thus attack of H$_2$O on the fully protonated species (pKa = 3.95) does not approach 13 sec^{-1} and the OH$^-$ promoted reactions of either the protonated or neutral molecule would require a high

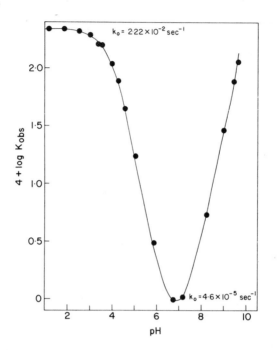

Fig. 17. *pH-Rate profile for the hydrolysis of acetylimi-dazole at 25° and μ = 1.0 M (NaClO₄).*

"effective" concentration of hydroxide in order to approach the enzymic rate.

Coordination of acetylimidazole does not help matters much either

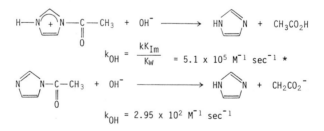

$$k_{OH} = \frac{kK_{Im}}{K_w} = 5.1 \times 10^5 \ M^{-1} \ sec^{-1} \ *$$

$$k_{OH} = 2.95 \times 10^2 \ M^{-1} \ sec^{-1}$$

---

*
The similar reaction for the N-methyl acetyl imidazole cation has a rate constant of 1.5 × 10⁵ M⁻¹ sec⁻¹ at 25°, μ = 0.2, and so it is reasonable to ascribe the "water" reaction to the attack of OH⁻ on the protonated species.

The $(NH_3)_5Co^{3+}$ moiety causes an acceleration of only $\sim$20-fold over that of the neutral species, and it is a little less effective than H^+,

$$(NH_3)_5Co-N\underset{\diagdown\diagup}{\diagup\diagdown}N-\underset{\underset{O}{\|}}{C}-CH_3 + OH^- \longrightarrow (NH_3)_5CoIm + CH_3CO_2^-$$

$$k_{OH} = 7 \times 10^3 \text{ M}^{-1} \text{ sec}^{-1} \ (25^\circ, \ \mu = 1.0)$$

$$k_{obs} \ (\text{pH } 7.6) = 5 \times 10^{-3} \text{ sec}^{-1}$$

as might be anticipated from what has gone before. Our previous arguments suggest that Zn^{2+} will be somewhat less effective than Co(III), and on this basis we predict a rate constant for hydrolysis of the $Zn-AcIm^{2+}$ complex

$$Zn-N\underset{\diagdown\diagup}{\diagup\diagdown}N-\underset{\underset{O}{\|}}{C}-CH_3 + OH^- \xrightarrow{k_{OH}} Zn-Im^{2+} + CH_3CO_2^-$$

of $\sim$1 $\times$ 10^3 M^{-1} sec^{-1}. Thus $k_{obs} < 10^{-3}$ sec^{-1} at the pH of maximum enzyme efficiency (7.6). Clearly this is only fractionally faster than for uncoordinated acetylimidazole, and for both species substantial catalysis is required, not only for the turnover number to be realized, but also to correspond to the inability to detect the intermediate. In this context it should be pointed out that the inability to trans-esterify 4-NPA by addition of MeOH[53(b)], and indeed the apparent inhibition of hydrolysis afforded by MeOH binding (probably) to the metal, is another argument against the direct involvement of "free" solvent OH$^-$.

5.3 PROTEIN IMIDAZOLE/ZnOH MECHANISM - The two most important pieces of experimental evidence to be accommodated when considering possible mechanisms for CA catalysis are (1) the pKa of 6.8 - 7.6 for the working enzyme and (2) the well-documented ability of anions and some neutral species (MeOH, aniline) to inhibit the enzymic process by displacing the base titratable group. The latter is probably coordinated to the Zn atom. These two observations most simply add up to the Zn-OH mechanism discussed previously, but as an alternative for ester hydrolysis Scheme 5 can be considered. This proposal has several attractive features, the most obvious being that the pKa of an uncoordinated imidazole group ($\sim$7.0) agrees with that of the enzyme (6.9 - 7.6, depending on function). Thus, there is no anomalously low pKa to consider and this contrasts with the $Zn-OH^+$ and $Zn-Im^+$ proposals given previously. Also, as we have already seen, reaction of the un-coordinated imidazole species with the ester is likely to be somewhat faster ($k_{Im} = 0.6$ sec^{-1} at pH 7.6, C = 1.0 M) than the similar reaction of $Zn-OH^+$ ($k_{ZnOH} \simeq 3 \times 10^{-3}$ sec^{-1}, pH 7.6, C = 1.0 M) or $Zn-Im^+$ ($k_{ZnIm} \simeq 4 \times 10^{-2}$ sec^{-1}, pH 7.6, C = 1.0 M) species, and this rate is not too different from the observed turnover number for 4-NPA hydrolysis ($k_{cat} = 1.3$ sec^{-1}, pH 7.6). So far, so good, but we now run into difficulties with hydrolysis of the acetylimidazole intermediate.

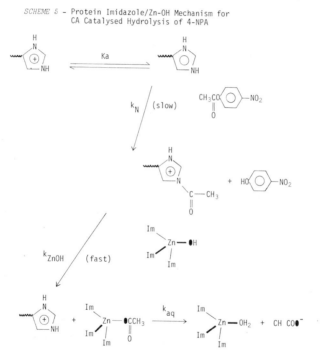

SCHEME 5 - Protein Imidazole/Zn-OH Mechanism for
CA Catalysed Hydrolysis of 4-NPA

NOTES:

1. *pKa (Im) = 7.1; pKa (enzyme) = 6.9 - 7.6 (depending on function).*

2. $k_N = 0.6$ *M*$^{-1}$ *sec*$^{-1}$ *(≈15× faster than Zn-Im$^+$ of similar pKa;*
    ≈200× faster than Zn-OH$^+$); $k_{enz} = 2 \times 10^2$ M^{-1} sec^{-1}.

3. $k_{ZnOH} = 10$ *M*$^{-1}$ *sec*$^{-1}$ *for Zn-OH$^+$ of pKa = 9.1* } *somewhat faster than*
    $k_{ZnOH} = 10^2$ *M*$^{-1}$ *sec*$^{-1}$ *for Zn-OH$^+$ of pKa = 7.6* } *acetylation step*

4. $k_{aq} \approx 10^6$ *sec*$^{-1}$ *(rate determining for CO$_2$ hydration only).*

In the previous section we pointed out the none-too-rapid hydrolysis of acetylimidazole by solvent species (Fig. 17) and suggested that this might be accelerated at neutral pH's by a Zn-OH nucleophile. This is obviously a round-about way to hydrolyse the ester substrate and clearly it needs a rapid subsequent reaction (> 13 sec^{-1} at pH 7.6) to account for the inability to detect an initial burst of 4-NP production by the enzyme. However, it does directly implicate the Zn-OH species, and it could account for the anion inhibition studies - displacement of a coordinated solvent molecule would prevent subsequent hydrolysis. However, for the titration data for the enzyme to agree with anion displacement at the metal center requires a coincidence of pKa values, the first governing k_{cat} and associated with the free imidazole residue in the present mechanism, and the second associated with the anion displacement of an acid titratable group coordinated (or closely associated) with the metal atom - this would be either the water molecule (most likely) or a coordinated imidazole residue. Thus it is required that the coordinated ligand also has a pKa equivalent to that of the catalytic group in the range 6.8 - 7.6. Alternatively, although less likely, the anion titratable group and the free uncoordinated imidazole residue are one and the same with the coordinated hydroxide complex being unaffected by pH changes and anion competition in neutral or alkaline conditions. If this were the case, the marked spectroscopic charges resulting from titrations of the Co(II) enzyme (visible spectrum, CD/ORD, epr) would require the metal to be drastically affected by adjacent changes in protein structure, rather than by direct substitution or titration of coordinated groups.

A second difficulty with this mechanism is that, although a more effective concentration of OH$^-$ can be realized in the Zn-OH$^+$ species, studies we have carried out show that metal-bound OH reacts only with the protonated form of acetylimidazole; no detectable reaction occurs with the neutral molecule, ($k_{OH} \leqslant 10^{-4}$ M^{-1} sec^{-1} for $(NH_3)_5CoOH^{2+}$). Since the pKa of AcImH$^+$ is 3.95 ($25°$, $\mu = 1.0$),

$$(NH_3)_5CoOH^{2+} + H-N \overset{\frown}{\underset{\smile}{+}} N-\overset{O}{\underset{\|}{C}}-CH_3 \longrightarrow (NH_3)_5CoOCOCH_3^{2+} + Im$$

$$(k_{CoOH} = 14.9 \text{ M}^{-1} \text{ sec}^{-1}, 25° \mu = 1.0)$$

a rapid reaction would require that the active site situation both lower the pKa of the Zn-OH$_2$ species and also raise the pKa

of an adjacent AcImH$^+$ residue - a situation difficult to visualize. The limited rate data for these reactions is plotted against
pKa in Fig. 18 and it is clear that a Brønsted slope of ≈0.5 is
involved. This suggests that the similar reaction with Zn-OH$^+$
of pKa 7.6 would have a rate constant of ≈50 sec^{-1} (at 1 M AcImH$^+$
concentration), which is certainly larger than the rate constant
for the initial acylation step (provided the above pKa modifications are realized) and would account for the absence of a
detectable concentration of acetylimidazole intermediate.

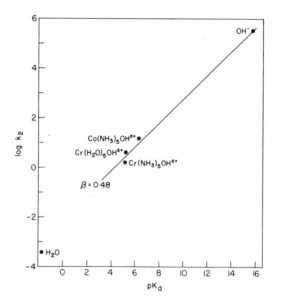

Fig. 18. Plot of log k_2 against basicity of the attacking
reagent for the hydrolysis of acetylimidazole in its protonated form (AcImH$^+$) at 25° and μ = 1.0 M (NaClO$_4$). A statis-
tical correction has been applied to the pK$_a$ of the Cr(H$_2$O)$_6$$^{3+}$
ion to take into account the six equivalent water molecules
(pK$_a$(obs) = 4.32).

Finally, product analyses studies on the model systems
(Co(III) and Cr(III)) show that the M-OH reactions are largely
(>93%), if not entirely, nucleophilic. No detectable general
base reaction was found. This suggests formation of a

$$\gtrless Zn-OCOCH_3^+$$

species in the enzyme and this would have to rapidly regenerate the active $Zn-OH^+$ species. As we have previously pointed out, such processes have rate constants of $\sim10^6$ sec^{-1} which is sufficiently fast to accommodate this step for ester hydrolysis.

5.4 CONCLUSIONS FOR CARBONIC ANHYDRASE - I have spent some time developing my ideas on this enzyme, since I thought it would be useful to air the views of an inorganic chemist to a biochemically motivated audience. Although the conclusions which I summarize below are based on studies on Co(III) systems, I believe they have something to say in the enzyme situation:

(1) $(Im)_3Zn-OH^+$ mechanism favored over "imidazole" mechanisms.
(2) $Zn-OH^+$ less reactive as a nucleophile than $Zn-Im^+$ (by ~10) but more effective at pH 7.5 than OH^-/H_2O (by $\sim10^3-10^5$ depending on substrate).
(3) pKa for isolated $Zn-OH_2^{2+}$ complex would have to be reduced by 1-2 units and this is possible (but not ~5 pKa units for a $Zn-ImH^{2+}$ species).
(4) Uncoordinated imidazole (His-63) is more nucleophilic than $Zn-OH^+$ or $Zn-Im^+$.
(5) General base mechanisms are unlikely.
(6) Rate determining step may be different for CO_2 hydration (dissociation of $Zn-OCO_2H^+$) and ester hydrolysis (acylation by $Zn-OH^+$).
(7) Co(III) enzyme is inactive.

ACKNOWLEDGEMENTS. The new results described in this lecture would not have been possible without the fine experimental skills of the following people: Dr. E. Baraniak, Mr. C. Boreham, Dr. C. Clark, Dr. L. Engelhardt, Dr. R. Keene and Dr. A. Zanella.

REFERENCES

1. P.D. Boyer, "The Enzymes", Vol. 1, Academic Press, New York (1971).
2. M.F. Dunn, "Structure and Bonding", Vol. 23, p. 61, Springer-Verloy (1975); A.S. Mildvan and J.E. Engle, "Methods in Enzymology", S.P. Colowick and N.O. Kaplan, Eds., Academic Press.
3. J.E. Coleman, "Inorganic Biochemistry", G.L. Eichhorn, Ed., Vol. 1, Chap. 16, p. 488, Elsevier (1973); also in "Progress in Bioorganic Chemistry", E.J. Kaiser and F.J. Kezdy, Eds., Vol. 1, p. 159, Wiley-Interscience (1971).
4. S. Lindskoy, "Structure and Bonding", Vol. 8, p. 160, Springer-Verloy (1970).

5.   E. Baraniak, D. Buckingham and A.M. Sargeson, unpublished results.
6.   D.A. Buckingham, D.M. Foster and A.M. Sargeson, *J. Am. Chem. Soc.*,**90**, 6032 (1968).
7.   D.A. Buckingham, J. MacB. Harrowfield and A.M. Sargeson, *J. Am. Chem. Soc.*,**96**, 1726 (1974).
8.   M.L. Bender and B.W. Turnquest, *J. Am. Chem. Soc.*,**79**, 1889 (1957).
9.   D. Pinnell, G.B. Wright and R.B. Jordan, *J. Am. Chem. Soc.*, **94**, 6104 (1972).
10. H. Shinar and G. Navon, *Biochem. Biophys. Acta,***334**, 471 (1974).
11. R.A. Anderson and B.L. Vallee, *Proc. Nat. Acad. Sci. U.S.A.*, **72**, 394 (1975).
12. E.P. Kang, C.B. Storm and F.W. Carson, *Biochem. Biophys. Res. Commun.*,**49**, 621 (1972); *idem. J. Am. Chem. Soc.*,**97**, 6723 (1975).
13. R.K. Murmann and H. Taube, *J. Am. Chem. Soc.*,**78**, 4886 (1956); D.E. Klimek, B. Grossman and A. Haim, *Inorg. Chem.*,**11**, 2382 (1972).
14. A.M. Sargeson and H. Taube, *Inorg. Chem.*,**5**, 1094 (1966).
15. D. Stranks, private communication.
16. A. Haim, *Inorg. Chem.*,**11**, 3147 (1972).
17. G. Sykes, personal communication to J. Edwards, 1976.
18. E. Lenz and R.K. Murmann, *Inorg. Chem.*,**7**, 1880 (1968).
19. T.A. Beech, N.C. Lawrence and S.F. Lincoln, *Aust. J. Chem.*, **26**, 1877 (1973); R.K. Wharton, R.S. Taylor and A.G. Sykes, *Inorg. Chem.*,**14**, 33 (1975).
20. D.A. Buckingham, J. McB. Harrowfield and A.M. Sargeson, *J. Am. Chem. Soc.*,**95**, 7281 (1973).
21. E. Chattee, T.P. Dasgupta and G.M. Harris, *J. Am. Chem. Soc.*, **95**, 4169 (1973).
22. D.A. Buckingham and L.M. Engelhardt, *J. Am. Chem. Soc.*,**97**, 5915 (1975).
23. D.A. Palmer and G.M. Harris, *Inorg. Chem.*,**13**, 965 (1975).
24. E. Breslow, in "The Biochemistry of Copper", J. Peisach, P. Aisen and W.E. Blumberg, Eds., p. 149, Academic Press, New York (1966).
25. P. Woolley, manuscript submitted to J.C.S. (Dalton).
26. A.J. Parker, *Chem. Revs.*,**69**, 1 (1969).
27. W.P. Jencks and M. Gilchrist, *J. Am. Chem. Soc.*,**90**, 2622 (1968).
28. S.F. Lincoln and D.R. Stranks, *Aust. J. Chem.*,**21**, 37 (1968) *idem., ibid.*,**21**, 57 (1968).
29. G.M. Harris, Personall communication to D.R. Stranks. $\Delta V^{\neq}$ studies support the intramolecular attack of a coordinated

aqua group on the carboxylate moiety.

30. D.A. Buckingham, J. MacB. Harrowifle and A.M. Sargeson, *J. Am. Chem. Soc.*,**95**, 7281 (1973).
31. C. Boreham, D.A. Buckingham, D. Francis, L.G. Warner and A.M. Sargeson, unpublished results.
32. Calculated rate constant from data of C. O'Connor and D.R. Llewellyn, *J. Chem. Soc.*, 2669 (1965).
33. D.A. Buckingham, D.M. Foster and A.M. Sargeson, *J. Am. Chem. Soc.*,**91**, 4102 (1969).
34. C. Boreham, D.A. Buckingham, F.R. Keene and A.M. Sargeson, unpublished results; a preliminary report has appeared, D.A. Buckingham, F.R. Keene and A.M. Sargeson, *J. Am. Chem. Soc.*, **96**, 4981 (1974).
35. W.P. Jencks, *J. Am. Chem. Soc.*,**94**, 4731 (1972).
36. D.M. Blow, J.J. Birktoft and B.S. Hartley, *Nature,* **221**, 337 (1969).
37. D.A. Buckingham, C.E. Davis, D.M. Foster and A.M. Sargeson, *J. Am. Chem. Soc.*,**92**, 5571 (1970); and D.A. Buckingham, F.R. Keene and C. Boreham, unpublished results.
38. S. Lindskog, L.E. Henderson, K.K. Kanna, A. Liljas, P.O. Nyman and B. Strandberg, "The Enzymes", P.D. Boyer, Ed., Vol. 5, Chap. 21, p. 587, Academic Press, New York (1971).
39. P. Woolley, *Nature,* **258**, 677 (1975).
40. "Stability Constants of Metal-Ion Complexes", L.G. Sillen and A.E. Martel, Eds., Chem. Soc., Spec. Pub. No. 17 (1964).
41. P.D. Perrin, *J. Chem. Soc.*, 4500 (1962).
42. D.R. Buckingham, C. Clark and L.M. Engelhardt, unpublished results.
43. D.L. Robenstein and G. Blakney, *Inorg. Chem.*,**12**, 128 (1973).
44. S. Chaberek, R.C. Courtney and A.E. Martell, *J. Am. Chem. Soc.*,**74**, 5057 (1952).
45. P.J. Morris and R.B. Martin, *J. Am. Chem. Soc.*,**92**, 1543 (1970).
46. T.P. Pitner, E.W. Wilson and R.B. Martin, *Inorg. Chem.*,**11**, 738 (1972).
47. J. Harrowfield, V. Norris and A.M. Sargeson, unpublished data.
48. Reference 28, p. 660 (Ref. 127).
49. W.P. Jencks in "Catalysis in Chemistry and Enzymology", Chap. 2, McGraw-Hill (1969).
50. P.L. Yeagle, C.H. Lochmüller and R.W. Henkens, *Proc. Nat. Acad. Sci. U.S.A.*,**72**, 454 (1975).
51. R.W. Henkens, private communication (manuscript in preparation).
52. C.K. Tu and D.N. Silverman, *J. Am. Chem. Soc.*,**97**, 5935 (1975); D.N. Silverman and C.K. Tu, *J. Am. Chem. Soc.*,**98**, 978 (1976).

53. (a) D.W. Appleton and B. Sarkar, *Proc. Nat. Acad. Sci. U.S.A.*, **71**, 1686 (1974);
    (b) *idem.*, *Bioinorg. Chem.*, **4**, 309 (1975).
54. J.M. Pesando, *Biochem.*, **14**, 675 (1975).
55. R.K. Gupta and J.M. Pesando, *J. Biol. Chem.*, **250**, 2630 (1975).
56. J.H. Saunders and K.C. Frisch, "Polyurethanes", Part I, p. 180, Interscience, New York (1965).
57. D.A. Buckingham, D.J. Francis and A.M. Sargeson, *Inorg. Chem.*, **13**, 2630 (1974).
58. P.C. Ford, *Inorg. Chem.*, **10**, 2153 (1971).
59. R.B. Martin, *Proc. Nat. Acad. Sci. U.S.A.*, **71**, 4346 (1974).
60. Calculated from data given in T.C. Bruice and G.L. Schmir, *J. Am. Chem. Soc.*, **80**, 148 (1958).

dinitrogen fixation in protic media (comparison of biological dinitrogen fixation with its chemical analogues)

A.E. SHILOV

Institute of Chemical Physics,
Academy of Sciences of U.S.S.R.,
Moscowskoja Oblast, U.S.S.R.

(cont'd)

1. INTRODUCTION

The primary process of molecular nitrogen fixation in indus-
try is the catalytic synthesis of ammonia from dinitrogen and
dihydrogen:

$$N_2 + 3H_2 \xrightarrow[\text{(Fe)}]{} 2NH_3$$

There is also the natural process of dinitrogen fixation,
which is presently the main source of bound nitrogen on our
planet, and is carried out by a number of bacteria and algae; in
this case, the first product of dinitrogen transformation is also
ammonia and this first stage can be schematically presented as
taking place with the participation of the protons of water:

$$N_2 + 6H^+ + 6\bar{e} \longrightarrow 2NH_3$$

Despite all the obvious outward resemblance of these two
reactions, chemists have always been amazed by the differences in
the conditions under which they proceed. Due to the well-known
chemical inertness of dinitrogen, industrial synthesis of ammonia
must be conducted at high temperatures and pressures, whereas the
biological fixation proceeds effectively directly from the air,
under natural environmental conditions. The existence of the
natural process has led to numerous attempts of implementing the
reduction of dinitrogen in the laboratory, in the same conditions
under which the biological reaction takes place. Until recently
such attempts had ended in failure. Reports appearing from time
to time in the literature on the formation of small amounts of
ammonia from dinitrogen in aqueous solutions were not subsequently verified.

It is only in the last six years that reproducible results
have been obtained demonstrating the reduction of dinitrogen to

hydrazine and ammonia in protic media. The present lecture reviews the results obtained in this area.

The way for the discovery of N_2 reduction in protic media was paved by the results, obtained mainly in the sixties (see below), on the biological fixation of dinitrogen, and the results on the chemical reduction of dinitrogen in aprotic media. The latter can be briefly summarized as follows (for details see Ref. *1*).

Starting with the work of Volpin and Shur in 1964 *(2)*, it became known that dinitrogen could be reduced to ammonia derivatives (usually nitrides) in the presence of compounds of transition metals (Ti, Zr, V, Nb, Cr, Mo, W, Mn, Fe, Co) under the action of such reducing agents as free metals (Li, Mg, Al), hydrides of metals (LiAlH₄), and various organometallic compounds. The role of the transition metals at their low oxidation states in these systems is, by all indications, associated with their forming dinitrogen complexes in which the N_2 molecule is activated. In 1965 Allen and Senoff discovered the first such N_2 complex, which was a Ru(II) compound *(3)*. The complex was obtained in an indirect way by the action of hydrazine on $RuCl_3$. Then in 1966, the work of Shilov, Shilova and Borod'ko *(4)* demonstrated the capability of similar complexes being formed directly from dinitrogen.

A large number of N_2 complexes is now known (see Refs. *5-7*), and their formulae are: L_nMN_2, $L_nMN_2ML_n$, or $L_nM(N_2)_2$, where M is the transition metal, and L, a ligand. They usually have a linear fragment MNN or MNNM. At first, investigators failed to observe complexes with dinitrogen, in which the N_2 would enter into chemical reactions (*e.g.*, those involving its reduction). Starting from 1969 *(8)*, however, such complexes were found and, in some cases, subsequently isolated for some systems involving titanium, iron and zirconium. All the complexes were formed in aprotic media with the participation of strong reducing agents, but it was noted that, in the iron N_2 complexes *(9,10)*, the reduction of N_2 to hydrazine took place only under the action of acid protons (in the presence of alcohol, the dinitrogen in the complex is fully converted into free dinitrogen). This was explained by protonation of N_2 in the coordination sphere of the metal with resulting facilitation of further reduction.

Recently (since the discovery of systems fixing dinitrogen in protic media) Chatt and his coworkers *(11)* succeeded in showing that, with the protonation of nitrogen in mononuclear bis-dinitrogen complexes of W(0) and Mo(0), one N_2 molecule was being reduced finally to hydrazine and ammonia.

2. THEORETICAL CONSIDERATIONS

2.1 ON THE POSSIBILITY OF ONE-STEP MULTIELECTRONIC REDUCTION OF DINITROGEN - Reduction of dinitrogen to hydrazine and ammonia requires the transfer of several electrons to the N_2 molecule and, in principle, this can proceed in several stages. In this connection, we shall examine the thermodynamics of the successive reduction of the N_2 molecule. We shall, first of all, note that the reduction of dinitrogen leads to the increase of electron density on N atoms and, correspondingly, to the enhancement of its basic properties. This results in ammonia stabilization, as well as stabilization of the intermediate products of N_2 reduction as compared with the gaseous phase or aprotic media, due to the formation of hydrogen bonds with the protons of the solvent. For example, the enthalpy of hydrazine formation from H_2 and N_2 is +22.8 kcal/mole in the gaseous phase and only +8.2 kcal/mole in aqueous solution. Although the entropy of N_2H_4 formation is also reduced (because of the "freezing" of the solvent molecules close to N_2H_4), the overall effect of the change in free energy of formation, with the transfer of N_2H_4 from the gaseous phase to the aqueous solution at 25°, amounts to -7.5 kcal/mole, which corresponds to the equilibrium constant of reaction

$$2 \ H_2 + N_2 \rightleftharpoons N_2H_4$$

increasing 3×10^5 times. It can be assumed that the rate constant for N_2 reduction in complexes of transition metal compounds will also increase in protic media as compared with aprotic solutions, all other conditions being equal, due to hydrogen bonds with the solvent being formed in the transition state. Thus, with the same reducing agent strength, protic media conditions are more favorable for reduction to take place.

Standard E_0 reduction potentials* for the formation of the products from one-, two-, four- and six-electron reduction of dinitrogen in aqueous solution at pH = 0 are presented below:

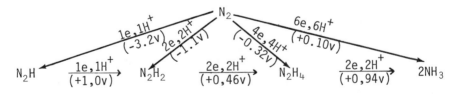

Reduction potentials will always be related to the standard H_2 electrode at pH = 0. A negative reduction potential means that the reducing agent is able to form H_2 from H^+ (at pH = 0).

The dependence of E_0 on pH for all the above redox pairs will be expressed by the formula: $E_0 = E_0^0 - 0.059$ pH. Energy expenditures among the stages of successive reduction of N_2 are seen to be distributed in an extremely non-uniform way. One can easily see that, if the reaction proceeds through N_2H, N_2H_2 or N_2H_4, a reducing agent stronger than H_2 is required at the first stage, which for aqueous or aqueous-alcohol media means that the reaction should be conducted in conditions when H_2 evolution is more advantageous. This results from an especially high energy required to split the first bond in N_2; this energy (~ 130 kcal/mole) amounts to more than half the energy of the $N{\equiv}N$ triple bond (225 kcal/mole). This is why both a one- and two-electron reduction of N_2 with the formation of N_2H_2, require a very strong reducing agent.

This distinguishes dinitrogen from the other compounds with multiple bonds, *e.g.*, acetylene, where, for the two-electron reduction to ethylene, a reducing agent with the potential of +0.37v is sufficient. Thus, in this case, dihydrogen is already a sufficiently strong reducing agent.

The high reduction potentials of the one- and two-electron reduction of N_2 explain why it is so difficult to reduce dinitrogen in protic media, although as noted above, the energy requirements for N_2 reduction in protic media are smaller than in aprotic media. For reduction to be accomplished, a reducing agent with a potential greater than -1.1v at pH 0, -1.5v at pH 7, or -1.9 at pH 14, must be stable enough in the protic media (free metals are among the reducers of such strength, but their inertness in water is associated with the formation of a protective film of oxides which will also interfere, evidently, with the interaction with dinitrogen).

At the same time, the existence of biological dinitrogen fixation with a moderate value of the reduction potential indicates that a mechanism can be implemented with this thermodynamic difficulty being avoided. The study of the properties of transition metals N_2 complexes made it possible, as early as 1969, to propose a mechanism in which the N_2H_2 molecule did not appear at an intermediate stage. The reaction could proceed with the transformation of coordinated dinitrogen directly into a derivative of hydrazine, for which a much less strong reducing agent is needed *(5)*. Various specific cases of this mechanism are considered in a review *(1)*. At present, the mechanism considered by the author to be the most probable for protic media, can be summarized in the following way:

$$M^{n+} \ldots N{\equiv}N \ldots M^{n+} \longrightarrow M{\diagup}^{N}_{\diagdown}{}_{N}{\diagup}^{}M \text{ (or } M{\diagup}^{N{-}N}{\diagdown}M) \xrightarrow{H^+} N_2H_4 + 2M^{n+2}$$

or with the prior participation of the solvent protons (taking into account the enhancement of the basic properties of nitrogen in the intermediate complex):

$$M^{n+}...N{\equiv}N ... M^{n+} \xrightarrow{H^+} M{\diagup}^{N}_{\diagdown}\underset{NH}{\overset{+}{\diagup}}M \xrightarrow{H^+} N_2H_4 + 2 \ M^{n+2}$$

In this process the oxidation level of every M must in the end increase by 2 units. If we assume that there is in the system next to every M at least one M_R ion capable of donating one electron to an M ion simultaneously with the reduction of di-nitrogen, then the formation of hydrazine in the presence of such a four-nuclei catalyst leads to an increase in the oxidation level of every M and M_R by 1 unit:

$$M_R^{m+} ... \ M^{n+}...N{\equiv}N...M^n \ ... \ M_R^m \xrightarrow{H^+} M_R...M{\diagup}^{N^+}_{\diagdown}\underset{H}{N}{\diagdown}^{M}_{\diagup} \ ... \ M_R \xrightarrow{H^+}$$

$$\longrightarrow N_2H_4 + 2 \ M^{n+1} + 2 \ M_R^{m+1}$$

Such a mechanism obviously makes it possible to utilize, in a catalytic process, a one-electron reducing agent (Red) (in a specific case Red = M_R), for which the reduction potential of the process

$$Red \longrightarrow Ox + e^-$$

is only slightly higher than the potential of the reaction

$$N_2 + 4e^- + 4H^+ \longrightarrow N_2H_4$$

Thus a reducing agent which is much weaker than one required for the formation of diimide can be used.
    We should note that for an optimum four-nuclei catalyst $(M^n)_2$ $(M_R^m)_2$, the mean oxidation-reduction potential of the processes

$$M^{n+} \longrightarrow M^{(n+1)+} + e^- \qquad \text{and}$$

$$M_R^m \longrightarrow M_R^{(m+1)+} + e^-$$

must lie between E_0 for Red and E_0 for N_2H_4. In the case when a

polynuclear catalyst includes six and more metal ions one can
visualize a further reaction with complete disruption of the N-N
bond and the formation of ammonia derivatives:

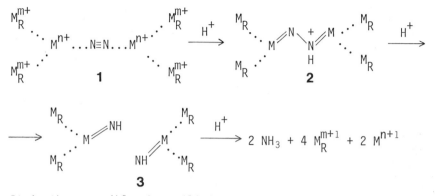

It is thus possible, by utilizing this mechanism, to exclude
intermediate formation not only of diimide but also of hydrazine
and conduct the reaction directly to derivatives of ammonia using
a reducing agent which is even weaker than hydrazine. In this
case the intermediate or transitional forms must certainly be
stabilized in an appropriate way; this seems to be assured by the
conjugation in the intermediate complexes **1** and **2**. We should
note that, if every M or M_R is capable of entering into two-electron reactions (with the formation of M^{n+2} or M_R^{m+2}, respectively),
the number of M or M_R atoms in the catalyst molecule can be reduced. It is, however, easy to see that in the general case this
will result again in an increase of the required reduction potential of Red, which, for protic media, is naturally undesirable.

*2.2   ON THE NATURE OF THE INTERMEDIATE COMPLEX IN THE REDUCTION
OF DINITROGEN IN PROTIC MEDIA* – Thermodynamic difficulties encountered in the reduction of dinitrogen in protic media, even if
the possibility of multielectron reduction of N_2 is taken into
account, require special properties for the intermediate complex
with dinitrogen. In fact, the requirement of a certain reduction
potential for the M complexing agent is supplemented by the requirements of a certain stability and a certain structure for the
intermediate complex.
    The stability must be optimal (it is felt that excessive
stability will impede further reactions of coordinated dinitrogen), and it is obviously useful for the intermediate complex to
be as close as possible to the reaction products.
    At present we have numerous data on various complexes of N_2

with transition metal compounds and can draw some conclusions on the nature of the optimum catalyst for the reduction of dinitrogen (certainly taking into account the already available experimental data on reactions in protic media).

As already noted, both the mono- and the binuclear complexes, whose structures have been determined, usually have a linear fragment M—N≡N or M—N≡N—M. It is only for a complex of low-valence nickel that a non-linear structure with a

fragment has been shown *(12,13)*. The linearity of the M—N≡N—M fragment is evidently determined by the possibility of maximum "conjugation of bonds" (*i.e.*, maximum overlapping of the d-orbitals of M and the p-orbitals of nitrogen). The transition into hydrazine derivatives corresponds to the formation of the planar configurations

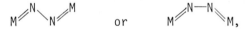

which is again determined by the best overlap of d and p orbitals. The order of the filling of molecular orbitals in a linear binuclear complex has been considered *(14,15)*. For the same M in a binuclear complex made up of two octahedra, a simplified picture of the molecular orbitals can be represented as:

$$1e_u < 1e_g < b_{2g} \approx b_{1u} < 2e_u < 2e_g$$

(Fig. 1). The bonding of dinitrogen in the M—N≡N—M complex is mainly effected by the $1e_g$ orbital, constructed from the d_{xy} and d_{yz} orbitals of M and the antibonding $1\pi_g$ orbital of N_2 (the lower doubly-degenerate $1e_u$ orbital accommodates 4 electrons of the π bonds of the N_2 molecule). The filling of the $1e_g$ orbital ensures the electron transfer from M on to the N_2 molecule and the weakening of the N—N bond. Addition of four electrons to this doubly-degenerate orbital, *i.e.*, with a change in electronic configuration from d^0 to d^2, will, all other conditions being equal, enhance the weakening of the N—N bond. Further increase

of the number of d electrons on M will first lead to the filling
of the non-bonding orbitals b_{2g} and b_{1u} (d^3 and d^4); then the
$2e_u$ orbitals (d^5 and d^6) will be filled which will weaken the
M—N bonds and stabilize the N—N bond; and finally, the filling
of the $2e_g$ orbital (d^7 and d^8) will correspond to the weakening
of both the M—N bonds and the N—N bond.

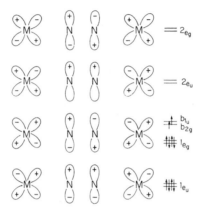

Fig. 1.   The molecular orbitals for a linear binuclear
$L_n M$—N≡N—ML_n *complex (D_{4h} symmetry).*

On the whole, the experimental data for binuclear complexes
conform to this picture.

For example, in linear M—N≡N—M binuclear complexes the N—N
bond for a d^6 metal system is no weaker than in the mononuclear
complexes, whose stability turns out to be greater than that of
binuclear complexes. For complexes with M (d^2), binuclear species are more stable than mononuclear species, and it is the binuclear complexes in which the N—N bond is characterized by considerable weakening. Following from the previous considerations,
it can be assumed that the catalytic properties of M in binuclear
complexes must be more optimal than in mononuclear complexes[*];
further, in a series of M_2N_2 complexes with the same reducing
properties for M, the optimum conditions for catalysis in trans-
forming a complex with linear configuration into one with planar

---

[*]*Since less change is necessary in the oxidation state of M during the process of*
N_2 *reduction.*

configuration will be realized when the N—N bond is maximally weakened and the charge of N atoms is maximally negative.

From this, it follows that in the case of the same oxidation-reduction potential for M and a linear intermediate complex M—N≡N—M, the catalytic properties of M will improve from d^0 to d^2 and deteriorate from d^5 to d^{10}.

Electronic configurations d^3 and d^4 deserve special consideration. The filling of each of the non-bonding b_{2g} and b_{1u} orbitals with one electron will prevent the possibility of electron transfer from the $1e_g$ orbital to the b_{2g} and b_{1u} orbitals (with a small difference in levels, this transition, for a d^2 configuration complying with Hund's rule, will decrease the weakening of the N—N bond and the negative charge on the N atoms.

The change from a d^2 to d^3 configuration will thus apparently improve the catalytic properties of M. The further filling of b_{2g} and b_{1u} orbitals with two electrons (to configuration d^4) is, from the point of view of creating optimum catalytic properties, undesirable, as these very orbitals will be the highest filled in the complex and, with equal reduction potentials for M (d^4) and M (d^3) (or equal ionization potentials), for a d^4 system, electron transfer onto the bonded N_2 molecule will be reduced. The same conclusion can be arrived at if we consider that in octahedral complexes with d^4 configuration, due to the ligand-field splitting, one electron is at a higher energy level, and this must lead to a greater difference in reduction potentials when there is a change from one- to two-electron processes (*e.g.*, Cr^{2+}, while being a strong one-electron reducing agent, is far less capable of transferring two electrons). For equal reduction potentials as regards a one-electron process, the transfer of 2 electrons to the N_2 molecule in a complex must therefore be greater for a d^3 configuration than for a d^4.

Thus in the series of transition elements, the d^3 configuration corresponds to the optimum properties for the catalytic reduction of dinitrogen in protic media following the mechanism discussed above, which seems to correspond to the most economical energy expenditure. For systems with the participation of strong reducing agents (*e.g.*, for complexes obtained in aprotic media), the linear binuclear M—N≡N—M complex may not be a necessary intermediate in the reduction of N_2. The participation of M with d^5 -d^{10} electronic configurations can then prove suitable for the activation of dinitrogen. As shown by the results of Chatt and coworkers *(11)*, the reduction of N_2 can take place via protonation of coordinated N_2 molecules in previously prepared mononuclear W(0) and Mo(0) complexes. Another possibility is to form a non-linear binuclear complex of the type with a strongly reduced N—N bond order, which can become referable with a high number of d electrons on M when the ionization potential of M is strongly reduced (*i.e.*, the donor properties are enhanced), as observed for a nickel complex *(12,13)*.

3. SYSTEMS FIXING NITROGEN IN PROTIC MEDIA

3.1 NITROGENASE - In the enzymatic reduction of dinitrogen, the protons of water take part in the formation of ammonia, which is the first detectable product. Taking into account that hydrophobic areas exist in enzymes, it is not known to what extent, in the strict sense of the word, one can speak of a "protic media" as a solvent in which the enzymatic process of dinitrogen fixation is developing. In any case, however, the active centre proves to be accessible to the water protons which, in the absence of N_2, are reduced to dihydrogen. It is thus necessary to consider here all the advantages and limitations resulting from the reaction being conducted with the participation of protons, which justifies our considering this process along with the non-biological reactions for the reduction of dinitrogen in protic media.

The structure and mechanism of action of nitrogenase have been studied with special intensity from the early sixties when crude extracts fixing dinitrogen, and then the enzyme itself, were successfully isolated.

Various aspects of the biological fixation of dinitrogen have been considered in a number of reviews (*e.g.*, see Refs. *16,17*), and we shall formulate only briefly here the principal results. The enzyme isolated from microorganisms is a combination of two proteins, one of which ("Mo-Fe protein") contains two molybdenum atoms and 24-32 iron atoms, as well as sulphide and thiol groups; the other ("Fe protein") contains 4 iron and 4 sulphur atoms.

The reaction of dinitrogen reduction *in vitro* can take place under the action of specially added sodium dithionite. No intermediate products (*e.g.*, N_2H_4) have been observed in the process of N_2 fixation.

A great role in the reaction mechanism belongs to ATP, whose hydrolysis, coupled with electron transfer from the reducing agent to the dinitrogen molecule bonded in the active centre, evidently raises the potential of the reduction system. The reduction of iron seems to take place first in the Fe protein, then the electrons pass from the Fe protein to the Mo-Fe protein, and it is at this stage that ATP takes part. It is felt *(1)* that the mechanism of ATP action includes the non-equilibrium protonation of one of the Mo-Fe protein ligands during the process of ATP hydrolysis, which imparts the capacity to accept electrons from the Fe protein. With subsequent equilibrium deprotonation of the ligand, the Mo-Fe protein potential increases above the initial potential of the reduced Fe protein; this actually means a transfer of the electron from the Fe to the Mo-Fe protein against the thermodynamic potential. According to the data of Likhtenstein and his coworkers *(18)*, it is essential that all iron atoms in the enzyme are bonded with each other, evidently forming a cluster capable of transferring electrons to the active centre. Electron transfer conjunct with ATP hydrolysis is the rate-controlling stage of the process. The reactions of NH_3 formation from N_2, and of H_2 formation from the protons of water, compete for the electrons of the reducing agent, the yield of NH_3 increasing and that of H_2 decreasing with the growth of dinitrogen pressure from zero to atmospheric. However, even at elevated dinitrogen pressures, the yield of H_2 does not drop below 25% of the used electrons of the reducing agent, 75% accounting for the ammonia formation from dinitrogen. The stoichiometry of the reaction for dinitrogen at sufficiently high pressures of N_2 can thus be represented:

$$N_2 + 8\,\bar{e} + 8\,H^+ \longrightarrow 2\,NH_3 + H_2$$

It is obvious that, along with the independent reaction of H_2 formation which is retarded by the introduction of dinitrogen, there exists a coupled reaction of hydrogen formation, which consumes 25% of the electrons at any dinitrogen pressures. Besides dinitrogen, other molecules with a triple bond, such as acetylene, CN^-, N_3^-, N_2O, CH_3CN, and CH_3NC (which, like dinitrogen, are capable of being reduced in the presence of nitrogenase), fully suppress dihydrogen evolution.

Carbon monoxide inhibits N_2 reduction, but at the same time has no effect on dihydrogen formation. It is of interest that acetylene is reduced only to ethylene, which is not reduced further to ethane and does not inhibit either the reduction of dinitrogen or the evolution of dihydrogen. The reduction of C_2D_2 produces cis-deuteroethylene.

The reduction rate of dinitrogen amounts to approximately 0.1-1 molecule of N_2 per one active centre in 1 sec.; the activation energy (pertaining, as mentioned, to the process of electron transfer) is 14 kcal at temperatures above $20^\circ C$.

Molybdenum evidently plays a significant part in the activation of N_2 and the other substrates of nitrogenase, and seems to participate in the formation of the intermediate complex.

It has been suggested recently *(19)* that in all molybdenum-containing enzymes, including nitrogenase, the same or a similar molybdenum-containing peptide is present; this is thought to be used in oxidation-reduction reactions catalyzed by the molybdenum. In this connection it is of interest to note that the molybdenum-containing peptide separated from xanthine oxidase *(20)* does not contain S–H bonds in the amino acids which form the peptide, and the molybdenum atom is bonded to the peptide by Mo–O bonds *(21)*.

3.2 HYDROXIDES OF Ti(III), Cr(II), and V(II), WITH Mo(III) PARTICIPATION - These systems were discovered in 1970 *(22-25)* based on previous assumptions for the mechanism of dinitrogen reduction *(5)*. They were the first systems that reproducibly reduced molecular nitrogen in aqueous alkaline and aqueous alcohol media with appreciable yields of the products, hydrazine and ammonia. Addition of magnesium salts produced an activating effect. The behaviour of these systems was subsequently described in a number of publications (see Ref. *1*).

Table 1 presents some of the results obtained.

TABLE 1: N_2 REDUCTION IN $Ti(OH)_3$-Mo(III) AND $Cr(OH)_2$-Mo(III) SYSTEMS				
Mo/Red	PN$_2$ (atm)	T$^{\circ}$C	Time (min)	N_2H_4/Mo
Red = Ti(OH)$_3$	Medium, CH$_3$OH + H$_2$O (mole ratio 6.5:1)			KOH = 1 M
2.0×10^{-5}	100	85	15	87.5
2.7×10^{-4}	120	95	15	46.2
1×10^{-2}	100	85	10	2
Red = Cr(OH)$_2$	Medium, CH$_3$OH + H$_2$O (mole ratio 5.5:1)			KOH - 0.4 M
1×10^{-3}·	40	90	15	0.8
1×10^{-3}	40	24	180	0.35

$Ti(OH)_3$ and $Cr(OH)_2$ are practically inactive toward dinitrogen without the Mo(III) under the experimental conditions. Mo(III) also activates vanadium (II) hydroxide preserved in the absence of N_2 (reduction of dinitrogen with freshly prepared vanadium (II)-magnesium hydroxide is described later). A Mo(III) compound is formed when the initially added Mo compound (*e.g.*, $MoOCl_3$ or MoO_4^{2-}) is reduced by a reducing agent present. $Mo(OH)_3$ does not by itself reduce dinitrogen, and thus in the case of $Ti(OH)_3$ and $Cr(OH)_2$, only the combined action of the reducing agent and Mo(III) ensures the N_2 reduction. In the case of $Ti(OH)_3$-Mo(III), at elevated temperatures (above $80°C$) catalytic yields of N_2H_4 and NH_3 with respect to molybdenum are formed, and the stoichiometry of the reaction for N_2H_4 formation can be written as follows:

$$4\ Ti(OH)_3 + N_2 \xrightarrow{\quad Mo(III) \quad} N_2H_4 + 4\ TiO_2 + 4\ H_2O$$

In the absence of dinitrogen, the system forms dihydrogen from water, this reaction proceeding even without Mo(III) and being retarded by additions of Mg salts. The parallel reaction of dihydrogen evolution is also partially suppressed when Mo(III) and N_2 are jointly present (26). Analysing this result, the authors concluded (26) that the catalytic action of the system was associated with the transfer of electrons along a polymer chain of Ti^{3+} ions in the hydroxide surrounding the complex, to a N_2 complex of Mo(III), which is an electron trap. Investigation into the electrical conductivity of dried Ti(III) hydroxide, which is a semiconductor, supports this conclusion; additions of molybdenum compounds decrease the electrical conductivity of the system, and this explains the decrease of the catalytic effect of Mo with increase of its concentration in the hydroxide. In contrast to $Ti(OH)_3$, the hydroxides of Cr(II) and V(II), which apparently have no pronounced electrical conductivity, do not give catalytic yields of products. However, the activity of these systems with respect to dinitrogen remains sufficiently high and, for example, hydrazine is observed to form with an appreciable rate in the $Cr(OH)_2$-Mo(III) system even at room temperature.

Carbon monoxide inhibits the reduction of dinitrogen in the presence of Mo(III). In a CO atmosphere the same yellow carbonyl complex of Mo passes into solution in all three systems (27), this complex undergoing equilibrium dissociation in the solution and forming free CO.

The complex, separated from the solution after the removal of one CO molecule, is also a carbonyl derivative of Mo and apparently contains several Mo atoms. This serves as indirect confirmation of at least a dimeric nature for the active Mo(III) complex in the hydroxide systems with respect to fixing dinitrogen. The carbonyl complex formed is again capable of catalyzing hydrazine formation when added to $Ti(OH)_3$ or $Cr(OH)_2$.

$Ti(OH)_3$ in alkaline media is a stronger reducing agent than H_2, thus the $Ti(OH)_3$-Mo(III) system cannot be a catalyst for hydrogenation of N_2 with N_2H_4 formation. However, the Ti(IV) formed can again be reduced by H_2 (*e.g.*, with participation of Pt catalyst) after addition of acid. Therefore H_2 can be used as a reducing agent for N_2, with accumulation of N_2H_4 successively changing the pH; in alkaline medium N_2H_4 is formed from N_2, while in acid medium Ti(III) is again regenerated from Ti(IV). The energy needed by the H_2 in order to form N_2H_4 from N_2 and H_2 is evidently provided by the neutralization reaction. The authors of this work *(28,29)* suggest that the acid action in this case is in effect similar to that of ATP in biological N_2 fixation, though naturally "acidification" in enzymatic ATP hydrolysis proceeds specifically only in the reaction centre, and this is so far unattainable in model experiments.

Study of the kinetics of dinitrogen reduction by the $Ti(OH)_3$-$Mg(OH)_2$-Mo(III) system *(30)* made it possible to assess the heat of formation of the intermediate complex with nitrogen (7 kcal/mole) and the activation energy of the subsequent reaction for the transformation of activated nitrogen into hydrazine (20 kcal/mole). The effective activation energy, when the active centres are filled to a small extent with dinitrogen, is accordingly equal to 13 kcal/mole.

Other substrates of nitrogenase do not require the joint presence of the reducing agent and Mo(III) for the reduction to take place. Acetylene, for example, is reduced both under the action of $Ti(OH)_3$, $Cr(OH)_2$, or $V(OH)_2$, in the absence of Mo(III), and under the action of $Mo(OH)_3$ itself *(31)*, ethylene and ethane being observed in the reaction products. Since ethylene is not by itself reduced in these systems, ethane formation indicates that the reaction with the participation of 4 electrons of a reducing agent, *e.g.*,

$$4\ Cr(OH)_2 + C_2H_2 + 4\ H_2O \longrightarrow 4\ Cr(OH)_3 + C_2H_6\ ,$$

proceeds without release of free C_2H_4 from the coordination sphere of the complex.

3.3   THE $V(OH)_2$-$Mg(OH)_2$ SYSTEM - If dinitrogen is passed into an aqueous or aqueous-alcohol suspension of freshly prepared V(II)-Mg(II) hydroxides, formed by adding alkali to a solution of a mixture of VCl_2 and $MgCl_2$, hydrazine can be formed even without added molybdenum. In this simple system, discovered by us in

1970 *(23)*, the reaction proceeds at high rates, even at room and lower temperatures, resulting in considerable yields of hydrazine and ammonia (Table 2).

				Yield (%)	
Mg^2/V^2	$T°C$	P_{N_2} (atm)	Time h	N_2H_4	NH_3
20	-20	1	8.0	14.4	traces
6	-20	100	0.16	40	"
6	-20	100	2.0	62	"
20	+20	1	1.0	5	44

TABLE 2: N_2 REDUCTION VIA THE $V(OH)_2$-$Mg(OH)_2$ SYSTEM [a]

[a] *Aqueous CH_3OH (20%), $KOH = 6$ M*

The presence of Mg^{2+} ions in the hydroxide is highly essential since the reduction virtually does not take place without them. Depending on the temperature, dinitrogen pressure, and the solvent, the reaction can proceed mainly either to hydrazine (high pressure, water medium) or ammonia. The following reactions can be carried out:

$$4\ V(OH)_2 + N_2 + 4\ H_2O \xrightarrow{Mg(OH)_2} 4\ V(OH)_3 + N_2H_4$$

$$6\ V(OH)_2 + N_2 + 6\ H_2O \xrightarrow{Mg(OH)_2} 6\ V(OH)_3 + 2\ NH_3$$

Kinetic studies show that NH_3 in this system is formed as a result of further N_2H_4 reduction. Activity of the mixed vanadium-magnesium hydroxide toward dinitrogen is demonstrated by the fact that it reduces N_2 directly from the air, with the yield being about half that from pure dinitrogen; moreover, carbon monoxide does not, in this case, appreciably inhibit the reaction *(32)*.

Acetylene is reduced by the hydroxide of divalent vanadium to ethylene and ethane, additions of magnesium salts not being necessary in this case. The mixed vanadium-magnesium hydroxide, as distinct from pure $V(OH)_2$, acquires the capability of also reducing ethylene to ethane *(31)*.

A quantitative study of the inhibition of dinitrogen reduction by adding V^{3+} salts led to the conclusion *(33)* that four V^{2+}

ions are present in the active centre (Fig. 2). The role of magnesium ions becomes somewhat clearer as a result of studying the
magnetic susceptibility of the system; in the presence of Mg^{2+}
ions the effective magnetic moment of the hydroxide increases,
which indicates that the spin exchange among V^{2+} ions forming
they hydroxide decreases (34).

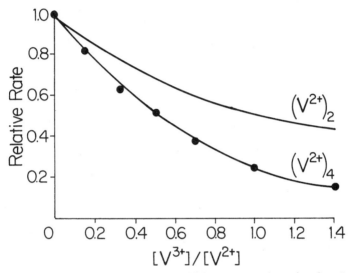

Fig. 2. *Inhibiting effect of V(III) additions on N_2H_4 formation from N_2 in
the $V(OH)_2$-$Mg(OH)_2$ system. Curves 1 and 2 are calculated for $(V^{2+})_2$ and
$(V^{2+})_4$- systems, respectively, on the grounds of equal probability of V^2 and
V^3 entering the system. The points correspond to experimental data.*

The following mechanism of N_2 reduction via the V(II)-Mg(II)
mixed hydroxide can be suggested:

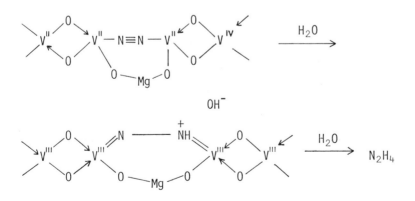

*3.4 AMALGAMS OF ALKALINE METALS AS REDUCING AGENTS OF DINITRO-
GEN* - Following the mechanism for the reduction of dinitrogen in
the $Ti(OH)_3$-$Mo(III)$ system, it was assumed *(35)* that, if electron
transfer to the activated dinitrogen molecule were conducted by
the lattice of titanium hydroxide, the use of a stronger reducing
agent, *e.g.*, sodium or potassium amalgam, should make it possible
to transform the $Ti(OH)_3$-$Mo(III)$ system into a catalyst for dinitrogen reduction under the action of the amalgam.

The presence of alkaline metal amalgam does, in fact, sharply
increase the activity of the $Ti(OH)_3$-$Mo(III)$ system toward dinitrogen, and considerable yields of hydrazine are observed even
at room temperature.

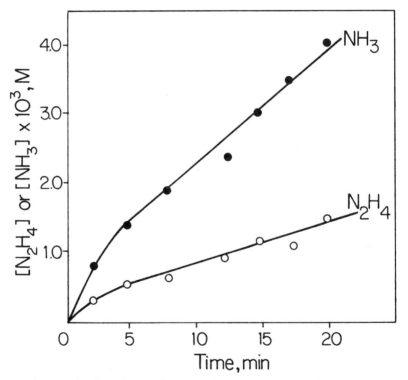

Fig. 3. *Kinetics of NH_3 and N_2H_4 formation in the $Na(Hg)$-$Ti(OH)_3$-$Mo(III)$
system in aq. CH_3OH; $T = 20°C$, 100 atm N_2.*

An interesting feature of this system is the concurrent formation of hydrazine and ammonia (Fig. 3), which shows that free
hydrazine is not the intermediate product in the formation of the

majority of the ammonia *(36)*. This conclusion was verified with the help of adding labelled hydrazine $^{15}N_2H_4$ to the system. Both in the absence and the presence of dinitrogen, hydrazine is slowly reduced at room temperature but its reduction rate cannot account for the observed rates of ammonia formation. This fact suggested that two types of active centres exist *(36)*. On one centre dinitrogen is reduced to hydrazine, while on the other type of active centre the reduction proceeds directly to ammonia without any release of hydrazine from the coordination sphere of the catalytic complex.

If titanium(III) hydroxide is the carrier of electrons from the alkaline metal amalgam to the Mo(III)-activated dinitrogen molecule, the reduction of N_2 should proceed without $Ti(OH)_3$ by direct contact of the alkaline metal amalgam with the Mo(III) complex, and the catalytic reduction of dinitrogen by sodium amalgam in the presence of Mo complexes in methanol was actually observed *(27,37)*. The reaction product was hydrazine, the reaction proceeding at room temperature with every Mo complex catalyzing formation of a large number of hydrazine molecules. The reaction, however, has to be carried out with very small concentrations of molybdenum (10^{-5} - 10^{-7} M), because with higher concentrations the catalytic formation of dihydrogen proceeds very rapidly. In this system, full reproducibility of the results could not be obtained, and the yields of hydrazine varied from one experimental series to another. The reason for the results being unreproducible seems to lie in the participation of impurities at the amalgam surface in contact with the molybdenum complex, since the Mo(III) complex by itself is incapable of forming even a stoichiometric quantity of hydrazine; the role of the amalgam is not only to reduce the molybdenum compound to Mo(III) but also to transfer electrons to the Mo(III)-activated N_2 molecule.

In the Na(Hg)-Mo(III) system, azide is catalytically reduced to N_2 and NH_3, acetylene to ethylene and ethane, and ethylene to ethane. These results, especially the capacity to reduce ethylene, are evidently the consequence of the high reduction potential of the amalgam (E_0 = -1.85 v).

3.5 ELECTROCHEMICAL REDUCTION OF DINITROGEN IN PROTIC MEDIA –
Only preliminary results have been obtained in this direction as yet. Small quantities of hydrazine can be obtained by the reduction of dinitrogen at a mercury cathode in the presence of molybdenum complexes *(38)*. The results, however, vary greatly from one experiment to another, which is obviously again associated with the participation of impurities at the contact of the

molybdenum complex with the surface of the cathode. The possibility of using amalgams of alkaline metals as reducing agents
for dinitrogen allows for application of electrochemical production of the amalgam, and considerable quantities of hydrazine and
ammonia were observed to form in the $Na(Hg)-Ti(OH)_3-Mo(III)$
system with the amalgam obtained from the electrolysis of an
aqueous-alcohol suspension of $Ti(OH)_3-Mo(III)$ on a mercury cathode at room temperature and 30 atm pressure of N_2 *(39)*.

Acetylene is readily reduced electrochemically with good
yields to ethylene and ethane in the presence of various dissolved Mo(III) complexes *(40,42)*. Ethane, for the most part, is
again formed when acetylene is reduced in the coordination sphere
of Mo(III) without intermediate formation of free ethylene. The
Mo(III) complexes can reduce acetylene catalytically only in
contact with the cathode which is the source of electrons; the
complexes alone practically do not react with acetylene.

3.6 SOLUBLE COMPLEXES OF V(II) WITH CATECHOL AND ITS AMALGAMS -
In 1972 and later, it was shown that dinitrogen can be reduced
with good yields to ammonia in homogeneous aqueous and alcoholic
solutions of V(II) complexes with catechol and some other aromatic diols *(43,44,46)*. In aqueous media for catechol, the reaction proceeds in the range of pH 8.5 - 13.5, the maximum yields
being observed at pH$\approx$10. The reduction of N_2 takes place at
room and lower temperatures. In alcoholic solutions at room
temperature and atmospheric pressure of dinitrogen, the ammonia
yield reached 60% with respect to the V(II) expenditure; in
aqueous solutions such yields are observed only at elevated pressures. In the absence of dinitrogen, a parallel oxidation reaction of V(II) by the solvent proceeds with H_2 evolution. The
yields of NH_3 increase with increase of dinitrogen pressure due
to the inhibition of the competing reaction of dihydrogen formation, but at no pressure does the yield of ammonia exceed 75% of
the reducing agent (Fig. 4); 25% of the V(II) electrons account
for the concurrently formed hydrogen. This residual process of
hydrogen evolution is coupled with the reduction of N_2, since
kinetic measurements show that both the rate of dihydrogen formation (at dinitrogen pressures corresponding to the limiting
yields of ammonia) and the rate of ammonia formation grow in
proportion to the dinitrogen pressure. Thus, in a manner similar
to the biological fixation of dinitrogen, the stoichiometry of
the reaction corresponds to the following equation:

$$8 \ V^{2+} + N_2 + 8 \ H_2O \longrightarrow 8 \ V^{3+} + 2 \ NH_3 + H_2 + 8 \ OH^-$$

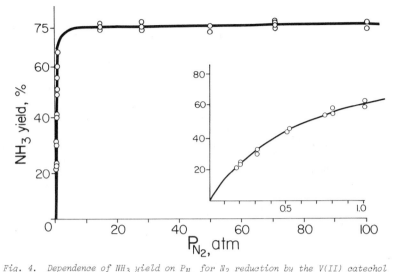

Fig. 4. *Dependence of NH₃ yield on P_{N_2} for N₂ reduction by the V(II) catechol complex in methanol.*

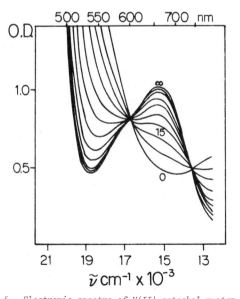

Fig. 5. *Electronic spectra of V(II) catechol system under N₂ in methanol. Figures on the curves correspond to time of the reaction in minutes.*

Again, as in biological dinitrogen fixation, carbon monoxide inhibits the formation of ammonia but not the formation of hydrogen; in fact, the rate of H_2 evolution is even increased in the presence of CO *(45)*. Acetylene is quantitatively reduced to ethylene and, in the case of C_2D_2 in H_2O, cis-deuteroethylene is formed. The range of pH for the acetylene reduction is much broader than that for dinitrogen (from pH 5 to concentrated alkali).

Kinetic studies on the oxidation of the V(II)-catechol complex in the dinitrogen atmosphere (Fig. 5) showed *(46)* the rate equation to have the following form:

$$- \frac{d\ [V(II)]}{dt} = k_1\ [V(II)]^2\ [N_2] + k_2\ [V(II)]^{1/2}$$

The first term pertains to the reduction of dinitrogen to ammonia, and the second term to the parallel (and independent) reaction of H_2 evolution from the solvent protons. Hydrazine, which could be the intermediate reaction product, was found to be easily reduced to ammonia in the V(II)-catechol solutions. When the reaction of dinitrogen reduction is abruptly stopped in its initial stages by adding excess acid or oxidant ($VOSO_4$), a small quantity of hydrazine is found, which may confirm its intermediate formation *(47)*. A kinetic study of the hydrazine reduction with the help of a similar procedure, the reaction being stopped by adding acid or V(IV), has, however, shown the rate constant to be at least two orders of magnitude smaller than it would have been had hydrazine, found in the system when dinitrogen is being reduced, been the only intermediate product during the ammonia formation. From this it can be inferred that if the main pathway of the reaction proceeds *via* hydrazine, this molecule must be in a different state compared to free hydrazine. This implies that the reducible hydrazine does not leave the coordination sphere of a vanadium complex.

The second order term in vanadium for the N_2 reduction, and the corresponding 0.5 order for the dihydrogen formation, are indicative of a polynuclear structure for the transition state during the reduction of dinitrogen. The H_2 formation from the solvent protons requires two electrons. A comparison of the kinetic orders shows that 8 electrons are available for the reduction of dinitrogen in the transition state, and it is reasonable to assume that the stoichiometry of the equation above reflects the reaction mechanism. It is possible, for example, to assume that at pH 8.5-13.5, a particle containing 4 vanadium

atoms is present in the solution. This particle forms a complex
with N_2 and then undergoes an equilibrium transformation into a
hydrazine derivative of the

type. Irreversible reduction to ammonia becomes possible during
collision with another tetranuclear particle. The N—N bond is
in this case disrupted, and an ammonia derivative is formed which
is then hydrolyzed to NH_3. The two "extra" electrons are at the
same time utilized for the formation of the H_2 molecule. An
alternative possibility is the reaction of two binuclear parti-
cles with a $V^{2+} \xrightarrow{-2e} V^{4+}$ transfer occurring when dinitrogen is
reduced to ammonia and H_2 is formed. A subsequent reaction of
V^{4+} with excess V^{2+} to give 2 V^{3+} will result in the formation of
the final V(III) complex observed. Kinetic data, therefore, undoubtedly indicate that at least a 4-nuclear, and taking into
account the energy requirements, more probably an 8-nuclear,
transition state is responsible for N_2 reduction.

The 0.5 order for the dihydrogen formation can be explained
by an equilibrium dissociation of the complex into two reactive
particles (each of them probably containing 2 V^{2+} ions). An
undetectable reaction rate for the undissociated state indicates
that the active centre is inaccessible for the solvent molecules
due to the organic ligands surrounding the V^{2+} ions. This appears to be an important factor as to why dinitrogen is a more
reactive substrate than the proton-containing molecules of the
solvent.

The catalytic role of CO in H_2 formation, and its inhibition
of N_2 reduction, can be explained by assuming that reaction of
the tetramer with CO gives a complex which can readily react
with H_2O (possibly via dissociation into dimers).

The complex nature of the transition state is confirmed by
the data obtained with ligands other than catechol *(48)*. Phenols
containing two *ortho* OH groups for complexing to the V(II) are
active with respect to dinitrogen reduction. When one or two OH
groups are substituted with SH, V(II) complexes are still formed,
but they are inactive toward N_2, although they are able to form
H_2 from the solvent and reduce acetylene.

V(II) complexes with pyrogallol and gallic acid are capable
of reducing dinitrogen at much higher pH values (≥ 13) than in
the case of catechol complexes, and the complex with protocatechuic acid (similar to gallic acid) is altogether inactive
with respect to N_2. These data, as well as the narrow pH range
within which the V(II) catechol complexes form ammonia from dinitrogen, are indicative of the necessity of the presence of only

certain definite types of complexes evidently in an associated form.

3.7 OTHER SYSTEMS REDUCING DINITROGEN IN PROTIC MEDIA; SCHRAU-ZER'S MODEL OF NITROGENASE - The systems described above, discovered in the author's laboratory and studied for the last six years, are presently the most effective in terms of the non-biological reduction of dinitrogen in protic media. At the same time, as already mentioned, a number of systems have been described in the literature which are thought capable of reducing dinitrogen with formation of small quantities of ammonia. A review *(1)* includes both the reports on these results, as well as later results, usually obtained with labelled nitrogen $^{15}N_2$, from which it follows that in these systems the ammonia is formed not from dinitrogen but from nitrogen-containing impurities in the reagents. On the other hand, there is no doubt that other substrates of nitrogenase: N_2O, CN^-, CH_3CN, CH_3NC, and C_2H_2, are effectively reduced in aqueous solutions under the action of a large number of reducing systems that are inactive or only slightly active with respect to dinitrogen *(1)*. Systems of this kind have already been partially considered above.

Systems discovered and studied by Schrauzer and coworkers *(49)*[*] should be noted specifically. These systems involve complexes which include molybdenum and compounds containing the SH group (cysteine, glutathione, thioglycerol); in the presence of $NaBH_4$, they are capable of reducing acetylene, N_2O, CN^-, RCN, and N_3^-. Acetylene is reduced to ethylene and partially to ethane. As distinct from nitrogenase, ethylene can also be reduced to ethane. Utilization of $NaBD_4$ shows that the reduction proceeds with the participation of hydrides *(51)*. In Schrauzer's opinion, mononuclear Mo(IV) complexes display the catalytic activity, and these systems are considered a close model of nitrogenase. A distinctive feature of these systems is the activating role of ATP; however, the mechanism of its action differs from that in the reaction of enzymatic reduction.

Thus it was shown *(52)* that ATP hydrolysis essentially does not take place, and so the energy of the macroergic bond is not utilized, which differs in principle from the ATP action in a biological system. It is significant that, if the pH value in the added ATP solution is brought to that of the reaction medium, ATP does not display any activating effect. Moreover, other protic acids, such as ADP and even sulphuric acid, also have a strong activating effect *(51)*, and the acid effect is possibly

---

[*]*The first information on molybdenum-thiol complexes appeared in 1970 (50), only a few months after the first report of hydroxide systems with molybdenum participation (22).*

associated with protonation of the ligands in the molybdenum complex, which can facilitate H^- transfer from $NaBH_4$ *(51)*.   In later work Schrauzer and coworkers *(53)* also assume a similar mechanism, but suggest that ATP, ADP, and phosphoric acid anions have an assisting effect.*

As for the possibility of dinitrogen being reduced in Schrauzer's systems, the question remains unclear.   In one of Schrauzer's early works *(54)* data are presented showing very small NH_3 yields (0.06-0.1% with respect to Mo, in 5 days with a N_2 pressure of 140 atm).   Elsewhere, a yield of NH_3 is reported corresponding to 4% of the Mo in 116 hr at 1 atm N_2 *(55)*.

It was reported *(56)* that the yield of NH_3 in systems similar to Schrauzer's can even reach 1.5 moles of ammonia per mole of Mo compound, but later studies did not confirm this result *(15)*. Negative results have been reported in attempts to reduce N_2 with these systems *(57,58)*, labelled dinitrogen also being used *(58)*.

In his latest reports, Schrauzer claims that the primary product of dinitrogen reduction is N_2H_2 "being accumulated during the first 40 min" *(59)*.   Taking into consideration the well-known reactivity of N_2H_2, this assertion cannot but raise doubt. Also, in Schrauzer's opinion, this reactive compound by itself is incapable of further reduction and is assumed to disproportionate with the formation of hydrazine, which is thought to undergo further reduction to ammonia.   Some other contradictions in the studies of Schrauzer and coworkers have already been discussed *(1)*.   In any case, it is obvious that the activity of Schrauzer's systems with respect to dinitrogen (if it does exist) is very small and incomparable with the activity displayed by the systems described in the preceding sections.

4. CONCLUSIONS

The analysis of quite a large amount of experimental data obtained by studying dinitrogen reduction in protic media makes it possible to conclude that, in all the cases presently known, the reduction proceeds in the coordination sphere of a catalytic complex without intermediate formation of free N_2H and N_2H_2, and in some cases, including nitrogenase, without the intermediate formation of free hydrazine.   For some substrates, there are

---

*This assumption is difficult to reconcile with the fact that anions of ATP and ADP do not affect the rate, if the pH of the added solution is equalized with the pH of the reaction mixture.   Differences in behaviour of different acids may be due to differences in pK's of the anions, such as $H_2PO_4^-$ (pK_A = 7.2) and HSO_4^- (pK_a = 2).

undoubtedly no intermediate one- and two-electron reduction products (*e.g.*, ethane is formed from acetylene without the intermediate formation of ethylene).

For dinitrogen, as already mentioned, it was shown in two cases that N_2 can be directly reduced to ammonia without the intermediate formation of free hydrazine. It seems clear that under these circumstances the intermediate formation of diimide, especially unfavourable from the thermodynamic point of view, appears to be virtually impossible, especially for such comparatively mild reducing agents (including nitrogenase) that are incapable to reduce ethylene. Moreover, there seems to be no reason to suppose, as done in some papers, that a diimide moiety is formed, and bonded and stabilized by the active centre of the catalyst without ever leaving the coordination sphere. And indeed, if the elementary process of N_2 reduction consists in a rearrangement of the following type,

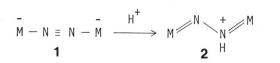

it follows that certainly neither the first nor the second complex is a diimide derivative (by analogy with organic chemistry, one could consider the first complex to be an analogue of CH—C≡C—CH (butadiyne), and the second an analogue of

$$CH_2 \diagup^{CH} \diagdown_{CH} \diagup^{CH_2} \text{ (butadiene).}$$

At present it is not clear in what sequence the protonation and disruption of the N—N and M—N bonds takes place, but it seems likely that formation of the hydrazine derivative (**2**) will be stabilized as a result of the conjugation of double bonds; this can correspond to a lowering of the required reduction potential for the metal compounds forming the N_2 complex.

The results of studying dinitrogen reduction in protic media, as well as numerous negative results obtained with various reduction systems, make it possible to draw the conclusion that dinitrogen can only be reduced when several conditions are jointly met. These conditions can be formulated as follows:

1) Dinitrogen must be activated in a complex with a metal of d^3 or d^2 electronic configuration, the most suitable being Mo(III) and V(II). It is becoming more and more probable that Mo(III), first found in model systems, is also an active Mo form in nitrogenase. These data are in agreement with the conclusion on the optimum properties of the d^3 configuration for the intermediate linear binuclear complexes which were discussed above.

2) The reduction potential of the reducing agent present in the system must be sufficiently high and for most cases presumably not lower than that required for the reaction $N_2 + 4 e + 4 H^+$ $\longrightarrow N_2H_4$. Table 3 shows the reduction potentials for dinitrogen-fixing systems at certain pH values. It is seen that all of them satisfy the formulated condition, except for nitrogenase where the potential of the reducing agent (ferredoxin or dithionite) is lower. Taking into account, however, the possibility of increasing the potential at the active centre by hydrolyzing ATP, this condition may be also satisfied in the enzymatic process. The stabilization of a hydrazine derivative, which was described above, seems to be small and can only ensure a small decrease in the potential as against that required for hydrazine production.

3) The system must be multinuclear and have a possibility to transfer several electrons to the centre which activates dinitrogen. In the case of V(II) in hydroxide, functioning both as activator and reducing agent of dinitrogen to hydrazine, no less than four V(II) species participate in the active centre, changing their valency in the process of N_2 reduction by unity and being oxidized to V(III). It is felt that the octavanadium transition state cluster, in the case of the V(II)-catechol complex, carries out the reduction of dinitrogen to ammonia with a simultaneous reduction of two protons to hydrogen. Systems with the participation of Mo(III) (mixed hydroxides and nitrogenase) include no less than two Mo(III) ions bonded with two or more atoms of a stronger reducing agent. Mo(III) by itself seems to be an insufficiently strong reducing agent for dinitrogen, and in principle may remain in the same oxidation state in the process of reduction, if the electrons are supplied from the surrounding ions.

One can visualize the following mechanism for nitrogenase by analogy with the V(II) catechol system where no intermediate hydrazine is formed and where the stoichiometry is similar to that of the nitrogenase reaction (25% of electrons are expended for H_2 formation). There are several (at least 6) Fe atoms together with 2 atoms of Mo in the active centre of nitrogenase capable of accepting electrons from a reducing agent. The number of required electrons to be transferred to the active centre is different for different substrates, and therefore the reduction potential of the acting enzyme is also different. A maximum number of electrons (presumably eight) is necessary for N_2 reduction and the potential is increased to its maximum value, two "extra" electrons being used for H_2 formation after N_2 reduction

TABLE 3: REDUCTION POTENTIALS OF DINITROGEN-FIXING SYSTEMS

Reaction	$E = E_{pH0} - a\ pH$	Minimum pH for N_2 reduction	E(v) of reducing system	$E(v)^a$ N_2H_4
$S_2O_4^{2-} + 4OH^- - 2e \rightleftharpoons$ $2SO_3^{2-} + 2H_2O$ $(4ATP \rightarrow 4ADP + 4P_i)$	$+0.56 - 0.118\ pH$	7	$-0.27^b\ (-1.12)^c$	-0.73
$V^{2+}(cat) + 2OH^- - e \rightleftharpoons$ $VO^+(cat) + H_2O$	$+0.10 - 0.118\ pH$	8.5	-0.903	-0.82
$2Cr(OH)_2 + 2OH^- - 2e \xrightarrow{Mo(III)}$ $Cr_2O_3 + 3H_2O$	$-0.56 - 0.059\ pH$	8.5	-1.06	-0.82
$2V(OH)_2 + 2OH^- - 2e \rightleftharpoons$ $V_2O_3 + 3H_2O$	$-0.55 - 0.059\ pH$	13	-1.32	-1.08
$Ti(OH)_3 + OH^- - e \xrightarrow{Mo(III)}$ $TiO_2 + 2H_2O$	$-0.80 - 0.059\ pH$	10	-1.40	-0.91
$Na(Hg) - e \rightleftharpoons Na^+$	-1.85		-1.85	

$^a E_{N_2H_4}$ is the reduction potential for the process $N_2 + 4H_2O + 4e \rightleftharpoons N_2H_4 + 4OH^-$

$^b E_{max} = 1.12v$ is the potential calculated on the assumption that a minimum 4 ATP molecules are hydrolyzed per one $S_2O_4^{2-}$ ion, increasing correspondingly the reduction potential (61).

$^c E_{min} = 0.27v$ is the potential of $S_2O_4^{2-}$ without ATP.

to 2 NH_3 molecules. Other substrates start their reduction with a smaller number of electrons in the active centre, and coupled H_2 formation does not proceed. Though the potential of the active centre for N_2 reduction is higher than that for other substrates, it is probably lower than for some known non-biological systems, as indicated by nitrogenase inactivity towards C_2H_4.

Thus with all the three conditions mentioned above being satisfied optimally, *i.e.*, the system is well "organized" (with suitable interatomic distances to Mo(III), and the possibility of easy electron transfer among the ions of the cluster), dinitrogen can be easily reduced in protic media, and can turn out to be a more reactive substrate than some other chemically rather active compounds (*e.g.*, ethylene).

When at least one of these conditions is not met, it seems generally that dinitrogen is essentially not reduced in protic media. These conditions are only applicable to dinitrogen and not to any other unsaturated molecules, *e.g.*, other substrates of nitrogenase. When we consider, for example, the reduction of acetylene into ethylene, or of nitrous oxide into N_2 and H_2O, none of the three conditions are essential, primarily because the processes are two-electron reactions. For acetylene to be reduced directly into ethane, or cyanide into methane and ammonia, it is essential that a polynuclear system should participate, as these reactions require correspondingly four and six electrons. However, in these cases, the participation of Mo(III) or V(II) in the activation of the substrate is not so essential, and the reduction properties of the system need not be so strongly manifested.

The poor (or zero) activity of Schrauzer's systems (Mo-thiol complexes + $NaBH_4$) with respect to dinitrogen is possibly associated with the insufficiently strong reducing properties for the N_2 to be reduced to hydrazine and/or the insufficiently good "organization" for the intermediate Mo—N≡N—Mo complex to be formed with direct reduction to ammonia.

Thus it seems that the better the preliminary "organization" of the polynuclear system, the weaker is the reducing agent necessary for carrying out N_2 reduction. For such a generalization, it is no doubt necessary to study a greater number of systems than those presently known that are capable, like nitrogenase, of reducing dinitrogen directly to ammonia in protic media. It is hoped that such new catalytic systems will be found in the future.

REFERENCES

1. A.E. Shilov, *Usp. Khim.*,**43**, 863 (1974).
2. M.E. Volpin and V.B. Shur, *Dokl. Akad. Nauk SSSR*,**156**, 1102 (1964).
3. A.D. Allen and C.V. Senoff, *Chem. Commun.*, 621 (1965).
4. A.E. Shilov, A.K. Shilova and Yu. G. Borod'ko, *Kinetika i kataliz*,**7**, 768 (1966).
5. Yu.G. Borod'ko and A.E. Shilov, *Usp. Khim.*,**38**, 761 (1969).
6. J. Chatt and R.L. Richards, "The Chemistry and Biochemistry of Nitrogen Fixation", Plenum Press, London, p. 57 (1971).
7. A.D. Allen, R.O. Harris, B.R. Loescher, J.R. Stevens and R.N. Whitely, *Chem. Rev.*,**73**, 11 (1973).
8. A.E. Shilov, A.K. Shilova and E.F. Kvashina, *Kinetika i kataliz*,**10**, 1402 (1969).
9. M.O. Broitman, N.T. Denisov, N.I. Shuvalova and A.E. Shilov, *Kinetika i kataliz*,**13**, 61 (1972).
10. Yu.G. Borod'ko, M.O. Broitman, L.M. Kachapina, A.E. Shilov and L.Yu. Ukhin, *Chem. Commun.*, 1185 (1971).
11. J. Chatt, *J. Organomet. Chem.*,**100**, 17 (1975).
12. C. Krüger and Y-H Tsay, *Angew. Chem.*,**85**, 1051 (1973).
13. K. Jonas, D.J. Brauer, C. Krüger, P.J. Roberts and Y-H Tsay, *J. Am. Chem. Soc.*,**98**, 74 (1976).
14. I.M. Treifel, M.T. Flood, R.E. Marsh and H.B. Gray, *J. Am. Chem. Soc.*,**91**, 6512 (1969).
15. J. Chatt and G.J. Leigh, *Chem. Soc. Rev.*,**1**, 121 (1972).
16. A.E. Shilov and G.I. Likhtenstein, *Izv. Akad. Nauk SSSR, ser. biol.*, 518 (1971).
17. R.W.F. Hardy, R.C. Burns and G.W. Parshall, *Adv. Chem. Ser.*, **100**, 219 (1971).
18. R.I. Gvozdev, A.P. Sadkov, A.I. Kotelnikov and G.I. Likhtenstein, *Izv. Akad. Nauk SSSR, ser. biol.*, 488 (1973).
19. P.A. Ketchum, H.Y. Cambier, W.A. Frazier, C.H. Madansky and A. Nason, *Proc. Nat. Acad. Sci. U.S.A.*,**66**, 1016 (1970).
20. G. MacKennz, N.P. L'vov, V.L. Ganelin, N.S. Sergeev and V.L. Kretovich, *Dokl. Akad. Nauk SSSR*,**217**, 228 (1974).
21. N.P. L'vov, V.L. Ganelin, Z. Alikulov and V.L. Kretovich, *Izv. Akad. Nauk SSSR, ser. biol.*, 371 (1975).
22. N.T. Denisov, V.F. Shuvalov, N.I. Shuvalova, A.K. Shilova and A.E. Shilov, *Kinetika i kataliz*,**11**, 813 (1970).
23. N.T. Denisov, O.N. Efimov, N.I. Shuvalova, A.K. Shilova and A.E. Shilov, *Zh. Fiz. Khim.*,**44**, 2694 (1970).
24. N.T. Denisov, V.F. Shuvalov, N.I. Shuvalova, A.K. Shilova and A.E. Shilov, *Dokl. Akad. Nauk SSSR*,**195**, 879 (1970).
25. A.E. Shilov, N.T. Denisov, O.N. Efimov, N.F. Shuvalov, N.T. Shuvalova and A.K. Shilova, *Nature (London)*,**231**, 460 (1971).

26. N.T. Denisov, A.E. Shilov, N.I. Shuvalova and T.P. Panova, *React. Kinet. and Catal. Lett.*,**2**, 237 (1975).
27. A.E. Shilov, A.K. Shilova and T.A. Vorontsova, *React. Kinet. and Catal. Lett.*,**3**, 143 (1975).
28. V.V. Abalyaeva, N.T. Denisov, M.L. Khidekel and A.E. Shilov, *Izv. Akad. Nauk SSSR, ser. khim.*, 196 (1973).
29. V.V. Abalyaeva, N.T. Denisov, M.L. Khidekel and A.E. Shilov, *Izv. Akad. Nauk SSSR, ser. khim.*,2638 (1975).
30. N.T. Denisov and N.I. Shuvalova, *React. Kinet. and Catal. Lett.*, **5**, 431 (1976).
31. A.G. Ovcharenko, A.E. Shilov and L.A. Nikonova, *Izv. Akad. Nauk SSSR, ser. khim.*, 534 (1975).
32. N.T. Denisov, E.I. Rudshtein, N.I. Shuvalova and A.E. Shilov, *Dokl. Akad. Nauk SSSR,***202**, 623 (1972).
33. N.T. Denisov, N.I. Shuvalova and A.E. Shilov, *Kinetika i kataliz,***14**, 1325 (1973).
34. N.T. Denisov, N.I. Shuvalova, N.N. Ivleva and A.E. Shilov, *Zh. Fiz. Khim,***48**, 2238 (1974).
35. G.V. Nikolaeva, O.N. Efimov, A.A. Brikenshtein and N.T. Denisov, *Zh. Fiz. Khim.*,**50**, 3030 (1976).
36. G.V. Nikolaeva, O.N. Efimov, A.A. Brikenshtein and A.E. Shilov, *React. Kinet. and Catal. Lett.*, in press.
37. L.P. Didenko, A.G. Ovcharenko, A.E. Shilov, and A.K. Shilova, *Kinet. Katal.*, in press.
38. A.F. Zueva, O.N. Efimov, A.D. Stirkas and A.E. Shilov, *Zh. Fiz. Khim.*,**46**, 760 (1972).
39. V.N. Tsarev, O.N. Efimov and A.A. Brikenshtein, *Zh. Fiz. Khim.*,**51**, in press (1977).
40. M. Ichikawa and S. Mesitsuka, *J. Am. Chem. Soc.*,**95**, 3411 (1973).
41. D.A. Ledwith and F.A. Schultz, *J. Am. Chem. Soc.*,**97**, 6591 (1975).
42. O.N. Efimov, A.F. Zueva, G.N. Petrova and A.E. Shilov, *Koord. Khim.*, **2**, 62 (1976).
43. L.A. Nikonova, A.G. Ovcharenko , O.N. Efimov, V.A. Avilov and A.E. Shilov, *Kinet. Katal.*,**13**, 1602 (1972).
44. L.A. Nikonova, N.I. Pershikova, M.V. Bodeyko, G.L. Oleinik, D.N. Sokolov and A.E. Shilov, *Dokl. Akad. Nauk SSSR,***216**, 140 (1974).
45. L.A. Nikonova, S.A. Isaeva, N.I. Pershikova and A.E. Shilov, *J. Mol. Catal.*,**1**, 367 (1975/76).
46. N.P. Luneva, L.A. Nikonova and A.E. Shilov, *React. Kinet. and Catal. Lett.*, **5**, 149 (1976).
47. N.P. Luneva, L.A. Nikonova and A.E. Shilov, *Kinetika i kataliz,***18**, in press (1977).
48. S.A. Isaeva and L.A. Nikonova, *Izv. Akad. Nauk SSSR, ser.*

khim., in press.
49. G.N. Schrauzer, *Angew. Chem. Int. Ed. Engl.*,**14**, 514 (1975).
50. G.N. Schrauzer and G. Schlesinger, *J. Am. Chem. Soc.*,**92**, 1808 (1970).
51. A.P. Khrushch, T.A. Vorontsova and A.E. Shilov, *J. Am. Chem. Soc.*,**96**, 4987 (1974).
52. T.A. Vorontsova and A.E. Shilov, *Kinetika i kataliz*,**14**, 1326 (1973).
53. G.N. Schrauzer, G.W. Kiefer, K. Tano and P.R. Robinson, *J. Am. Chem. Soc.*,**97**, 6088 (1975).
54. G.N. Schrauzer, G. Schlesinger and P.A. Doemeny, *J. Am. Chem. Soc.*,**93**, 1803 (1971).
55. G.N. Schrauzer, G.W. Kiefer, K. Tano and P.A. Doemeny, *J. Am. Chem. Soc.*,**96**, 641 (1974).
56. R.E.E. Hill and R.L. Richards, *Nature (London)*,**233**, 114 (1971).
57. D. Werner, S.A. Rullel and H.J. Evans, *Proc. Nat. Acad. Sci. U.S.A.*,**70**, 339 (1973).
58. T.A. Vorontsova, A.P. Khrushch and A.E. Shilov, *Kinetika i kataliz*,**16**, 1618 (1975).
59. G.N. Schrauzer, G.W. Kiefer, K. Tano and P.A. Doemeny, *J. Am. Chem. Soc.*,**96**, 641 (1974).
60. V.V. Streletz, O.N. Efimov, L.A. Nikonova and Ya.M. Zolotov-itzkii, *Zh. Fiz. Khim.*,**50**, 1019 (1976).
61. G.D. Watt, W.A. Bulen, A. Burns and K. LaMont Hadfield, *Bio-chemistry*,**14**, 4266 (1975).

the activation of molecular nitrogen

J. CHATT

Unit of Nitrogen Fixation, School of Molecular Sciences,
University of Sussex, Brighton, E. Sussex, BN1 9QJ, U.K.

1. INTRODUCTION

Molecular nitrogen, or dinitrogen as it is now called, is the most inert diatomic molecule. It owes its inactivity to the large energy difference between its filled and vacant molecular orbitals. Its filled molecular orbitals are low in energy ($\leq$-15.6ev) and its vacant orbitals are high ($\geq$-7ev). It is thus very difficult to put the electrons into the nitrogen molecule or to remove them from it, in the ground state. Dinitrogen, at room temperature, is inert to such highly oxidising substances as difluorine and such strongly reducing ones as sodium metal. Nevertheless, some electropositive metals which form stable nitrides, such as lithium, calcium and titanium, react with dinitrogen at or a little above room temperature, and more important, the metal centres in some strongly reduced transition metal complexes take up dinitrogen as a terminal or bridging ligand under normal temperatures and pressures *(1-4)*. This they do by the synergic effect of interacting with both the low energy electron-donor ($3\sigma_g$) and high energy electron-acceptor ($1\pi_g$) orbitals of the dinitrogen molecule.

2. NITROGENASE

Chemists have been intrigued for decades by the fact that certain bacteria can assimilate molecular nitrogen to build up their own protein. The first recognised stage of this reaction is the reduction of dinitrogen to ammonia. It is thought to take place on a molybdenum centre in the enzyme nitrogenase, and this idea has inspired much chemical research into the reaction of dinitrogen with transition metal complexes under various conditions. In little more than a decade an extensive complex chemistry of dinitrogen as a ligand has been discovered. Generally it is found that dinitrogen in its complexes is little more reactive than in the molecular state. However, in some of its complex compounds with titanium, zirconium, molybdenum and tungsten the dinitrogen reacts directly with protic acids to give hydrazine and/or ammonia. Some dinitrogen complexes of molybdenum and tungsten are also known to react with organic compounds to give organonitrogen complexes *(5)*. Since you have asked me to speak I assume that you wish me to discuss recent work in this field from my own laboratory.

It is useful to mention first some of the gross properties of nitrogenase as they affect a chemist's ideas on the subject. Nitrogenase is a brown molybdo-iron protein which separates on a 'Sephadex' column into two metallo-proteins which are essential for nitrogenase action *(6)*. These are a large brown protein (M = 200,000 - 220,000) which contains ideally two atoms of molybdenum per mole but more usually 1.3 - 1.8 atoms, and a smaller yellow protein (M = 65,000) which contains no molybdenum but 4 atoms of iron. The large protein also contains around 24 atoms of iron, but 18 to 34 atoms have been reported. Both proteins appear to contain sulphide ions equal in number to the iron atoms. The iron is present mainly as Fe_4S_4 cubane clusters which provide a system of electron storage. Both proteins are sensitive to air. The large protein will withstand brief exposure to air and be reduced again to the active form; the small protein is destroyed irreversibly by the briefest exposure to air.

The two proteins together catalyse the reduction of dinitrogen to ammonia in neutral aqueous solution in the presence of the monomagnesium salt of ATP (adenosine triphosphate) and a suitable reducing agent such as sodium dithionite. Twelve molecules of ATP are hydrolyzed to ADP (adenosine diphosphate) and orthophosphate in the reduction of one molecule of dinitrogen to ammonia. Evidently the natural process is an energy intensive one. The small protein acts as a specific electron carrier to the large one with which it forms a complex. The transfer of electrons from the small to the large protein occurs only in the presence of the monomagnesium salt of ATP and neither ATP nor its dimagnesium salt appear to effect the electron transfer. It is thought that the dinitrogen molecule attaches itself to one or both molybdenum atoms in the enzyme and is protonated to ammonia with simultaneous take-up of electrons, *via* the molybdenum atom, from the iron-sulphur electron-storage system.

The oxidation state or states of the molybdenum in the enzyme is uncertain. It gives no ESR spectrum which can be attributed to molybdenum, and so the molybdenum is not in oxidation state V unless the molybdenum atoms are in pairs. The large protein can be separated into two pairs of identical sub-units, and since the molybdenum atoms must occupy separate sub-units they are probably not present in pairs. Nevertheless the possibility that they could lie as a pair in connection between matching sub-units must not be overlooked. The nitrogenase activity of the large protein is roughly proportional to its essential molybdenum content, this again suggests that the molybdenum atoms behave independently, but does not prove it.

Until very recently there has been no spectroscopic method of detecting the molybdenum in the active enzyme, however, in March this year scientists at Stanford University published their observation of the MoK X-ray absorption edge at 20,015.5 ± 0.5ev in the photoelectron absorption spectrum of nitrogenase isolated from *Clostridium pasteurianum (7)*. This value in octahedral molybdenum complexes covers the range of oxidation states IV, V and VI depending on the mix of N, O and S ligand atoms. No definite information concerning the nature of the ligand atoms is available but sulphur is thought to be a likely candidate because nitrogenase is rich in sulphide ion. This conjecture that sulphur ligates the molybdenum now appears to be receiving confirmation from the fine structure at the above absorption edge, which shows an amplitude corresponding to ligating sulphur rather than to lighter ligand atoms such as oxygen or nitrogen *(8)*.

It was the consideration that the molybdenum atoms in nitrogenase probably react independently with dinitrogen which inspired the special study of mononuclear molybdenum and tungsten complexes by my group at Sussex. We have examined particularly their reactions with dinitrogen, and the reactions of dinitrogen in such complexes.

3.   CHEMISTRY OF THE REDUCTION OF DINITROGEN TO AMMONIA

In the nitrogenase reaction there are three parts of the cyclic process to be considered:
1.   The uptake of dinitrogen by the enzyme.
2.   The reduction of the dinitrogen to ammonia.
3.   The liberation of the ammonia.

There have been no really new developments in chemistry concerning stages 1 and 3 during the past few years. The only well established example of a reaction in which dinitrogen reacts with a transition metal salt in aqueous solution to give stable dinitrogen complexes has been known for almost a decade (Reaction 1).

$$[Ru(NH_3)_5(H_2O)]^{2+} \begin{array}{c} \xrightarrow{N_2, \; 1 \; Atm.} [(NH_3)_5Ru-N\equiv N-Ru(NH_3)_5]^{4+} \\ \xrightarrow{N_2, \; 100 \; Atm.} [Ru(NH_3)_5(N_2)]^{2+} \end{array} \qquad (1)$$

Reaction 1 demonstrates the remarkable ease with which dinitrogen will enter the coordination sphere of a complex compound when the electronic conditions are right. The reaction leads to either bridging or mononuclear dinitrogen complexes in aqueous solution, depending in this case on the pressure. It very probably provides the best model for the uptake of dinitrogen by the molybdenum site in nitrogenase, if indeed it is taken up by a molybdenum site.

Dinitrogen complexes of molybdenum have not been produced in aqueous solution, but they are readily produced at room temperature and pressure by Reaction 2.

$$[MoCl_4(thf)_2] + PR_3 + N_2 + Mg \xrightarrow{thf} [Mo(N_2)_2(PR_3)_4] \qquad (2)$$

$$(PR_3 = PMe_2Ph, \; PMePh_2, \; \tfrac{1}{2}Ph_2PCH_2CH_2PPh_2 \; etc.;$$
$$thf = tetrahydrofuran).$$

Reaction 2 is typical of the method of preparation of a great number of dinitrogen complexes. Representative complexes are listed in Table 1, which also gives the frequencies of the N-N stretching vibration. This vibration gives rise to a very strong band in the infra-red spectra of all mononuclear dinitrogen complexes, suggesting that the dinitrogen has become electrically strongly asymmetrical in its complexes. Dinitrogen in the rhenium (I) complex listed in the table is so asymmetrical that its atoms are resolved in the X-ray photoelectron spectrum and show a separation of 2 ± 0.5ev.

TABLE 1: *SOME CHARACTERISTIC DINITROGEN COMPLEXES AND DINITROGEN STRETCHING FREQUENCIES,* [$\nu(N_2)$]

Complex	$\nu(N_2)$, cm^{-1}
trans-[Mo(N$_2$)$_2$(dpe)$_2$]^a	1970
cis-[W(N$_2$)$_2$(PMe$_2$Ph)$_4$]	1931, 1998
trans-[ReCl(N$_2$)(PMe$_2$Ph)$_4$]	1925
[FeH$_2$(N$_2$)(PEtPh$_2$)$_3$]	2057
[Ru(NH$_3$)$_5$(N$_2$)]Cl$_2$	2105
mer-[OsCl$_2$(N$_2$)(PMe$_2$Ph)$_3$]	2082
[CoH(N$_2$)(PPh$_3$)$_3$]	2082
trans-[IrCl(N$_2$)(PPh$_3$)$_2$]	2095

^{a}dpe = Ph$_2$PCH$_2$CH$_2$PPh$_2$

TABLE 2: *BASIC STRENGTHS OF COMPLEXES B DETERMINED FROM THE FOLLOWING EQUILIBRIUM ESTABLISHED IN BENZENE SOLUTION:* B + Et$_2$OAlMe$_3$ $\rightleftharpoons$ BAlMe$_3$ + Et$_2$O *(EQUILIBRIUM CONSTANT K)*

B	K
Tetrahydrofuran	70
[ReCl(N$_2$)(PMe$_2$Ph)$_4$]	20.6
[Mo(N$_2$)$_2$(dpe)$_2$]*	33 (16.5)^a
[W(N$_2$)$_2$(dpe)$_2$]*	15 (7.5)^a
[ReCl(N$_2$)(PMe$_2$Ph)$_2$$\{$P(OMe)$_3$$\}_2$]	5.5
[ReCl(CO)(PMe$_2$Ph)$_4$]	3.3
Diethyl ether	1
[OsCl$_2$(N$_2$)(PEt$_2$Ph)$_3$]	0.3

^{a}dpe = Ph$_2$PCH$_2$CH$_2$PPh$_2$; in parentheses, half values of K for bis (dinitrogen)-complexes to provide strict comparison of basic strengths with mono(dinitrogen)-complexes.

233

The dinitrogen in its stable mononuclear complexes is bonded in much the same way as carbon monoxide in its complexes. There is electron donation from a σ-bonding orbital and back donation into the anti-bonding orbitals of the dinitrogen molecule. The low values of $\nu(N_2)$ in the complexes (Table 1) as compared with that of dinitrogen itself (2331 cm^{-1}) is mainly attributable to this back donation, and a very low value of $\nu(N_2)$ as in [ReCl(N$_2$) (PMe$_2$Ph)$_4$] is evidence of a strong electron drift from the metal into the π^* orbitals of the ligating dinitrogen molecule. This back donation strengthens the Re-N bond at the expense of the N$\equiv$N bond, and it also renders basic the terminal nitrogen atom. We have tested the basicity of a number of stable dinitrogen complexes against trimethylaluminium relative to diethyl ether. These are listed in Table 2. It will be seen that the rhenium (I) complex with its very low $\nu(N_2)$ has a basic strength lying between that of diethyl ether and tetrahydrofuran; also the carbonyl, [ReCl(CO)(PMe$_2$Ph)$_4$], is a much weaker base than the corresponding dinitrogen complex. The rhenium (I) dinitrogen complex can be attached through the terminal nitrogen atom to all sorts of Lewis acids to form bridged dinitrogen complexes, but not to the proton.

Protic acids usually react with the dinitrogen complexes to form hydride complexes with oxidation of the metal ion and loss of dinitrogen. If the hydride complexes react with protic acid, which is usual, dihydrogen is evolved. The normal products of protic attack on dinitrogen complexes are thus dinitrogen and dihydrogen, and not hydrides of nitrogen (*e.g.* Reaction 3); sometimes the intermediate dinitrogenhydrido-complex is stable (*e.g.* Reaction 4).

$$[CoH(N_2)(PPh_3)_3] \xrightarrow[\text{fast}]{\text{HCl}} [CoCl(PPh_3)_3] + N_2 + H_2$$

$$\text{HCl} \downarrow \text{slow}$$

$$[CoCl_2(PPh_3)_2] + \tfrac{1}{2}H_2 + PPh_3 \qquad (3)$$

$$[ReCl(N_2)(PMe_2Ph)_4] \xrightarrow[\text{(2) NaBPh}_4]{\text{(1) CF}_3\text{COOH}} [ReClH(N_2)(PMe_2Ph)_4]BPh_4 \qquad (4)$$

The only exceptions amongst mononuclear dinitrogen complexes to the general Reactions 3 and 4 are provided by the reactions of certain tungsten and molybdenum dinitrogen complexes of the general formula [M(N$_2$)$_2$(PR$_3$)$_4$] (M = Mo or W; PR$_3$ = PMe$_2$Ph, PMePh$_2$, or $\tfrac{1}{2}$ Ph$_2$PCH$_2$CH$_2$PPh$_2$). When these bis(dinitrogen)-complexes contain monophosphines, they react directly with protic acids of oxo anions to give ammonia. The yields are best in a protic medium such as methanol, and sulphuric acid in methanol yields up to 90% ammonia from certain tungsten complexes according to Reaction 5 *(9)*.

$$[W(N_2)_2(PR_3)_4] + H_2SO_4 \xrightarrow[20^\circ]{MeOH} 2NH_3 + N_2 + W^{VI} \text{ products} + PR_3 \quad (5)$$

The tungsten is in oxidation state zero and to pass to its normal oxidation state of 6 it must lost 6 electrons, just the number required to reduce one dinitrogen molecule to ammonia. Reaction 5 demonstrates that 6 electrons can pass from a metal atom into ligating dinitrogen, smoothly and easily in a protic medium to produce ammonia. It may possibly represent the type of reaction which occurs during the reduction of dinitrogen on molybdenum in nitrogenase. However, in nitrogenase the electrons would be provided from the iron-sulphur cluster electron-storage system and would not involve the molybdenum atom changing oxidation state by 6 units; such change seems highly unlikely.

The acid is not necessary for the reduction of dinitrogen to ammonia; methanol alone is sufficiently protic and the reaction by the tungsten dinitrogen complex to give ammonia occurs just as efficiently, but much more slowly, at 60° in methanol. Under tungsten filament irradiation the reaction goes slowly in methanol alone at room temperature.

The molybdenum complex corresponding to the above tungsten complex reacts much less efficiently with acids to give ammonia, and the yields obtained are only about 35%. Perhaps the greater lability of molybdenum complexes, as compared with tungsten, allows the dinitrogen molecules to escape more easily. Then the oxidation of the molybdenum occurs by passage of electrons directly to the protons to produce a hydride complex which reacts further with acid to form dihydrogen. A small amount of hydrazine is produced in such reactions as 5, and it may reach a few percent from the tungsten complexes. It is formed only in minute traces when the central atom is molybdenum. The mechanism of the reduction of the molybdenum dinitrogen complex may be different from that of the tungsten ones.

The degree of reaction to produce ammonia according to Reaction 5 and the ratio of hydrazine to ammonia produced depends very much upon the solvent. The more protic the solvent the more ammonia is produced, the less protic the more hydrazine. Halogen acids are less effective at producing ammonia than the oxo-acids and it is essential that the phosphine is lost from the metal as the complex oxidises during the reduction of the ligating dinitrogen to ammonia. Thus when the complexes contain the more strongly bound diphosphine, $Ph_2PCH_2CH_2PPh_2$ (dpe), the reduction stops at the N_2H_2 stage (Reaction 6).

$$[W(N_2)_2(dpe)_2] \xrightarrow{\quad HCl \quad} [WCl_2(N_2H_2)(dpe)_2] \qquad (6)$$

To take Reaction 6 beyond the N_2H_2 stage the complex must be heated with acid in a suitable high boiling donor solvent to cause the removal of a molecule of the diphosphine.

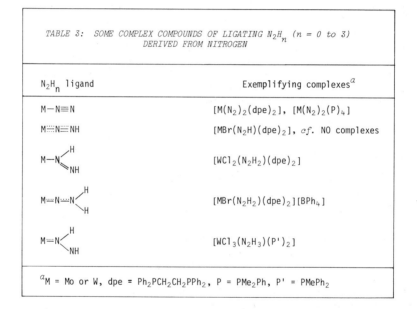

TABLE 3: *SOME COMPLEX COMPOUNDS OF LIGATING N_2H_n (n = 0 to 3) DERIVED FROM NITROGEN*

N_2H_n ligand	Exemplifying complexes[a]
$M\!-\!N\!\equiv\!N$	$[M(N_2)_2(dpe)_2]$, $[M(N_2)_2(P)_4]$
$M\cdots N\equiv NH$	$[MBr(N_2H)(dpe)_2]$, *cf.* NO complexes
$M\!-\!N\diagdown{}^{H}_{NH}$	$[WCl_2(N_2H_2)(dpe)_2]$
$M\cdots N\cdots N\diagdown{}^{H}_{H}$	$[MBr(N_2H_2)(dpe)_2][BPh_4]$
$M\cdots N\diagdown{}^{H}_{NH}$	$[WCl_3(N_2H_3)(P')_2]$

[a]M = Mo or W, dpe = $Ph_2PCH_2CH_2PPh_2$, P = PMe_2Ph, P' = $PMePh_2$

By the use of different tertiary phosphines and acids, representative complexes of various stages in the reduction process from N_2 to N_2H_3 have been isolated and the N_2H_2 ligand has been found in the diazene (NH=NH) and the hydrazide (2-)(N-NH_2) forms. Some typical examples of intermediate complexes are listed in Table 3 *(10)*. Particularly note the first isolation of complexes of the imminonitrosyl or diazenido-ligand and the di-imide or diazene ligand. The diazenido-ligand is stable only in complexes containing the diphosphine and the only complex which we have been able to isolate at the N_2H_3 stage is that formulated in the Table. All complexes containing the N_2H_n (n = 1-3) ligand, provided they contain monodentate phosphines, react with methanolic sulphuric acid to give yields of ammonia which parallel very closely those obtained from the parent bis(dinitrogen)-complexes, which suggests that they may be intermediates on the way to ammonia.

The series of compounds listed in Table 3 show how the degradation of the triply bonded dinitrogen molecule occurs by the transfer of multiple bonding from the N≡N bond into the M-N bond with each successive protonation until there is essentially a single bond between the two nitrogen atoms at the third protonation. The breaking of the N-N bond probably occurs as a consequence of the fourth protonation because hydrazine is only a very minor product. The breaking of the bond between the nitrogen atoms may occur simply as illustrated in Reaction 7, because if the fourth protonation produced the M-NH$_2$-NH$_2$ grouping hydrazine should be a major product.

$$M\equiv N{\overset{H}{\underset{NH_2}{\diagup}}} + H^+ \longrightarrow M\equiv N{\overset{H}{\underset{\overset{+}{N}H_3}{\diagup}}} \longrightarrow M{\overset{+}{=}}NH + NH_3 \qquad (7)$$

Reaction 7 yields one molecule of ammonia and an imido-complex; the latter will readily hydrolyze to ammonia. We have not yet been able to isolate an imido-complex from this reaction. However, the related nitrido-complex, [MoCl$_2$N(PMePh$_2$)$_2$], gives ammonia quantitatively with methanolic sulphuric acid *(11)*. Also, in the reaction of [Mo(N$_2$)$_2$(PMePh$_2$)$_4$] with an excess of hydrogen iodide a product has been obtained which appears to contain a bridging amido-group. We have tentatively formulated this product as [P$_2$I$_3$Mo(NH$_2$)MoI$_2$(thf)P$_2$]I (P = PMePh$_2$). It would appear that the single nitrogen species has been captured and stabilised by its bridging to another molybdenum species *(10)*.

Although one can see a logical sequence of single protonation steps leading from ligating dinitrogen to ammonia, it is not clear what starts the protonation of dinitrogen in the first place. Indeed, simple protonation of the metal has been observed when the complex [W(N$_2$)$_2$(dpe)$_2$] is treated with 2 mols of hydrogen chloride in tetrahydrofuran (Reaction 8).

$$[W(N_2)_2(dpe)_2] + 2HCl \longrightarrow [WH(N_2)_2(dpe)_2][HCl_2] \qquad (8)$$

$$[W(N_2)_2(dpe)_2] + 12HCl \longrightarrow [WCl_2(N_2H_2)(dpe)_2] + N_2 \qquad (9)$$

The resulting hydride complex is stable and remains unchanged on treatment with an excess (greater than 12 mols) of hydrogen chloride, whereas the precursor dinitrogen complex forms an N_2H_2 complex under the same reaction conditions (Reaction 9). It appears that the reduction of the dinitrogen ligand does not occur by hydriding of the metal and migration of the ligating hydride ion to the ligating dinitrogen molecule.

Other possibilities are that direct protonation of one ligating N_2 molecule does occur, in contrast to what happens in most dinitrogen complexes. Since this is effectively an oxidation of the metal the remaining dinitrogen molecule is then readily replaced by the anion of the acid to give the intermediate $[MA(N_2H)P_4]$ according to Scheme 1. Alternatively the anion may displace one dinitrogen ligand which triggers off the protonation of the second. This can be readily understood in terms of the displacement of the π-acceptor dinitrogen ligand by a π-donor oxygen ligand, which would force more negative charge on to the remaining ligating dinitrogen molecule, especially if the oxygen ligand is anionic. These alternatives are illustrated in Scheme 1.

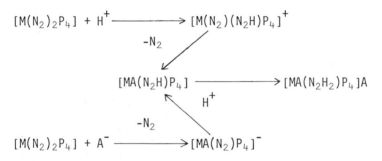

$$[M(N_2)_2P_4] + H^+ \longrightarrow [M(N_2)(N_2H)P_4]^+$$

$$-N_2$$

$$[MA(N_2H)P_4] \longrightarrow [MA(N_2H_2)P_4]A$$

$$H^+$$

$$-N_2$$

$$[M(N_2)_2P_4] + A^- \longrightarrow [MA(N_2)P_4]^-$$

(M = Mo or W; P = monotertiary phosphine; HA = protic acid)

Scheme 1. Alternative proposals for the initiation of the reduction of ligating dinitrogen.

When the bis(dinitrogen)-complex contains two molecules of the diphosphine, protonation of the dinitrogen stops at the N_2H_2 stage at room temperature, but when the complex contains four monophosphine ligands reaction continues through to ammonia. The continuation of the reaction in the second case probably depends upon the fact that as the reduction of the ligating dinitrogen molecule proceeds, at the expense of the central metal atom which is oxidised, the phosphine molecules become less strongly bound. A ligating monophosphine molecule is readily displaced at the N_2H_2 stage, but not a diphosphine. The phosphine molecule which

is lost will be replaced by an oxygen donor most probably HSO_4^-.
We thus have the replacement of a neutral weakly π-acceptor ligand
by a negatively charged π-donor which will tend to force more
electrons from the metal into the N_2H_2 ligand in the $N-NH_2$ form,
and increase its susceptibility to protonation. The monophosphine
complexes appear to provide very nicely balanced systems whereby
in the early stages of protonation electron pressures within the
complexes are not so high as to pass electrons directly into the
protons of the medium, but only through the dinitrogen molecule.
As the reaction proceeds the phosphine molecules are gradually
lost and replaced by the π-donor oxygen of the solvent or the
acid, so raising the electron density on the metal and increasingly forcing electrons into the dinitrogen moiety. There they are
neutralised by protons until the dinitrogen has been degraded
completely into molecules of ammonia. This suggests that an oxygen ligand may play a similar crucial role in the nitrogenase
reaction *(12)*.

The above considerations lead one to ask whether it is likely
that the molybdenum in nitrogenase is in an oxygen rather than a
sulphur environment. The above reaction shows that electrons can
be passed efficiently from a transition metal into a ligating dinitrogen molecule in a protic medium to produce ammonia. In our
complexes above the electrons are already stored in the complex
and the oxidation state of the metal, certainly in the case of
tungsten, changes from zero to six, but such changes of oxidation
state are unlikely in the enzyme. It seems probable therefore
that molybdenum operates in some intermediate oxidation state in
the enzyme, a state which takes up N_2 but does not hold on to the
ammonia which is produced, probably preferring water or hydride
ion in the absence of dinitrogen. The hydride ion would be produced by reduction of protons in the environment.

All the stable dinitrogen complexes contain the metals in low
oxidation states of closed shell structure, but there is some
evidence that the early transition metals in the presence of
oxygen ligands catalyse the reduction of dinitrogen to ammonia.
The most striking example was provided by Professor Shilov's
group. They showed that dinitrogen is reduced to hydrazine by
the mixed hydroxides of vanadium(II) and magnesium, obtained by
co-precipitation of vanadium(II) and magnesium chlorides with
potassium hydroxide. Almost quantitative yields were obtained at
10 atm. pressure. The yields were calculated on the reducing
capacity of the system (Reaction 10) *(4)*.

$$VCl_2 + MgCl_2 + N_2 \ (10 \text{ atm.}) \xrightarrow[\text{pH10}]{\text{KOH}} V(OH)_3 + Mg(OH)_2 + N_2H_4 + KCl \qquad (10)$$

If molybdenum interacts with dinitrogen when it is in an aqueous or oxygen environment it is unlikely to have an oxidation state lower than III. Recent work has indicated that molybdenum in oxidation states IV and V has a strong affinity for dinitrogen as it is modified in the complex $[ReCl(N_2)(PMe_2Ph)_4]$. In this complex the dinitrogen molecule has been strongly polarised, but otherwise changed very little because, admittedly in a difficult structure, its N-N distance is indistinguishable from that in molecular dinitrogen *(13)*. However, in the stable adducts $[(PMe_2Ph)_4ClRe-N\equiv N-MoCl_4-N\equiv N-ReCl(PMe_2Ph)_4]$ *(14)* and $[(PMe_2Ph)_4$ $ClRe-N\equiv N-MoCl_4(OMe)]$ *(15)* the N-N bond distances are 1.28 and 1.21 Å respectively as compared with 1.12 Å in most mononuclear dinitrogen complexes and 1.10 Å in molecular nitrogen. Since slightly modified dinitrogen in the rhenium complex has a high affinity for molybdenum (IV) and molybdenum (V) it may be that a molybdenum atom, as modified by the enzyme, and in one of those higher oxidation states, could have an affinity for normal molecular nitrogen. Before I saw Dr. Cramer's poster session *(8)* I was inclined to the view that the chemistry of dinitrogen was pointing strongly to molybdenum in an oxygen or oxygen-nitrogen environment as the most likely centre for the reduction of dinitrogen is nitrogenase. Dr. Cramer's results are tentative; so is my view.

4.   REACTIONS OF LIGATING DINITROGEN WITH ORGANIC COMPOUNDS

Since we are concerned here mainly with reactions of biological significance I have tended to concentrate on the reduction of dinitrogen to ammonia. However, the dinitrogen in the bis(dinitrogen)-complexes of molybdenum and tungsten reacts with certain organic compounds and the reactions are of intrinsic interest *(5)*. They have lent themselves to the production, for the first time, of complexes containing aliphatic organodiazenido-(RN=N-), organohydrazido(2-)-(RR'N-N=) and aliphatic diazo-(RN_2) ligands.

Aroyl, acyl, and alkyl halides react readily in benzene solution upon irradiation with tungsten filament light, according to Reaction 11.

$$[M(N_2)_2(dpe)_2] \xrightarrow[h\nu, \ C_6H_6]{RX} [MX(N=NR)(dpe)_2] \qquad (11)$$

(M = Mo or W; R = *e.g.* PhCO, MeCO, Me, Et; X = halogen; dpe = $Ph_2PCH_2CH_2PPh_2$).

The product does not undergo a second addition of RX to give

$[MX_2RN=NR)(dpe)_2]$, or $[MX(N-NR_2)(dpe)_2]X$ except when M=W and R= Me; then the compound $[WBr(N-NMe_2)(dpe)_2]Br$ results. Neverthe- less, all of the above diazenido-complexes react immediately with acids, HX, to form organohydrazido(2-)-complexes of the type $[MX(N-NHR)(dpe)_2]X$. The addition of acid is readily reversed by reaction with aqueous potassium carbonate or other mild base.

When Reaction 11 is carried out with α,ω-dibromoalkanes, $Br(CH_2)_nBr$, the product may be "normal", *e.g.* $[MBr\{N(CH_2)_nBr\}$ $(dpe)_2]$, or a heterocyclic product, *e.g.* $[MBr\{N-N(CH_2)_{(n-2)}CH_2\}$ $(dpe)_2]$. When n=3 or 6-9 "normal" products result but when n=4, 5, 10 or 12, heterocyclic products predominate. When n=2 there is no identifiable product, and when n=1 and M=W a diazomethane complex, $[MBr(N-N=CH_2)(dpe)_2]Br$, the first ever recorded is formed, but when M = Mo, $[MoBr(N=NCH_2Br)(dpe)_2]$ results *(16)*. Nevertheless, 2,2'-dibromopropane with $[Mo(N_2)_2(dpe)_2]$ yields $[MoBr(N-N=CMe_2)(dpe)_2]Br$ *(17)*.

When Reaction 11 with RX = MeBr is attempted in tetrahydro- furan, another complex reaction takes place and the products are ω-diazobutanol complexes, $[MBr\{N-N=CH(CH_2)_2CH_2OH\}-(dpe)_2]Br$ (M = Mo or W) in which the four carbon atom chain and oxygen are derived from the solvent *(18)*. Since methane is evolved during the formation of the ω-diazobutanol complex it is evident that the mechanism of formation involves methyl-radicals from irradiation of the methyl bromide. These create a tetrahydro- furan radical which attaches itself to the terminal nitrogen of the ligand. This suggests that the above reactions with alkyl bromides, which either need irradiation or are greatly acceler- ated by it, result from the formation of organic radicals and bromine atoms which attack the bis(dinitrogen)-complexes *(19)*.

The structure of the ω-diazobutanol complex as determined by X-rays shows that in its complexes the N=N-C chain approaches the structure =N-N=CH- rather than $=N\equiv N=CH-$ *(18)*. The CH adjacent to the nitrogen is strongly electrophilic and takes hydride ion from lithium tetrahydroborate to form the corresponding diazenido- complex (Reaction 12).

$$[MBr(N-N=CHR)(dpe)_2]^+ + H^- \longrightarrow [MBr(N=N-CH_2R)(dpe)_2] \qquad (12)$$

It is evident that the organic chemistry of ligating dinitro- gen is only just beginning, and that there is a real possibility of the development of catalytic processes for the production of organo-nitrogen compounds directly from elementary nitrogen.

ACKNOWLEDGEMENT. The above work has involved myself with a number of co-workers whose names are on our joint papers. I wish to express my gratitude to them all for the effort and ideas they have contributed, and especially to Drs. G.J. Leigh, R.L. Richards and J.R. Dilworth who have collaborated with me throughout.

REFERENCES

(References 1-6 are reviews covering most of the early literature)

1. "The Chemistry and Biochemistry of Nitrogen Fixation", J.R. Postgate, Ed., Plenum Press, London and New York, 1971.
2. J. Chatt and G.J. Leigh, *Chem. Soc. Rev.*,**1**, 121 (1972).
3. D. Sellman, *Angew. Chem. Int. Ed. Engl.*,**13**, 639 (1974).
4. A.E. Shilov, *Russ. Chem. Rev.*,**43**, 378 (1974).
5. J. Chatt, *J. Organomet. Chem.*,**100**, 17 (1975).
6. R.R. Eady and J.R. Postgate, *Nature (London)*,**249**, 805 (1974); W.G. Zumft and L.E. Mortenson, *Biochim. Biophys. Acta.*,**416**, 1 (1975).
7. S.P. Cramer, T.K. Eccles, F.W. Kutzler, K.O. Hodgson, and L.E. Mortenson, *J. Am. Chem. Soc.*,**98**, 1287 (1976).
8. S.P. Cramer, K.O. Hodgson, L.E. Mortenson and J. Dawson, Poster No. 6 at symposium "Biological Aspects of Inorganic Chemistry", University of British Columbia, June, 1976.
9. J. Chatt, A.J. Pearman, and R.L. Richards, *Nature (London)*, **253**, 39 (1975).
10. J. Chatt, A.J. Pearman and R.L. Richards, *J. Organomet. Chem.*, **101**, C45 (1975) and continuing work.
11. J. Chatt and J.R. Dilworth, *Chem. Commun.*,517 (1974) and continuing work.
12. J. Chatt, A.J. Pearman, and R.L. Richards, *Nature (London)*, **259**, 204 (1976).
13. B.R. Davis and J.A. Ibers, *Inorg. Chem.*,**10**, 580 (1971).
14. P.D. Cradwick, J. Chatt, R.H. Crabtree, and R.L. Richards, *Chem. Commun.*, 351 (1975).
15. M. Mercer, R.H. Crabtree, and R.L. Richards, *Chem. Commun.* 808 (1973); M. Mercer, *J. Chem. Soc., Dalton Trans.* 1637 (1974).
16. R. Ben-Shoshan, J. Chatt, W. Hussain, and G.J. Leigh, *J. Organomet. Chem.*,**112**, C9 (1976).
17. J. Chatt, G.J. Leigh and C. Torreros, continuing work from Ref. 16.

18. P.C. Bevan, J. Chatt, R.A. Head, P.B. Hitchcock and G.J. Leigh, *Chem. Commun.*, 509 (1976).
19. H. Gray and co-workers have independently reached a similar conclusion (private communication at symposium of Ref. 8).

current status of the mechanism of action of B$_{12}$-coenzyme

ROBERT H. ABELES

Graduate Department of Biochemistry, Brandeis University, Waltham, Massachusetts 02154

1. INTRODUCTION

The investigation of the mechanism of action of B$_{12}$ coenzymes began approximately twenty years ago when Barker demonstrated the envolvement of a derivative of vitamin B$_{12}$ in an enzymatic reaction (Reaction 1, Table 1)[1]. The rearrangement described by Barker is a most remarkable reaction since, until very recently, no analogous chemical reaction was known. Shortly after the discovery of this reaction, the structure of the coenzyme (Fig. 1) was elucidated through the work of Barker and his collaborators and through the X-ray crystallographic analysis of Dorothy Hodgkins and her collaborators [1]. In the coenzyme form of vitamin B$_{12}$ an adenosyl moiety is bonded through its 5'-carbon to form a covalent bond to the central carbon atom of vitamin B$_{12}$. This is

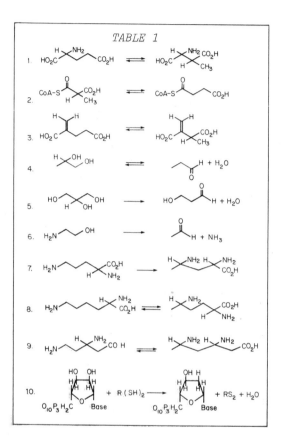

TABLE 1

the first example of a water-stable naturally occurring organometallic compound. Elucidation of the structure of the coenzyme
form of vitamin B_{12} did not clarify its mechanism. In fact, it
deepened the mystery. In a sense, the investigation of the
mechanism of action of vitamin B_{12} was unique and differed from
other investigations of the mechanism of action of enzymes. In
general, when the mechanism of an enzymatic reaction is investigated, organic chemistry provides a number of analogous reactions
and potential mechanisms by which the reaction can occur. It is
then the task of the investigator to decide which of the possible
organic mechanisms is applicable. In investigating the mechanism
of action of vitamin B_{12} coenzymes, no suitable organic precedent
seemed to exist. The current conclusions concerning the mechanism of action of vitamin B_{12} coenzymes are based entirely upon
evidence obtained from the study of the enzymic reaction itself.

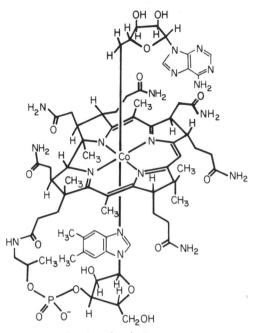

Fig. 1. Vitamin B₁₂ coenzyme.

It is fair to say, I believe, that at this point some concensus concerning the mechanism of some of the reactions involving vitamin B_{12} coenzyme has been reached and is agreed on by most, but not all, investigators in the area.

Table 1 lists all of the reactions known today which require B_{12} coenzyme. All of these reactions, with the exception of the reaction catalyzed by methylmalonyl-CoA-isomerase (Reaction 2, Table 1) occur in bacterial systems. The interconversion of methylmalonyl-CoA to succinyl-CoA is the only reaction which occurs in a mammalian system. The reactions in Table 1 involve unique carbon skeleton rearrangements, as well as group migrations. The general nature of the reactions can be summarized by Equation 1:

$$-\overset{|}{\underset{X}{C_1}}-\overset{|}{\underset{H}{C_2}}- \longrightarrow -\overset{|}{\underset{H}{C_1}}-\overset{|}{\underset{X}{C_2}}- \tag{1}$$

In all cases, some entity represented by X in Equation 1 migrates from one carbon to an adjacent carbon, and concomitantly, a hydrogen migrates in the opposite sense. A common feature of all of the reactions involving B_{12}-coenzyme, with the exception of ribonucleotide reductase (Reaction 10, Table 1) is that the hydrogen migrates without exchange with the protons of the solvent. The significance of this observation will become apparent when the reaction mechanism is discussed in more detail.

In this article, I shall primarily discuss the reaction catalyzed by the enzyme dioldehydrase (Reaction 4, Table 1) which catalyzes the following reactions:

$$CH_3-CH-CH_2 \longrightarrow CH_3-CH_2-CHO$$
$$||$$
$$OHOH$$

Other substrates: CH_2OH-CH_2OH, $CH_2F-CHOH-CH_2OH$

2.   THE HYDROGEN-TRANSFER STEP

The conversion of ethylene glycol or 1,2-propanediol to acetaldehyde or propionaldehyde *appears* to proceed with direct hydrogen transfer. When the reaction is carried out in D_2O, no deuterium is incorporated into the product, or when the reaction is carried out with C-1-di-deutero-1,2-propanediol, deuterium is transferred from C-1 to the α-carbon of the resulting propionaldehyde. This hydrogen transfer is highly stereoselective. With L-1,2-propanediol, the H_S hydrogen migrates. The replacement of the hydroxyl group by the migrating hydrogen occurs with inversion at the C-2 carbon.* Elegant experiments carried out in Arigoni's *(2)* laboratory have demonstrated that the reaction catalyzed by dioldehydrase proceeds as represented by Equation 2:

$$CH_3-CHOH-CH_2OH \longrightarrow CH_3-CH_2-CH(OH)_2 \longrightarrow CH_3-CH_2-CHO \qquad (2)$$

In the conversion of the 1,2-diol to the intermediate 1,1-gemdiol the C-2 OH group is transferred to the C-1 carbon. It should be noted that the enzyme also catalyzes the dehydration of the intermediate gemdiol, since the actual product of the reaction is the aldehyde.

In the early phases of the investigation of the mechanism of action of dioldehydrase, it became apparent that the reaction does not involve a 1,2-hydrogen shift. The coenzyme functions as an intermediate hydrogen carrier in these reactions. Since the

---

*   *Inversion is also observed with glutamate mutase (Reaction 1, Table 1). The reaction catalyzed by methylmalonyl-CoA mutase and ribonucleotide reductase (Reactions 2 and 10, Table 1) proceed with retention.*

experiments from which these conclusions are derived have been published and reviewed *(1)*, I shall only summarize these experiments: 1) When a mixture of C-1-H^3-1,2-propanediol and unlabeled ethylene glycol is allowed to react with the enzyme, α-H^3 acetaldehyde is isolated, *i.e.*, hydrogen is transferred from propanediol to the product derived from ethylene glycol. Hydrogen transfer, therefore, can be largely intermolecular. 2) When the enzyme-coenzyme complex is allowed to react with C-1-H^3 propanediol, and the coenzyme is re-isolated after termination of the reaction, the coenzyme contains tritium. When the tritiated coenzyme so obtained is added back to apoenzyme and unlabeled 1,2-propanediol, all of the tritium of the coenzyme is transferred to propionaldehyde. Chemical degradation of the tritiated coenzyme, established that ^{3}H is located exclusively on the C-5' carbon directly bonded to cobalt. 3) ^{3}H was introduced non-stereospecifically by chemical synthesis into the C-5' position of the coenzyme. When synthetically prepared 5'-^{3}H-coenzyme is added to apo-dioldehydrase and 1,2-propanediol, all of the tritium is transferred from coenzyme to reaction products with a single-rate constant. No distinction is made between the two nonequivalent tritium atoms.

These results suggested to us the following mechanism:

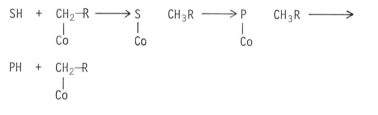

$$(3)$$

SH, PH = substrate and product

—CH$_2$—R = adenosyl

According to the mechanism of Equation 3, a ligand exchange takes place, between substrate and B₁₂-coenzyme which leads to the formation of 5'-deoxyadenosine and a new substrate-cobalamin adduct. In the ligand interchange, a new carbon-cobalt bond is formed and a hydrogen atom is transferred from the substrate to the C-5' carbon of the coenzyme. Next, the substrate-cobalamin adduct rearranges to form a product-cobalamin adduct. Finally, the product is formed through a second ligand exchange and the coenzyme is regenerated. Regeneration of the coenzyme again involves a transfer of hydrogen from the 5'carbon of 5'-deoxyadenosine to the product. The mechanism accounts for the transfer

of tritium from substrate to coenzyme and from coenzyme to substrate, as well as for the observed intermolecular hydrogen
transfer. The intermediate formation of 5'-deoxyadenosine is
also consistent with the observation that the two nonequivalent
hydrogens of the coenzyme are transferred to the product at equal
rates. Breaking of the carbon-cobalt bond and the formation of
5'-deoxyadenosine leads to loss of chirality of the 5'-hydrogens.

The participation of 5'-deoxyadenosine in the reaction sequence is a central feature of the proposed mechanism and a considerable body of evidence has accumulated in support of its
participation in the catalytic process. I shall cite here two,
of many experiments, in support of the intermediate involved of
5'-deoxyadenosine. The first experiment was a kinetic experiment
and was carried out prior to the actual isolation of 5'-deoxy-
adenosine, which shall be described later.

The experiment which led to the conclusion that an intermediate with at least three equivalent hydrogen atoms is involved
was based on the following considerations. Let us assume that an
intermediate exists which contains three hydrogen atoms, two
derived from the coenzyme and one from the substrate. We will
designate this intermediate by: $\begin{bmatrix} & H \\ X & H \\ & H \end{bmatrix}$. When the product is
formed, any one of the hydrogen atoms can be transferred to the
product. Now consider an intermediate containing hydrogen and
tritium $\begin{bmatrix} & T \\ X & H \\ & H \end{bmatrix}$. What factors decide whether hydrogen or tritium
is transferred to the reaction product? Steric factors and iso-
tope effects will determine whether hydrogen or tritium is transferred. Clearly, there will be discrimination against tritium
due to isotope effects. From the intermediate $\begin{bmatrix} & T \\ X & H \\ & D \end{bmatrix}$ the
probability of tritium transfer to the product will be greater
than from $\begin{bmatrix} & T \\ X & H \\ & H \end{bmatrix}$. The probability of tritium transfer to the

product will be even further enhanced from the intermediate
$\begin{bmatrix} T \\ X\ D \\ D \end{bmatrix}$. Now consider experiments in which enzyme -H^3-B_{12} coenzyme is allowed to act on CH_3—$CHOH$—CD_2OH (**1**) and CH_2OH—CH_2OH
(**2**). In the first set of experiments (**2**) will be in excess over
(**1**).

According to our model, when (**1**) reacts with enzyme-coenzyme
the intermediate $\begin{bmatrix} T \\ X\ H \\ D \end{bmatrix}$ will be formed and (**2**) will give rise to
$\begin{bmatrix} T \\ X\ H \\ H \end{bmatrix}$. Therefore, it is predicted that the product derived from
(**1**), *i.e.*, the deuterated substrate, will have a higher probabi-
lity of taking ^3H from the coenzyme. This means that the specific activity of the product derived from the deuterated substrate
should be higher than that derived from the non-deuterated substrate. This prediction was experimentally verified. The specific activity of the product derived from the deuterated substrate was four times that of the product derived from the non-
deuterated substrate. These results establish that an intermediate exists in which hydrogen (or deuterium) provided by the
substrate competes with tritium provided by the coenzyme for
transfer to the reaction product. These results are consistent
with our model; however, they do not require an intermediate
containing three hydrogen atoms. An intermediate containing only
two hydrogen atoms would be consistent with these results. An
additional experiment was carried out which establishes that the

intermediate must contain at least three hydrogen atoms. The experiment described above was repeated with a large excess of (**1**) over (**2**). Under these conditions, the C-5' position of the enzyme bound coenzyme will gradually become deuterated. In the limiting case the two intermediates which provide ^{3}H for the product will be $\begin{bmatrix} T \\ X\ D \\ D \end{bmatrix}$ derived from the deuterated substrate and $\begin{bmatrix} T \\ X\ D \\ H \end{bmatrix}$ derived from the non-deuterated substrate. As stated above the probability of tritium transfer from $\begin{bmatrix} T \\ X\ D \\ D \end{bmatrix}$ to the reaction product is much higher than from $\begin{bmatrix} T \\ X\ D \\ H \end{bmatrix}$ or $\begin{bmatrix} T \\ X\ H \\ H \end{bmatrix}$ The experimental consequence is that as the ratio of (**1/2**) increases, the specific activity of the product from (**1**) should increase relative to that derived from (**2**). This was experimentally observed. At the lowest ratio of (**1/2**) the specific activity of the product derived from (**1**) was four times that of (**2**) and at the highest ratio of (**1/2**) the specific activity of the product derived from (**1**) was eight times higher than the specific activity of the product derived from (**2**). This increase in specific activity, as the fraction of deuterated substrate increases, cannot be explained by a two hydrogen intermediate and requires an intermediate containing at least three hydrogen atoms.

3. 5'-DEOXYADENOSINE

Evidence for the intermediate involvement of 5'-deoxyadenosine is not only based on kinetic arguments, but 5'-deoxyadenosine has been isolated under several conditions. If the enzyme-coenzyme complex is denatured in the presence of substrate, one finds no or little 5'-deoxyadenosine. We attribute this to the fact that, with the optimal substrates, the steady state concentration of intermediates is such that there is very little 5'-deoxyadenosine present. It is possible that with substrates, where the intermediate substrate-cobalamin adduct is unstable, 5'-deoxyadenosine will accumulate. We have found a number of substrates where 5'-deoxyadenosine accumulates. The most interesting of these is chloroacetaldehyde. Chloroacetaldehyde exchanges tritium with the enzyme-coenzyme complex. It is not converted to any other product, only an exchange reaction occurs. This exchange means that it must form the intermediate of interest, *i.e.*, a reaction must take place between substrate and coenzyme. Additions of chloroacetaldehyde to the enzyme-coenzyme complex produces a spectral change, suggesting the formation of $B_{12}(r)$, *i.e.*, cobalt in the plus two oxidation state. A similar change, but not nearly as intense, is seen upon addition of the normal substrate to the enzyme-coenzyme complex. Addition of chloroacetaldehyde also brings about loss of enzyme activity. This is in accordance with our expectations. If the intermediate substrate-cobalamin adduct decomposes, one expects to lose enzyme activity. As enzyme activity is lost, 5'-deoxyadenosine accumulates, which is exactly what we hoped. The rate of formation of 5'-deoxyadenosine and loss of enzyme activity are precisely identical. Similar results have been obtained for dioldehydrase with other substrate analogs and also with ethanolamine-ammonia lyase *(3)*.

These experiments therefore establish that the adenoxyl moiety of the coenzyme can be converted to 5'-deoxyadenosine. They are, however, subject to a serious criticism: in all cases cited so far, when 5'-deoxyadenosine has been isolated in substantial amounts, it was isolated from a catalytically inactive enzyme-coenzyme-substrate complex. Therefore, it can be argued that formation of 5'-deoxyadenosine leads to loss of catalytic activity. It was, therefore, essential to find a system in which 5'-deoxyadenosine could be isolated from a catalytically active enzyme-coenzyme complex. The enzyme, ethanolamine-ammonia lyase, which catalyzes Reaction 6, Table 1, was suitable for this purpose. This enzyme also catalyzes the conversion of 2-amino-1-propanol to propionaldehyde, although at a rate substantially slower than that observed with ethanolamine, the normal substrate. When 2-amino-1-propanol is added to the enzyme-coenzyme complex, and the complex is then rapidly denatured, less than 20% of the coenzyme is recovered. Instead 5'-deoxyadenosine, derived from the adenosyl moiety of the coenzyme can be isolated. The amount of 5'-deoxyadenosine isolated corresponds to approximately 80% of

the amount of coenzyme present. The complex, prior to denaturation, has full catalytic activity. When the same experiment is
carried out with $1-H^3-2$-amino-1-propanol, tritium is found in
5'-deoxyadenosine. 5'-Deoxyadenosine can thus be isolated from
a complex which has *not* lost catalytic activity.

Reversibility of 5'-deoxyadenosine formation was established
through the following experiment: 2-amino-1-propanol was added
to the enzyme-B_{12}-coenzyme complex. An aliquot of this reaction
mixture was denatured and, as mentioned above, 80% of the coenzyme originally present is recovered as 5'-deoxyadenosine.
Another aliquot was denatured after addition of ethanolamine.
Addition of ethanolamine should shift the steady state concentration of intermediates so that very little enzyme bound 5'-
deoxyadenosine should be present. After denaturation of the
enzyme, essentially no 5'-deoxyadenosine is found and only intact
B_{12}-coenzyme is recovered. These experiments therefore show that
enzyme species in which the adenosyl moiety of the coenzyme is
largely present as 5'-deoxyadenosine can be converted to a species containing primarily B_{12}-coenzyme. The reversible cleavage
of the carbon-cobalt bond and the reversible formation of 5'-
deoxyadenosine is therefore established.

It might be argued that 5'-deoxyadenosine is formed during
the denaturation process from an enzyme-bound adenosyl-radical.
This radical could, during the process of denaturation, abstract
hydrogen from the environment. To investigate this possibility
we added 2-amino-1-propanol-1-D_2 to enzyme-B_{12}-coenzyme. The
coenzyme employed in these experiments contained two deuterium
atoms in the C-5'-position. The complex was then denatured and
5'-deoxyadenosine was isolated and the isotopic composition of
its methyl group was examined. The methyl group was found to be
predominantly -CD_3 *(4)*. This is expected if the hydrogens of the
methyl group of 5'-deoxyadenosine are derived from the two C-5'
hydrogens of the coenzyme and one hydrogen contributed by the
substrate. These experiments put the intermediate involvement of
5'-deoxyadenosine in the reaction catalyzed by dioldehydrase as
well as ethanolamine ammonia lyase on a very firm basis. Evidence for the intermediate involvement of 5'-deoxyadenosine has
also been obtained for glutamate mutase and methylmalonyl CoA
isomerase *(5)*.

4.   THE LIGAND EXCHANGE REACTION

I shall now return to a consideration of the "ligand exchange
reaction." It has been known for some time that addition of substrate to the dioldehydrase coenzyme complex brings about a

change in the spectrum of the enzyme-bound coenzyme. Due to experimental difficulties, the precise nature of the spectral change was difficult to identify. However, it suggested one of two things: either the coenzyme was converted to $B_{12(r)}$ (Co^{+2}), which would imply that the carbon-cobalt bond has been broken by homolytic process, or the dimethylbenzimidazole ligand was no longer coordinated to the bottom possition of the corrin ring. The work of Babior and Gould *(6)* with ethanol ammonia lyase provided the first indication that the addition of substrate to a B_{12}-dependent enzyme (ethanolamine ammonia lyase) led to the formation of radical species. Two types of radical signals were obtained. One, which could be attributed to the formation of $B_{12(r)}$, and the other signal to the formation of organic radicals. Similar results were subsequently obtained with dioldehydrase. After addition of substrate or substrate analogs to dioldehydrase, ESR signals indicative of the formation of radical species can be detected. When the normal substrate is used, 40-50% of the enzyme-bound coenzyme is converted to $B_{12(r)}$. With the use of isotopically labeled substrate analogs it could be demonstrated that at least part of the signal is due to a substrate radical *(7)*. By measuring the rate of formation of the radical species after addition of substrate to dioldehydrase, it was established that the radical intermediates are kinetically competent. These observations, therefore, led to revision of the proposed mechanism. A modified mechanism is shown in Equation 4.

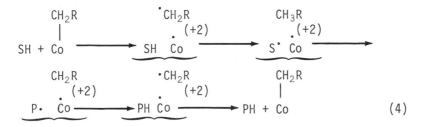

$$\text{(4)}$$

_____ *denotes an enzyme bound complex. Other symbols as in Equation 3.*

According to this mechanism, upon addition of substrate to the enzyme-B_{12} coenzyme complex, the carbon cobalt bond breaks homolytically to produce $B_{12(r)}$ and a deoxyadenosyl radical. The deoxyadenosyl radical abstracts a hydrogen atom from the substrate, which leads to the formation of a substrate radical and 5'-deoxyadenosine. The substrate radical then rearranges to a product radical by an as yet unspecified process. Finally, essentially through a reversal of the steps outlined above, the final product is formed and the coenzyme regenerated. The deoxy-adenosyl radical provides a satisfactory mechanism for the abstraction on an inactivated hydrogen which occurs in many of the reactions catalyzed by B_{12} coenzymes.

Further support for the intermediate involvement of a product derived radical is provided by the work carried out in Arigoni's laboratory *(8)*. Chirally labelled [2-^{2}H,^{3}H] ethanolamine was converted to acetaldehyde in the presence of ethanolamine-ammonia lyase. The acetaldehyde formed was racemic. This result is consistent with the intermediate formation of $\cdot CH_2$—$CHOH\ NH_2$.

The next question we must consider is, "How is the substrate radical converted to a product radical?" Fig. 2 summarizes several possible pathways. It is highly probable that any one enzyme will utilize only one pathway. It is possible, however, that different enzymes may operate by different mechanisms. It may well be one of the important properties of B_{12} coenzyme that it can be utilized in reactions involving different mechanisms.

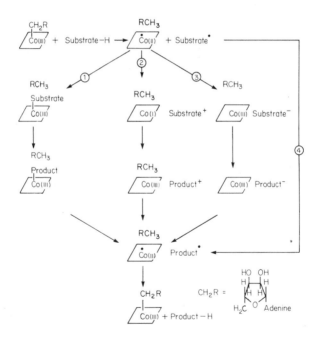

Fig. 2. Possible pathways for B_{12}-coenzyme mechanisms.

For none of the B_{12}-coenzyme-dependent reactions can a definitive mechanism for the rearrangement be stated at this time. For
the reaction catalyzed by dioldehydrase, we initially considered
Pathway 1 (Fig. 2) and have made many attempts to detect a covalent intermediate between substrate and coenzyme. All such
attempts have been unsuccessful. Although it is dangerous to
draw conclusions from negative results, it is our view, at this
time, that no such intermediates exist. We now favor a radical
rearrangement. The mechanism proposed *(9)* for the conversion of
ethylene glycol to glycol aldehyde and acetaldehyde in the presence of Fenton's reagent (Equation 5), may provide a non-enzyme
precedence for this type of radical rearrangement.

$$CH_2OH\text{—}CH_2OH + Fe^{+3} \longrightarrow CH_2OH\text{—}CHOH + H^+ + Fe^{+2}$$

$$CH_2OH\text{—}CHO + Fe^{+2} + H^+$$

$$\overset{+}{CH_2}\text{—}\overset{\cdot}{CHOH} + Fe^{+2}$$

$$CH_3\text{—}CHO + Fe^{+3}$$

$$\overset{\cdot}{CH_2}\text{—}\overset{+}{CHOH}$$

$$(5)$$

A similar mechanism appears attractive for the reaction catalyzed
by dioldehydrase. A disadvantage of the mechanism, as formulated
above, is that it does not lead to the formation of a 1,1-gemdiol
which is an intermediate in the reaction catalyzed by dioldehy-
drawe. A radical mechanism was proposed by Golding which does
not suffer from this advantage, but involves an intermediate protonated epoxide *(10)*. It is difficult to see how a protonated
species of this type could be formed on the enzyme. It has been
reported that the 4,1-dihydroxypentyl (pyridine) cobaloxime *(11)*
is converted in low yield from photolysis to pentanal. This
reaction may provide a model for the enzymic reaction. Dolphin
has suggested that a covalent adduct is formed between the substrate radical and $B_{12(r)}$ and that this adduct rearranges through
a π complex *(12)*. It has been shown that such rearrangements
occur in nonenzymic systems. On even less firm ground than the
mechanism of the rearrangement catalyzed by dioldehydrase is the
mechanism of the carbon-skeleton rearrangements catalyzed by B_{12}-
coenzymes. Until recently, no analogous organic reactions were
known. This situation has been remedied by Dowd, who has described nonenzymic reactions very similar to the enzyme-catalyzed

reactions (Reactions 2 and 3, Table 1) *(13)*. For instance, he
has observed the following rearrangement.

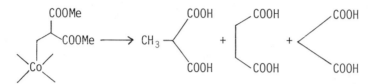

This reaction is very similar to the rearrangement catalyzed by
methylmalonyl-CoA isomerase. A nonenzymic model reaction for
this rearrangement also has been reported recently by Retey *(14)*.
The availability of these model reactions should be helpful in
the further elucidation of the enzymic mechanisms.

    In all discussions of the mechanism of action of vitamin B_{12}
coenzyme, the presence of the protein and the possible role of
the protein in the catalytic process, has been completely ignored.
This, of course, is a forgivable offense, since it is difficult
to consider the catalytic role of the protein when nothing is
known about the catalytic process. Now that some features of the
catalytic process are beginning to emerge, we have considered the
role of the apoprotein. In the early stage of the reaction,
according to the mechanism shown in Equation 4, the homolytic
cleavage of the carbon-cobalt bond takes place. It appears very
likely that the protein plays a major role in bringing about this
homolytic cleavage. How could this happen? Dissociation of the
carbon-cobalt could be facilitated by distortion of the carbon-
cobalt bond through interaction of the protein with the adenosyl
moiety of the coenzyme or the interaction of the protein with the
corrin ring, or perhaps, through both effects. It will be noted
that the coenzyme contains a number of amide groups around the
periphery of the corrin ring. The possibility occurred to us
that these amide groups might hydrogen bond to the protein and
that through this interaction, distortion of the corrin ring
might occur. In order to test this hypothesis, we removed one of
the amide groups (probably one of the propionamide groups) of the
corrin ring by hydrolysis *(15)*. The monocarboxylic analog of B_{12}
so obtained did not bind to the apoenzyme, whereas vitamin B_{12}
itself binds extremely tightly (irreversibly) to the apoenzyme.
Possibly the presence of a negative charge in the corrin ring
prevents reaction with the apoenzyme. We therefore prepared a B_{12}
analog in which the carboxylic acid group was esterified, *i.e.*,
one of the amide groups is replaced by a methoxy group. This vitamin B_{12} analog binds to the apoenzyme. It was then converted to
the coenzyme form and its interaction with the apoenzyme was

examined. This modified coenzyme differed from the normal coenzyme in several respects: the complex formed between apoenzyme and the modified coenzyme has only 5% the catalytic activity of the normal form of the coenzyme. When substrate is added to the modified-coenzyme-apoenzyme complex, an extensive spectral shift is observed, as is the case with the unmodified coenzyme. The spectrum resembles the spectrum of the coenzyme under acidic conditions. The modified coenzyme differs from the normal coenzyme in that no ESR signal can be detected. An attractive interpretation of this spectral change is provided by the NMR studies carried out in the laboratories of Hogenkamp *(16)* and Williams *(17)*. These studies indicate that the yellow shift, which has generally been associated with the formation of 5-coordinated cobalamine, can also be the result of conformational changes in the corrin ring. The spectral changes observed upon addition of the substrate to the enzyme-coenzyme complex could be due to a change in the conformation of the enzyme bound cobalamine. It is very likely that the reaction sequence up to this stage is qualitatively similar for the modified and unmodified coenzyme. With the unmodified coenzyme, the next step is the homolytic cleavage of the carbon-bond, and the appearance of an ESR signal. With the modified coenzyme, this cleavage must occur to a much lower extent since no ESR signal can be detected. These results are consistent with the hypothesis that interaction of the peripheral amide group of the corrin with the apoprotein is essential for the homolytic cleavage of the carbon-cobalt bond. This aspect of the catalytic action of dioldehydrase is now under investigation.

ACKNOWLEDGEMENT. This work was supported by grants from the National Institute of Health. I would like to thank my collaborators who have been involved in this work over the past years, and who have made invaluable contributions to this work.

REFERENCES

1. For recent reviews see, "Cobalamin Biochemistry and Pathophysiology," B.M. Babior, Ed., John Wiley and Sons, New York, 1975; R.H. Abeles and D. Dolphin, *Acc. Chem. Res.*,**9**, 114 (1976); B.M. Babior, *Acc. Chem. Res.*,**8**, 376 (1975).
2. J. Retey, A. Umani-Ronchi, J. Seibl, and D. Arigoni, *Experientia*,**22**, 502 (1966).
3. B.M. Babior, *J. Biol. Chem.*,**245**, 1755 (1970).
4. K. Sato, B.M. Babior, and R.H. Abeles, *J. Biol. Chem.*, in press.

5. J.H. Richards and W.W. Miller, *J. Am. Chem. Soc.*,**91**, 1948 (1969).
6. B.M. Babior and D.C. Gould, *Biochem. Biophys. Res. Commun.*, **34**, 441 (1969).
7. T.H. Finlay, J. Valinsky, A.S. Mildvan, and R.H. Abeles, *J. Biol. Chem.*,**248**, 1285 (1973).
8. J. Retey, C.J. Suckling, D. Arigoni, and B.M. Babior, *J. Biol. Chem.*,**249**, 6359 (1974).
9. C. Walling and R.A. Johnson, *J. Am. Chem. Soc.*,**97**, 2405 (1975) and references therein.
10. B.T. Golding and L. Radon, *Chem. Commun.*, 939 (1973).
11. B.T. Golding, T.J. Kemp, E. Nocchi, and W.P. Watson, *Angew. Chem. Int. Ed. Engl.*,**14**, 813 (1975).
12. R.B. Silverman and D. Dolphin, *J. Am. Chem. Soc.*,**94**, 4028 (1972).
13. P. Dowd, M. Shapiro, and K. Kang, *J. Am. Chem. Soc.*,**97**, 4754 (1975); P. Dowd and M. Shapiro, *J. Am. Chem. Soc.*,**98**, 3724 (1976).
14. J. Retey, G. Bidlingmaier, A. Flahi, M.E. Kempe, and T. Krebs, *Angew. Chem. Int. Ed. Engl.*,**14**, 822 (1975).
15. E.K. Krodel and R.H. Abeles, unpublished.
16. H.P.C. Hogenkamp, P.J. Vergamini and N.A. Matwiyoff, *J.C.S., Dalton Transactions*, 2628 (1975).
17. S.A. Cockle, O.D. Hensens, H.A.O. Hill and R.J.P. Williams, *J.C.S., Dalton Transactions*, 2633 (1975).

some bio-inorganic chemical reactions of environmental significance

J.M. WOOD, J.D. LIPSCOMB, L. QUE, JR., R.S. STEPHENS,
W.H. ORME-JOHNSON*, E. MÜNCK, W.P. RIDLEY, L. DIZIKES,
A. CHEH, M. FRANCIA, T. FRICK, R. ZIMMERMAN AND J. HOWARD

*Freshwater Biological Institute, University of Minnesota,
Navarre, Minnesota 55392, U.S.A.*

**Department of Biochemistry, University of Wisconson - Madison,
Madison, Wisconsin 53706, U.S.A.*

1. INTRODUCTION

Before the evolution of organic matter in living systems a number of inorganic equilibria existed. These equilibria were influenced by the various physical constraints on the earth. Upon this matrix of inorganic compounds evolution led to the synthesis of both coordination complexes and organometallic complexes. Initially, metal-ligand interactions occurred in a strictly anaerobic environment. Many primitive coordination complexes still exist; for example the ferredoxins represent electron transfer proteins designed on a system of solubilizing, by protein association, the very insoluble iron sulfide complex. Later, molecules of more adaptable design evolved which could function in a variety of catalytic processes (e.g. DNA-polymerase which has a zinc atom at its active site). A second example is the iron complex heme which initially may have evolved to facilitate the reduction of soluble inorganic ions such as sulfate and nitrate. The appearance of molecular oxygen in the biosphere caused the adaptation of heme complexes which could function in the reduction of oxygen to water. This reaction clearly evolved as a detoxification mechanism since "activated" molecular oxygen is exceedingly toxic to the anaerobic bacteria. The scene was set for the evolution of complexes which could deal with toxic elements so as to give organisms selective advantages. The environment is of critical importance to the survival of the cell. It is clear that the initial inorganic matrix in the earth's crust forced the evolution of natural biogeochemical cycles for all elements including toxic elements. In every case where we are dealing with natural biological cycles, we should not be surprised to find inorganic and organometallic equilibria for toxic elements in aqueous systems.

During evolution many toxic compounds appeared and threatened survival. Molecular oxygen represents a very important toxic compound in the evolutionary sense because it can be "activated" by iron-containing enzymes or flavoproteins to give superoxide anion (O_2^-) *(1)*. With the onset of aerobic systems superoxide dismutase is thought to have evolved as a defense mechanism to oxygen toxicity *(2)*. Anaerobic bacteria do not contain superoxide dismutase and therefore these organisms die when exposed to an environment which contains oxygen. Clearly the ability to cope with molecular oxygen became increasingly important in the evolution of higher forms of life. Oxygen "fixation" reactions, though neglected in Biochemical texts, are of prime importance to the synthesis and degradation of the majority of biomass on this planet *(3)*. Recently, it has been shown that a dioxygenase links the oxygen cycle to the carbon cycle in photosynthesis. Furthermore, the second most predominant biopolymer in nature (lignin) is synthesized in a series of oxygen insertion reactions (Fig. 1). Since this polymer was evolved in an O_2 atmosphere it appears that anaerobic bacteria never had to develop the ability to degrade these highly oxidized polymeric aromatic ethers. Therefore, lignin can only be degraded oxidatively and not reductively.

In what follows we shall discuss one of the mechanisms used to "activate" oxygen to facilitate the degradation of aromatic natural products to aliphatic acids.

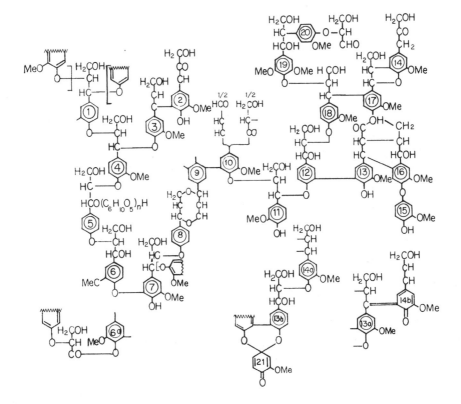

Fig. 1. Schematic presentation of the structure of lignin showing extent of oxygen insertion into the polymer.

2. REACTIONS WITH MOLECULAR OXYGEN

During the last decade a great deal of effort has been made to determine the structure and function of mixed function oxygenases. Elegant work has been done particularly by the Gunsalus and Coon groups on the cytochrome P450 system *(4)* as well as interesting model study work *(5)*. However, the majority of mixed function oxygenases which have been purified to homogeneity are flavoproteins *(6-11)*. Those enzymes responsible for hydroxylation of aromatic compounds have all been characterized as flavoproteins, *e.g.*, (Fig. 2):

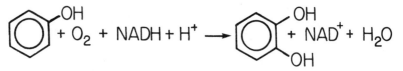

Fig. 2. Flavoprotein catalyzed mixed function oxygenation of phenol to catechol.

Hydroxylation of the benzene ring is a very important step in the biosynthesis of aromatic natural products as well as in their subsequent return to the carbon cycle by biodegradation. After the introduction of two hydroxyl groups, either *ortho* or *para* to each other, the aromatic nucleus is prepared for ring cleavage by a second group of enzymes called the dioxygenases. The majority of dioxygenases are non-heme iron proteins which convert aromatic compounds to aliphatic carboxylic acids capable of entering the tricarboxylic acid cycle.

In 1950 Hayaishi and Hashimoto *(12)* discovered that enzyme-catalyzed oxidative ring cleavage of the aromatic nucleus of catechol occurred to give *cis,cis*-muconic acid as the reaction product. In 1955 it was shown by $^{18}O_2$ studies that both atoms of the same oxygen molecule were incorporated into the product, indicating the direct addition of molecular oxygen to the aromatic substrate to give an aliphatic dicarboxylic acid as the product. Since these earlier studies a large number of dioxygenases have been isolated and classified on the basis of their specificity.

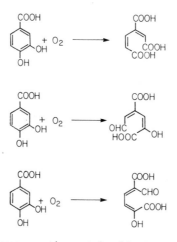

Fig. 3. Dioxygenase reactions catalyzed by 3,4 PCase, 4,5 PCase and 2,3 PCase respectively.

For example, three different enzymes have been isolated which cleave 3,4-dihydroxybenzoic acid (protocatechuic acid [PCA]) at different positions on the aromatic nucleus *(13,14,15)* (Fig. 3). The specificity of these three enzyme systems serves as an example of how natural systems have evolved which catalyze reactions hitherto unattainable in chemical model systems.

Enzymes which cleave catechol by extradiol and intradiol mechanisms (metapyrocatechase [2,3 CTase]) and (pyrocatechase [1,2 CTase]), as well as those which cleave protocatechuic acid (3,4 PCase and 4,5 PCase) have been purified to homogeneity *(16-23)*.

Other dioxygenases which cleave *ortho*-diphenols have been purified to homogeneity (3,4-dihydroxyphenylacetate 2,3-dioxygenase *(23)*, 2,3-dihydroxygenzoate dioxygenase *(24)*, and 7,8-dihydroxyflavone 8,9-dioxygenase *(25,26)*). The majority of dioxygenases purified to date cleave the benzene nucleus of *ortho*-diphenols. Relatively few dioxygenases that cleave *para*-diphenols have been purified. This is an unfortunate oversight, since this class of oxygenases appears to be ubiquitous among natural microbial populations *(27,28,29)*, and in mammalian tissues *(29)*. For example, homogentisate 1,2-dioxygenase (1,2 HGase) has been found in both mammalian tissues and in bacteria; this enzyme plays an important role in the metabolism of tyrosine *(28,30)* (Fig. 4).

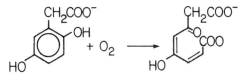

Fig. 4. *Cleavage of the para-diphenol homogentisate by 1,2 HGase.*

Protocatechuate 3,4-dioxygenase (3,4 PCase) from *Pseudomonas aeruginosa* (EC1.13.1.3) catalyzes the cleavage of protocatechuate to β-carboxy-*cis,cis*-muconate with the incorporation of two atoms of molecular oxygen (Fig. 5). In the past decade, this enzyme has been studied extensively by Hayaishi and his coworkers *(31)*.

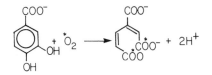

Fig. 5. *Incorporation of oxygen ($^{18}O_2$) by 3,4 PCase.*

Molecular weights of 700,000 and 660,000 have been estimated for
the pure enzyme by sedimentation equilibrium and total amino acid
composition, respectively *(31)*. Reportedly eight identical sub-
units, each containing one iron atom, comprise the holoenzyme
(32). Each subunit was shown to contain 1 atom of high spin Fe^{3+}
(33). The native enzyme is easily isolated and crystallized from
cell extracts of *Pseudomonas aeruginosa* grown with p-hydroxy-
benzoic acid as carbon source. We have been able to obtain 0.5
gr of crystalline enzyme from a 20 liter culture of this organism
routinely. The native enzyme is red in color, having a λ_{max} at
450 mμ with a broad absorption band between 400-650 mμ. When
protocatechuic acid is added to the enzyme in the absence of O_2
there is an increase in extinction of the chromophore accompanied
by a shift to the red (λ_{max} 480 mμ). By using stopped flow techniques Fujisawa *et al.* *(34)* have shown that the introduction of
oxygen gives a new spectral intermediate with a λ_{max} near 520 mμ.
This complex breaks down rapidly to give β-carboxy-*cis,cis*-muco-
nic acid and the native enzyme (450 mμ). Based on this kinetic
study, the following reaction pathway has been proposed by the
Japanese group *(34)*.

$$E + S \;\rightleftharpoons\; ES$$
$$ES + O_2 \;\rightarrow\; ES\text{-}O_2$$
$$ESO_2 \;\rightarrow\; E + P$$

E = 3,4 PCase
S = protocatechuic acid
P = β-carboxy-*cis,cis*-muconic acid

During the last six months we have re-investigated the subunit structure of 3,4 PCase and we have made a number of new
observations on subunit size and interactions.

Our studies on the native enzyme showed that it dissociates
into identical subunits (mol.wt. 76,000) in base. When these
subunits are treated with 1.0% sodium dodecylsulfate (SDS) at
pH 10.5, further dissociation occurs to give equal fractions of
two smaller units of molecular weights (15-16,500) and (18-20,500)
respectively. Sequence analysis of these nonidentical subunits
has yielded the following residues from the amino-terminal:
Peptide 1, Pro-Ile-Glu-Leu-and *Peptide 2* Pro-Ala-Gln-Asp-. At
the present time we are isolating large quantities of Peptides 1
and 2 for complete sequence analyses.

Resolution of 28 tryptic peptides together with the amino
acid composition and UV absorbance of the apo-enzyme suggest that

the molecular weight of the minimum repeating unit is about 38,000. The molecular weight of the base dissociable subunit determined by SDS gel electrophoresis and Biogel P-200 column fractionation is approximately 66,000. Iron reconstitution studies show that up to 10 active iron atoms and 10-14 inactive iron atoms can be retained by the enzyme in similar sites. These results suggest each base dissociable subunit is an $\alpha_2\beta_2$ structure with one active iron site and perhaps one inactive iron site *(35)*.

EPR spectroscopy has firmly established that the irons in the native enzyme are in a high-spin ferric state (S=5/2). 3,4 PCase exhibits a prominent feature near g = 4.3, typical of high-spin ferric iron in a "rhombic" environment. A low temperature EPR spectrum is presented in Fig. 6.

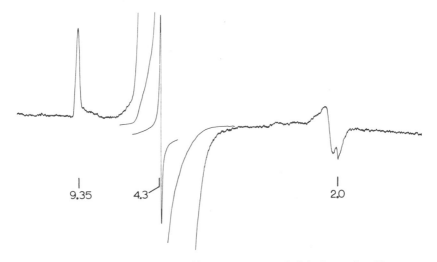

Fig. 6. EPR spectrum of polycrystalline 3,4 PCase, packed in its mother liquor. Conditions: T = 12°K; microwave frequency, 9.196 GHz; microwave power, 3 mW; modulation amplitude, 10 gauss; sweep rate, 1000 gauss/min; time constant, 0.3 sec; receiver gain, 3200 in the low field region and 320 and 32 in the g = 4.3 region, respectively. The feature in the g = 2 region is due to some adventitiously bound copper (≤ 0.1 spin/mole).

Variable temperature EPR spectroscopy on native 3,4 PCase has determined that the high-spin ferric iron exhibits a zero-field splitting parameter, D, of 1.6 cm^{-1} *(36)*. This is similar to that found for ferric rubredoxin *(36)*. 3,4 PCase and rubredoxin are unique among iron-proteins which exhibit a g = 4.3 resonance in that they exhibit values of D much larger than the other proteins. This similarity has led to the suggestion that 3,4 PCase

may have a rubredoxin-like environment, *i.e.*, a tetrahedral coordination of cysteines *(37)*. The saturation field (H$_{sat}$) obtained from the Mössbauer spectra on native 3,4 PCase suggests
otherwise (Fig. 7). Rubredoxin exhibits an H$_{sat}$ of -370 kG *(38,
39)* while 3,4 PCase exhibits an H$_{sat}$ of -525 kG *(40)*. This H$_{sat}$
further suggests that cysteine may not be coordinated to the
metal at all. Mössbauer data on iron complexes having one mer-
captide coordination show that these complexes all reflect fields
of ca. -450 kG *(41)*. Our Mössbauer data further eliminate an
octahedral oxygen coordination, since these complexes all exhibit
much smaller D values (ca. 0.5 cm^{-1})*(42)*.

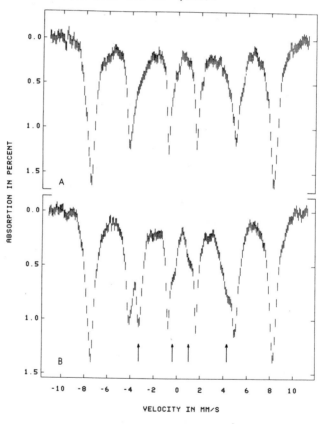

Fig. 7. *Mössbauer spectra of* ^{57}*Fe-enriched 3,4 PCase taken at
1.5°K(A) and 4.2°K(B), respectively. The spectra were taken in
a field of 600 gauss applied parallel to the Mössbauer radiation.
The absorption lines of the "5 = 4.3" Kramers doublet are indicated by arrows (for details see Ref. 40).*

Mössbauer spectra taken on dithionite reduced 3,4 PCase show that the irons are in a high-spin ferrous state. The data obtained for the reduced protein further substantiates the dissimilarity of 3,4 PCase with rubredoxin. The characteristic isomer shift for reduced rubredoxin (δ = 0.68 mm/s) reflects its more covalent, tetrasulfur environment while that of 3,4 PCase (δ = 1.21 mm/s) suggests more ionic, oxygen-nitrogen type of environment.

The Mössbauer parameters for reduced 3,4 PCase are quite similar to those of deoxyhemerythrin (43). In fact, there is good agreement in their isomer shifts (δ = 1.21 mm/s, 1.15 mm/s, respectively), suggesting that the irons may reside in similar environments. X-ray diffraction studies on myohemerythrin and hemerythrin implicate histidine and tyrosine residues for iron binding (44,45). The involvement of tyrosine in iron binding is also suggested by the optical spectrum of 3,4 PCase (46). The native enzyme exhibits an intense absorption centered near 450nm ($\varepsilon \sim 3mM^{-1}cm^{-1}$/Fe). The large extinction coefficient indicates a charge transfer transition rather than a d-d transition which would have a much smaller extinction coefficient (ca $50M^{-1}cm^{-1}$). With cysteine as an unlikely ligand, tyrosine is most probably responsible for this absorption. Similar optical properties have been observed for transferrin (λ_{max}, 480nm, ε 2.6mM$^{-1}cm^{-1}$) (47). Resonance Raman studies on transferrin have implicated tyrosine in iron coordination (48).

Optical absorption spectra for oxygenated intermediates of 3,4 PCase have been reported by Fujisawa, *et al.* (34). While the nature of such complexes has been extensively investigated for heme proteins, very little work has been published for such intermediates in dioxygenase reactions. We have characterized the ternary complex of 3,4 PCase with 3,4-dihydroxyphenylpropionate and molecular oxygen by EPR and Mössbauer spectroscopy. Our investigation reveals a high-spin ferric complex whose electronic structure gives rise to a large and negative zero-field splitting. To our knowledge, such parameters have not been observed for any other iron protein. Our data shows clearly that the electronic state of the iron in the enzyme is changed upon oxygen binding.

EPR spectra of the oxygenated complex are displayed in Fig. 8. The upper trace shows a spectrum of the oxygenated complex frozen after steady state conditions had been achieved.

The most prominent features in Fig. 8A are strong resonances at g = 6.7 and g = 5.3. Such resonances can result from a high-spin ferric ion in an environment of almost axial symmetry; they result commonly from the ground Kramers doublet of high-spin ferric heme proteins, for which D > 0. Fig. 9 shows a Mössbauer spectrum of an oxygenated sample prepared under the same conditions as the EPR sample for which the upper trace in Fig. 8 was obtained. The spectrum in Fig. 9 was taken at 4.2°K in a magnetic field of 600 gauss applied parallel to the Mössbauer radiation.

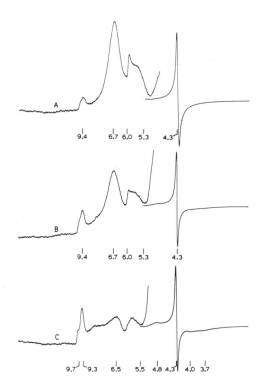

Fig. 8. *EPR spectra of the ternary complex of 3,4 PCase with 3,4-dihydroxy-phenylpropionate and O_2. Conditions: $T = 12.3°K$; 9.196 GHz, 1 mW; modulation amplitude, 10 gauss; sweep rate, 1000 gauss per min.; receiver gain, 3200 in the low-field region, 160 in the $g = 4.3$ region in (A) and 63 in (B) and (C). The magnetic field increases linearly to the right and selected values of the frequency-to-field ratio (g-value) are given on the abscissa. A. The sample was frozen after steady-state conditions had been achieved. B. The sample was frozen after the ternary complex had decayed for one half-life (4 min.) as monitored by optical spectroscopy. C. The sample was frozen after the ternary complex had decayed for four half-lives (for details see Ref. (40)). The decay of the $g = 6.7$ resonance parallels product formation and the decay of the oxy-genated complex as observed with optical spectroscopy.*

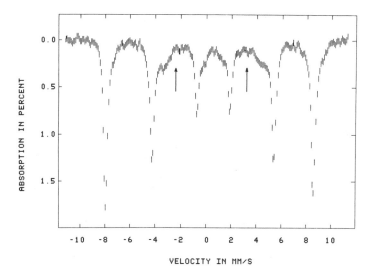

Fig. 9. Mössbauer spectrum of the ternary complex of 3,4 PCase with 3,4-di-hydroxyphenylpropionate and O_2 taken at $4.2°K$ in a 600 gauss parallel field. The prominent six-line pattern results from the ground Kramers doublet of the ternary complex.

The intensities of the six prominent lines did not change when the field was applied perpendicular to the Mössbauer radiation. This observation implies that this spectrum results from a Kramers doublet for which $g_x \simeq g_y \ll g_z$; such doublets yield either no EPR signal (if $g_x = g_y = 0$) or only extremely weak signals. On the other hand, a Kramers doublet with g-values at 6.7 and 5.3 has to yield a Mössbauer spectrum with intensities depending quite sensitively on the direction of the applied field. Thus the EPR and Mössbauer spectra result from two different Kramers doublets. We have shown that these observations can be reconciled by assuming a high-spin ferric system with almost axial symmetry and with a negative zero-field splitting *(40)*. By measuring the temperature dependence of the g = 6.7 resonance (which results from the upper Kramers doublet of the S = 5/2 system) we have determined the zero-field splitting parameter D $\cong$ $-2cm^{-1}$.

To date we do not fully understand the nature of enzyme substrate complexes. These are certainly in a high-spin ferric state characterized by *positive* zero-field splittings. What is puzzling is that we observe at least two species no matter what substrate is used. These species appear in the same ratio at low and high substrate concentrations. However rudimentary our under-standing of the enzyme-substrate complexes addition of oxygen causes dramatic changes of the electronic configuration of the iron. This, however, does not necessarily imply that the oxygen molecule is directly coordinated to the iron. The details of oxygen binding clearly need further elucidation.

The goal of this study is the elucidation of the catalytic mechanism. Clearly any mechanism proposed has to be consistent with the magnetic resonance data on the iron center. In addition, we have to determine the mode of substrate binding. Enzyme inhi-bition studies have been useful in determining the relative im-portance of various functional groups on substrate binding. Protocatechuic acid could coordinate to the iron through the carboxylate group, or one hydroxy group, or chelate to the iron with both hydroxy groups, or it may not coordinate at all. K_I determinations have been made for a large number of competitive inhibitors. These data are presented in Fig. 10.

Substrate

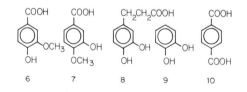

Fig. 10. *Kinetic studies with 3,4 PCase.*
K_m *for substrate* $= 3.0 \times 10^{-5} M$. *The following* K_I *values were determined for competitive substrate analogs:* **1** $= 1.4 \times 10^{-5} M$, **2** $= 2.5 \times 10^{-4} M$, **3** $= 4.5 \times 10^{-3} M$, **4** $= 1 \times 10^{-6} M$, **5** $= 8.0 \times 10^{-4} M$, **6** $= 3 \times 10^{-3} M$, **7** $= 2 \times 10^{-2} M$, **8** $= 1.8 \times 10^{-5} M$, **9** $= 7.0 \times 10^{-4} M$, **10** $=$ *not determined precisely.*

It is clear that the carboxylate group is important for binding (see **8** and **9**). Using protocatechuic aldehyde (**1**) we have preliminary evidence that the carboxylate binding site is an ε-amino group of lysine, because the aldehyde forms a Schiff's base which is reduced to an enamine derivative with sodium borohydride. The carboxylate group does not coordinate to the iron because similar spectral changes to those seen with the substrate occur when catechol (**9**) binds to the active site.

The inhibition data indicate that only the *para* hydroxy group is involved in iron binding, since there are large differences in inhibition between the compounds having p-OH and m-OH groups (compare **2** and **3**, **4** and **5**, **6** and **7**). Significant spectral changes occur when the *para* isomer is added to the enzyme, while only small perturbations of the original optical and EPR spectra are observed when the *meta* isomer is added.

On the basis of this kinetic study, together with our studies on the iron center, we can now propose a mechanism for 3,4 PCase (Fig. 11).

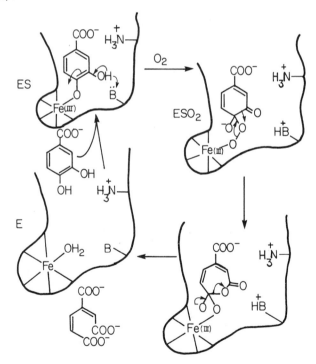

Fig. 11. Proposed reaction mechanism for 3,4 PCase.

In the proposed mechanism, the iron serves to bind substrate and promote the polarization of electrons toward C-4. The iron also catalyzes what, in the absence of the metal ion, would be a spin-forbidden reaction, the reaction of oxygen (a spin-triplet) to the substrate (a spin-singlet) through an ionic mechanism. The iron possibly binds the peroxide which is formed, though this remains to be demonstrated. This oxygenated intermediate then rearranges to give product.

Model studies on the oxidation of catechols tend to support this mechanism. Organic chemists have been able to isolate products from O_2 oxidation of catechols in alkaline media derived from the expected *cis,cis*-muconic acid *(49,50)*. In particular, Jack Baldwin's work in this area may shed light on the enzymatic mechanism. The catechol (**11**) was synthesized (Fig. 12). At -50°C (**11**) was shown to isomerize to (**12**) in the absence of O_2. Upon exposure to O_2, the corresponding *cis,cis*-muconic acid was formed. This experiment emphasizes that O_2 can add to an organic substrate and that a peroxy intermediate may indeed be formed. If this mechanism is correct then the role of the iron in 3,4 PCase is to catalyze the isomerization presented in Fig. 12. Therefore the catechol is activated so that it can react with molecular oxygen to give a peroxy intermediate. We believe that this enzyme provides the first evidence for substrate "activation", as a prerequisite to oxygen "activation."

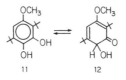

Fig. 12. *Isomerization of substituted catechol* (**11**) *to a substituted keto-alcohol* (**12**).

3. REACTIONS WITH TOXIC METALS

The biosynthesis of metal-alkyls is now well recognized as a microbiological response to the toxicity of inorganic heavy metals. Metal-alkyls effectively dilute local cellular concentrations of heavy metals, because small organometallic complexes such as methylmercury and dimethylmercury are dissipated by diffusion controlled processes.

Three methylating coenzymes are available for alkyltransfer reactions in biological systems:

(1) S-adenosylmethionine
(2) N^5-methyltetrahydrofolate derivatives
(3) Methylcorrinoid derivatives.

S-adenosylmethionine and N^5-methyltetrahydrofolate derivatives transfer their methyl groups to nucleophiles and this rules out methyl transfer to metals and to salts of metalloids in an aerobic environment. However, in 1964 Halpern and Maher [51] demonstrated that mercuric ion reacts with methylpentacyanocobaltate to give methylmercury as one of the products:

$$CH_3-Co^{III}(CN)_5 + Hg^{2+} + H_2O \rightarrow H_2O-Co^{III}(CN)_5 + CH_3Hg^+$$

From these pioneering studies it was apparent that electrophilic attack by metal ions on the Co-C bond occurs to yield products of extreme toxicity.

In 1968, Wood *et al.* showed that methylcobalamin would function very efficiently in this reaction to yield a mixture of methylmercury and dimethylmercury [52]. At the same time Jensen and Jernelöv (1968) [53] showed that aquarium sediments were capable of converting inorganic mercury salts to methylmercury and control experiments proved that this reaction was catalyzed by microorganisms. Since these independent discoveries, Hill *et al.* [54], Schrauzer *et al.* [55], Bertilsson and Neujahr [56], Imura *et al.* [57], and Adin and Espenson [58], have all studied the reaction between mercuric ion and the Co-C bond and each laboratory has confirmed Halpern and Maher's original observation.

Recent studies have elucidated the details of the mechanisms for the synthesis of both methylmercury and dimethylmercury from methyl-B_{12} compounds (Wood *et al.* [59]; DeSimone *et al.* [60]). The first reaction involves the displacement of benzimidazole from the coordination sphere of the cobalt atom to set up an equilibrium mixture of "base on" and "base off" methylcobalamin:

$$H_2O + Hg^{2+} + \quad \begin{array}{c} CH_3 \\ | \\ Co^{III} \\ \uparrow \\ Bz \end{array} \quad \overset{K}{\rightleftharpoons} \quad \begin{array}{c} CH_3 \\ | \\ Co^{III} \\ \uparrow \\ O \\ H \quad H \\ BzHg^+ \end{array}$$

"Base On"   "Base Off"

This reaction is first order in Hg^{2+} and in CH_3B_{12}. The equilibrium constant, K, is 70 ± 5 in favour of the "base off" species.

The second reaction involves electrophilic attack by Hg^{2+} on both "base on" and "base off" species to give methylmercury as the common reaction product:

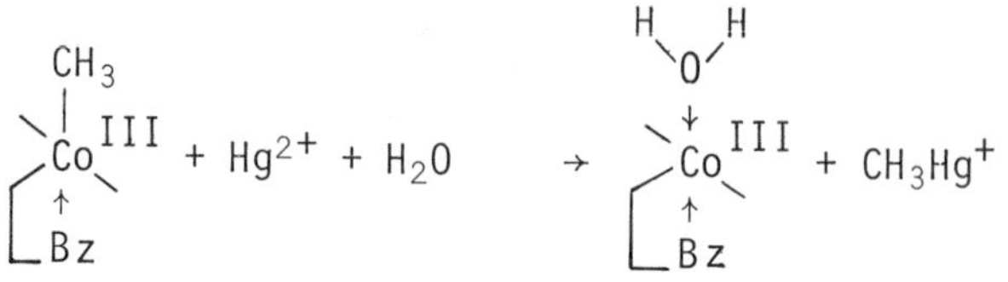

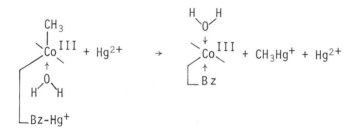

"Base On"

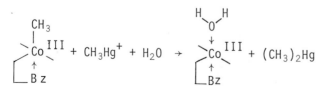

"Base Off"

Both reactions are first order in Hg^{2+}, but the reaction with the "base on" species is 1000 times faster than the reaction with the "base off" species. This is, presumably, because the co-ordination of benzimidazole to the cobalt atom increases the electron density on the cobalt atom, which facilitates the displacement of the methyl group as CH_3^-.

In addition to the sequence of reactions just described, methylmercuric ion reacts with methylcobalamin to give dimethyl-mercury as product, but this reaction is 6000 times slower than the reaction with mercuric ion:

After this initial discovery it was important to determine which other metals would react with methylcobalamin by a similar mechanism. Scoville *(61)* was able to show that palladium reacts by an identical mechanism to that reported for mercury, and we have confirmed the original observation made by the Oxford Group *(62)* that thallium follows the same pattern. It was then left to determine whether reactions would occur with tin, lead and platinum. Lead and tin are particularly important in the environmental context because large quantities of these metals are mined and used by advanced industrial societies. Diethyl lead dichloride was found to react *very slowly* with methyl-B_{12} (Fig. 13). The reaction rate could be enhanced by adding a chloride ion complexing agent such as silver acetate.

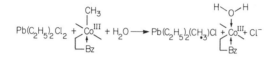

Fig. 13. Electrophilic attack of diethyl lead dichloride on the Co-C bond of methyl-B_{12}.

Recently we have shown that stannic salts do not react with methylcobalamin directly, but stannous salts react slowly to give a methyl-stannic complex as the product. At first glance the reaction appeared to progress by oxidative-addition (Fig. 14). Schrauzer and Kratel *(63)* have prepared alkyl-tin-cobalt derivatives for cobaloximes. Patmore and Graham *(64)* have demonstrated tin insertion into the cobalt-carbon bond for alkyl-cobalt tetracarbonyl complexes, and so insertion into the Co-C bond of methyl-B_{12} seemed reasonable. However, a detailed study of the reaction between stannous chloride and methylcobalamin shows that the role of stannous ion is to facilitate electrophilic attack by contaminating stannic ion. We can now report experiments which allow us to formulate a mechanism for the biomethylation of tin. Fig. 15 presents spectroscopic evidence that there is no change in the oxidation state of the cobalt atom during the demethylation reaction since "base off" methylcobalamin is converted to "base on" aquocobalamin.

Secondly, we have shown that the reaction rate is dependent on the concentration of Cl^- indicating that :$SnCl_3^-$ may have a labilizing effect by coordinating trans to the Co-C bond (Fig. 16). The initial reaction probably involves the reaction of chloride ion with stannous chloride to give the weak nucleophile $SnCl_3^-$, which coordinates to the cobalt atom to facilitate electrophilic attack by stannic chloride to give CH_3SnCl_3 as the product. Chloride ion is critical to this reaction because it establishes a reasonable concentration of $SnCl_3^-$.

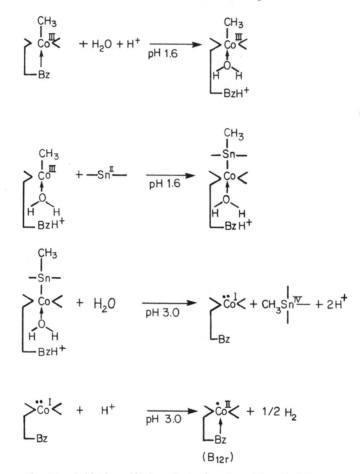

Fig. 14.   Oxidative-addition of stannous ion with methyl-B_{12}.

From this survey it appears that heterolytic cleavage of the Co-C bond is a general reaction for both strong and weak electrophiles. However, weak electrophiles may need assistance by the coordination of good electron donors transaxial to the methyl group.

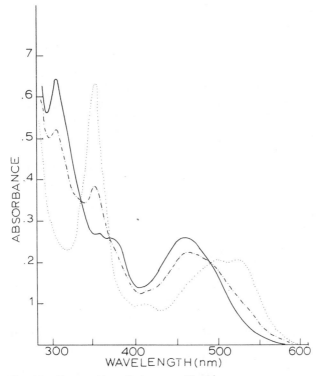

Fig. 15. *The reaction of stannous chloride.*
$4 \times 10^{-3}M$ *with methylcobalamin* 4×10^{-5} *at pH 1.2. Methyl-B$_{12}$
(base off)* (————), *Methyl-B$_{12}$ + stannous chloride (intermediate reaction time)* (—·——·—), *and aquocobalamin formed after methyl-transfer to stannic ion* (. . . . . .).

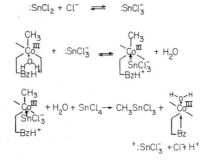

Fig. 16. *Proposed reaction mechanism for methyl transfer from methyl-B$_{12}$ to stannic chloride in the presence of stannous chloride plus chloride ions.*

279

4.   REACTIONS WITH METALLOIDS

Schrauzer has postulated that the coordination of thiols to methylcobalamin facilitates heterolytic cleavage of the Co-C bond by the following electrophiles: (1) H^+ to give methane, (2) carbon dioxide to give acetic acid, and (3) selenium salts to give alkyl selenium compounds *(65,66)*. Schrauzer predicts that the coordination of the thiol increases the electron density on the methyl group to generate a "krypto" carbanion which is readily displaced by a number of electrophiles (Fig. 17).

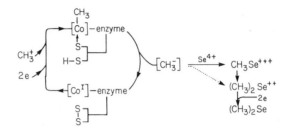

Fig. 17. Schrauzer's proposed mechanism for thiol-promoted electrophilic attack on the Co-C bond (65)(66).

Using this scheme Schrauzer suggests mechanisms for the cobalamin-dependent biosynthesis of methane, acetic acid, dimethylarsine and dimethylselenide. Although this mechanism is very appealing, it is important to determine whether thiols coordinate to methylcobalamin. Several laboratories have reported on the interaction of thiols to both cobalamins and cobaloximes *(67,68)*. In fact, we reported that reduced glutathione is capable of displacing benzimidazole and coordinating to the cobalt atom for the important B_{12}-coenzymes, methylcobalamin and 5'-deoxyadenosylcobalamin *(69)*. Our conclusions were based on ultraviolet-visible titrations and the pH dependence for the formation of reduced glutathione complex. We have re-examined this reaction using ^{13}C and ^1H NMR and we can find no evidence for the formation of a stable coordination complex between these corrinoids under conditions where we have a twenty-five fold excess of glutathione over cobalamin (Table 1). Furthermore, a low molecular weight thiol called Coenzyme M has been isolated recently and implicated as an intermediate methyl-acceptor from methylcobalamin in methane biosynthesis (Fig. 18). This rules out electrophilic attack on the Co-C bond by H^+ to give methane.

We have studied the non-enzymatic reaction between methylcobalamin and Coenzyme M to determine whether this non-enzymatic reaction proceeds by a similar route to that reported for the

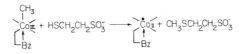

Fig. 18. Methyl-transfer from methyl-B_{12} to coenzyme M, after McBride and Wolfe (72). (N.B. One electron is unaccounted for).

enzymatic reaction. McBride and Wolfe *(70)* and Taylor and Wolfe *(71,72)* report that cob(II)alamin (B_{12r}) is the product of methyl-transfer to Coenzyme M. We find that the demethylation of methyl-cobalamin by Coenzyme M is best explained in terms of thiol pro-moted homolytic cleavage to give cob(II)alamin and the thioether as products. The reaction rate is not affected by pH in the range 7.0 to 14.2 (Fig. 19), and we can find no evidence for heterolytic cleavage of the Co-C bond even at pH 14.2.

TABLE 1: COUPLING CONSTANTS, $J^{13}C-^{1}H$ OF [^{13}C]METHYLCOBALAMIN SAMPLES					
SAMPLE	SOLVENT	$J^{13}C-^{2}H$ (Hz)	A (Hz)	B (Hz)	A-B (Hz)
[^{13}C]Methylcobalamin	0.1 M $K^{2}H_{2}PO_{4}$				
	$p^{2}H$ 7.0	138	69	69	0
	$p^{2}H$ 4.0	138	68	70	2
	$p^{2}H$ 1.5	146	73	73	0
[^{13}C]Methylcobalamin-glutathione	0.1 M $K^{2}H_{2}PO_{4}$				
	$p^{2}H$ 4.0	138	69	69	0

SAMPLE	^{13}C NMR		CHEMICAL SHIFT* (ppm)
[^{13}C]Methylcobalamin	(8.6 mM)	$p^{2}H$ 5.6	+7.55
	(30 mM)	$p^{2}H$ 2.8	+4.65
	(34 mM)	$p^{2}H$ 1.1	+0.02
[^{13}C]Methylcobalamin ($\approx$ 34 mM)			
- glutathione	(0.4 M)	$p^{2}H$ 3.6	+6.16
- glutathione	(0.8 M)	$p^{2}H$ 3.6	+6.27
- glutathione	(0.8 M)	$p^{2}H$ 2.8	+5.03
[^{13}C]Methylcobinamide	(30 mM)	$p^{2}H$ 7.0	+0.36
	(15 mM)	$p^{2}H$ 7.0, T = 20°C	- 0.14
[^{13}C]Methylcobinamide	(30 mM)		
- glutathione	(0.5 M)	$p^{2}H$ 7.0	+0.30

Chemical shift given in ppm relative to TMS at 0.00 ppm. Positive sign indicates shift to low field. All studies taken at 40°C unless otherwise noted.

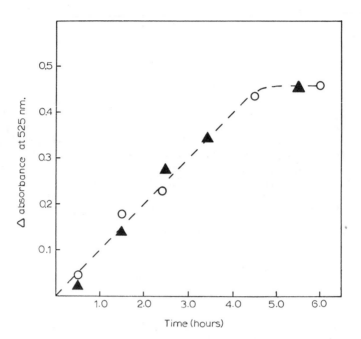

Fig. 19. pH independence of demethylation of methyl-B_{12} (1 × 10^{-4}M) by Coenzyme M (6 × 10^{-2}M) at pH 7.0 (0-0) and at pH 14.2 (▲-▲).

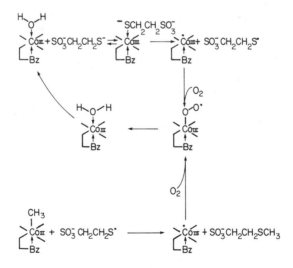

Fig. 20. Proposed mechanism for the alkylation of Coenzyme M.

These data suggest a mechanism where aquocobalamin is involved in a catalytic role to generate a sufficient concentration
of thiol radical to promote homolytic cleavage of the cobalt-
carbon bond (Fig. 20). A kinetic study of this reaction shows a
lag period during which a catalytic cycle is established before
the Co-C bond can be cleaved at any appreciable rate (Fig. 21).

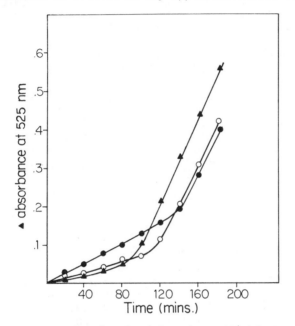

Fig. 21. Demethylation of methyl-B_{12} (3.4 × 10^{-4}M) by Coenzyme
M at 1.8 × 10^{-2}M (▲—▲), 3.6 × 10^{-2}M (●—●) and 5.4 × 10^{-2}M (○—○).
Reactions were performed in 0.1M KH_2PO_4 buffer at pH 7.2 and $40°C$.

Also the mechanism proposed in Fig. 20 is consistent with the
observation that the lag phase is inversely proportional to the
thiol concentration. This is expected due to thiol radical
coupling at high thiol concentrations. The rate enhancement
which we observe when the reaction is carried out under aerobic
conditions is to be expected because oxygen is required to oxi-
dize cob(II)alamin ($B_{12}r$) to aquocobalamin (Co(III)). In this
non-enzymatic system aquocobalamin is the electron acceptor from
Coenzyme M.

For the strictly anaerobic methane enzyme system obviously
oxygen is not involved, but a number of alternative electron

acceptors are present in the methane system. Therefore, radical attack on methylcobalamin to yield cob(II)alamin (B_{12}) and methyl-Coenzyme M is not an unreasonable enzyme mechanism, and such a reaction yields the same products which are found in the enzyme catalyzed reaction.

The most intriguing aspect of this study is the strong evidence for thiol promoted homolytic cleavage of the Co-C bond. It appears that cobalamin-dependent methyl-transfer reactions may proceed by a free radical mechanism not too dissimilar from that currently being proposed for the 5'-deoxyadenosylcobalamin catalyzed rearrangements *(73)*. Finally, it should be acknowledged that as early as 1935 Frederick Challenger demonstrated the synthesis of trimethylarsine by the bread mold *Scopulariopsis*. Using ^{14}C labelled methionine he demonstrated classical carbonium ion transfer from "activated" methionine (S-adenosylmethionine) *(74)*. Similar results have been obtained for the biomethylation of selenium *(75)*. Recently, we have implicated S-adenosylmethionine in methyl-transfer to mercury by studying methyl-mercury synthesis in a number of B_{12}-independent mutants *(76)*.

In this review we examine how biochemical systems evolved to cope with the introduction of toxic elements in the biosphere. It appears that living systems adapted to deal with molecular oxygen early in geological time. Closer examination of the biochemistry of molecular oxygen reveals that its intervention as terminal electron acceptor in respiration represents but one important function for this molecule. According to our Biochemistry textbooks, the electron transport chain is the only point in metabolism where oxygen is involved; there is no mention of the important role played by oxygen in both the biosynthesis and biodegradation of the majority of living matter on this planet. Couple this serious omission with the current treatment of metabolism, which is centered around mammalian liver and the fecal organism *Escherichia coli*, and we fail to teach the chemistry of biological systems in any meaningful fashion. Without a thorough understanding of important metabolic events in the biosphere we will always have difficulty in recognizing the extent of man-made perturbations. The most difficult problem to assess is the extent to which man's activities have perturbed natural cycles for toxic elements. The movement of toxic metals in advanced industrial societies not only poses a threat to those societies, but may upset the delicate balance which Mother Nature has established for all the elements on this planet.

Methyl-B_{12} appears to hold a central position in the biosynthesis of metal-alkyls in the environment. It has been shown to be involved in the methylation of mercury, tin, lead, palladium, gold, thallium, platinum, selenium, tellurium, sulfur, and arsenic. The majority of methylated products are much more toxic than the parent elements. Furthermore, some of the products bioaccumulate, and others are volatile so that they enter the atmosphere. We are just beginning to learn the great significance of the metabolic cycles for toxic elements and how they affect biogeochemical cycles.

ACKNOWLEDGEMENTS. We would like to express our gratitude to the Freshwater Biological Research Foundation for making the Institute and the research facilities available. In particular, we wish to acknowledge the efforts of Richard Gray, Sr., a private citizen, without whose efforts the Freshwater Biological Institute would not exist. We thank Mr. W. Hamilton for capable technical assistance with the EPR spectroscopy, Mr. R.L. Thrift for his valuable help in getting the computer facility and the Mössbauer spectrometer running and Mr. Francis Engle for assistance with biological preparations. This work was supported by U.S. Public Health Service Grants GM22701, GM17170 and AM18101, by a Research Career Development Award K04-GM70683 (E.M.), by National Science Foundation Grants BMS 14980 and PCM-17318, by a contract from the FDA, a grant from the International Lead Zinc Research Organization, and by the Graduate Research Committee and the College of Agricultural and Life Sciences of the University of Wisconsin. We are especially indebted to Dr. P.J. Chapman for providing us with the fluorocatechols used in this research project, and to Dr. J. Howard for his sequence analyses.

REFERENCES

1. S. Dagley, "A Biochemical Approach to Some Problems of Environmental Pollution", Essays in Biochemistry, Vol. 1976, pp. 81-137. Biochemical Society, P.N. Campbell, Ed.
2. J.M. McCord, B.B. Kelle and I. Fridovich, *Proc. Nat. Acad. Sci. U.S.A.*, **68**, 1024 (1971).
3. S. Dagley, *Am. Sci.*, **63**, 681 (1975).
4. I.C. Gunsalus, J.R. Meeks, J.D. Lipscomb, P. Debrunner and E. Münck, "Molecular Mechanisms of Oxygen Activation", O. Hayaishi, Ed., Academic Press, **14**, pp. 559-613 (1974).
5. See C.K. Chang and D. Dolphin in "A Survey of Contemporary Bioorganic Chemistry", E.E. Van Tamelen, Ed., Academic Press, N.Y., in press.
6. S. Strickland and V. Massey, *J. Biol. Chem.*, **248**, 2944 (1973).
7. S. Takemori, H. Yasau, K. Mihara, K. Suzuki and M. Katagiri, *Biochim. Biophys. Acta*, **191**, 58 (1969).

8.   R.H. White-Stevens and H. Kamin, *J. Biol. Chem.*,**247**, 2358 (1972).
9.   Y. Ohta and D.W. Ribbons, *Febs. Letters*,**11**, 189 (1970).
10. B. Hesp, M. Calvin and K. Hosokawa, *J. Biol. Chem.*,**244**, 5644 (1969).
11. L.G. Howell, T. Spector and V. Massey, *J. Biol. Chem.*,**247**, 4340 (1972).
12. O. Hayaishi and Z. Hashimoto, *J. Biochem.*, *Tokyo*,**37**, 371 (1950).
13. H. Fujisawa and O. Hayaishi, *J. Biol. Chem.*,**243**, 2673-2681 (1968).
14. R.M. Zabinski, E. Münck, P.M. Champion and J.M. Wood, *Biochemistry*,**11**, 3212-3221 (1972).
15. R.L. Crawford, *J. Bacteriol.*,**121**, 531-536 (1975).
16. M. Nozaki, S. Kotoni, K. Ono and S. Senoh, *Biochim. Biophys. Acta*, **220**, 213-223 (1970).
17. M. Nozaki, H. Kagamiyama and O. Hayaishi, *Biochem. Biophys. Res. Commun.*,**11**, 65-69 (1963).
18. M. Nozaki, K. Ono, T. Nakazawa, S. Kotoni and O. Hayaishi, *J. Biol. Chem.*,**243**, 2682-2690 (1968).
19. S. Takemoni, T. Komiyama and M. Katagiri, *Europ. J. Biochem.*, **23**, 178-184 (1971).
20. S. Dagley, P.J. Geary and J.M. Wood, *Biochem. J.*,**109**, 559-568 (1968).
21. Y, Kojima, H. Fujisawa, A. Nakazawa, T. Nakazawa, F. Konetsona, H. Tanuchi, M. Nozaki and O. Hayaishi, *J. Biol. Chem.*, **242**, 3270-3278 (1967).
22. T. Nakazawa, M. Nozaki, O. Hayaishi and T. Yamano, *J. Biol. Chem.*,**243**, 119-125 (1969).
23. H. Fujisawa and O. Hayaishi, *J. Biol. Chem.*,**243**, 2673-2681 (1968).
24. H. Kita, *J. Biochem.*, *Tokyo*,**58**, 116-122 (1965).
25. E. Schultz, F.E. Engle and J.M. Wood, *Biochemistry*,**13**, 1768-1776 (1974).
26. E. Schultz and J.M. Wood, *Biochim. Biophys. Acta* (in press) (1976).
27. R.L. Crawford, *Appl. Microbiol.*,**30**, 439-447 (1975).
28. D.J. Hopper and P.J. Chapman, *Biochem. J.*,**122**, 19 (1970).
29. K. Adachi, Y. Iwayama, H. Tanioka and Y. Takeda, *Biochim. Biophys. Acta*,**118**, 88 (1966).
30. W.G. Flamm and D.I. Crandall, *J. Biol. Chem.*,**238**, 389 (1963).
31. K. Bloch and O. Hayaishi, "Biological and Chemical Aspects of Oxygenases", Proceedings of a U.S.A.-Japan Symposium on Oxygenases, Maruzen Company Ltd., Tokyo, Japan (1966).
32. M. Nozaki, "Non-heme Iron Dioxygenase in Molecular Mechanisms of Oxygen Activation", O. Hayaishi, Ed., Academic Press, Chap. 4, 142-144 (1974).

33. H. Fujisawa, M. Uyeda, Y. Koiima, M. Nozaki and O. Hayaishi, *J. Biol. Chem.*,**247**, 4414-4421 (1972).
34. H. Fujisawa, K. Hiromi, M. Uyeda, M. Nozaki and O. Hayaishi, *J. Biol. Chem.*,**247**, 4422-4428 (1972).
35. J. Lipscomb, J. Howard, T. Lorsbach and J.M. Wood, "Federation Proceedings in Oxygenase Session" (1976).
36. W.R. Blumberg and J. Peisach, *Ann. N.Y. Acad. Sci.*,**222**, 539 (1973).
37. L.H. Jensen, in "Iron-Sulfur Proteins", W.E. Lovenberg, Ed., Vol. 2, Academic Press, New York, Chap. 4 (1973).
38. K.K. Rao, M.C.W. Evans, R. Cammack, D.O. Hall, C.L. Thompson, P.J. Jackson and C.E. Johnson, *Biochem. J.*,**129**, 1063 (1972).
39. P.G. Debrunner, E. Münck and L. Que in "Iron-Sulfur Proteins", W.E. Lovenberg, Ed., Vol. 3, Academic Press, N.Y., in press.
40. L. Que, Jr., J.D. Lipscomb, R. Zimmermann, E. Münck, N.R. Orme-Johnson, and W.H. Orme-Johnson. Submitted for publication.
41. S. Koch, R.H. Holm and R.B. Frankel, *J. Amer. Chem. Soc.*,**97**, 6714 (1975) 40.
42. W.T. Oosterhuis, *Structure and Bonding*,**20**, 59 (1974).
43. M.Y. Okamura and I.M. Klotz, in "Inorganic Biochemistry", G.L. Eichhorn, Ed., Elsevier, Amsterdam, Chap. 11 (1973).
44. W.A. Hendrickson, G.L. Klippenstein and K.B. Ward, *Proc. Nat. Acad. Sci. U.S.A.*,**72**, 2160 (1975).
45. R.E. Stenkamp, L.C. Sieker and L.H. Jensen, *Proc. Nat. Acad. Sci. U.S.A.*,**73**, 349 (1976).
46. H. Fujisawa and O. Hayaishi, *J. Biol. Chem.*, **243**, 2673 (1968).
47. P. Aisen, in "Inorganic Biochemistry", G.L. Eichhorn, Ed., Elsevier, Amsterdam, Chap. 9 (1973).
48. B.P. Gaber, V. Miskowski and T.G. Spiro, *J. Amer. Chem. Soc.*, **96**, 6868 (1974).
49. R.R. Grinstead, *Biochemistry*,**3**, 1308 (1964).
50. J. Baldwin, private communication.
51. J. Halpern and A. Maher, *J. Amer. Chem. Soc.*,**90**, 2311-2312 (1964).
52. J.M. Wood, F.S. Kennedy and C.G. Rosen, *Nature*,**220**, 173 (1968).
53. S. Jensen and A. Jernelöv, *Nordforsk*,**14**, 3-6 (1968).
54. H.A.O. Hill, J.M. Pratt, S. Ridsdale, F. Williams and R.J.P. Williams, *Chem. Commun.*,**6**, 341 (1970).
55. G.N. Schrauzer, J. Weber, T. Beckham and R. Ho, *Tetrahedron Letters*,**3**, 275-278 (1971).
56. L. Bertilsson and H.J. Neujahr, *Biochemistry*,**10**, 2805-2810 (1971).
57. N. Imura, S. Sakegawa, E. Pan, K. Nagao, J. Kim, T. Kwan and T. Ukita, *Science*,**172**, 1248-1251 (1971).

58. A. Adin and W. Espenson, *Chem. Commun.*,**13**, 653-654 (1971).
59. J.M. Wood, *Naturwissenschaften*,**62**, 357-364 (1975).
60. R.E. DeSimone, M.W. Penley, L. Charbonneau, S.G. Smith, J.M. Wood, H.A.O. Hill, J.M. Pratt, S. Ridsdale and R.J.P. Williams, *Biochim. Biophys. Acta*,**304**, 851-863 (1973).
61. W. Scoville, *J. Amer. Chem. Soc.*,**96**(11), 3451-3456 (1974).
62. G. Agnes, H.A.O. Hill, J.M. Pratt, S.C. Ridsdale, F.S. Kennedy and R.J.P. Williams, *Biochim. Biophys. Acta*,**252**, 207 (1971).
63. G.N. Schrauzer and G. Kratel, *Chem. Ber.*,**102**, 2392-2407 (1969).
64. D.J. Patmore and W.A.G. Graham, *Inorg. Chem.*,**7**, 771 (1968).
65. G.N. Schrauzer, J. Seck, R. Holland, T. Beckham, E. Rubin and J. Sibert, *Bio-inorg. Chem.*,**2**, 93-115 (1973).
66. G.N. Schrauzer, *Prog. Chem. Org. Nat. Prod.*,**31**, 563-587 (1974).
67. G.N. Schrauzer, *Acc. Chem. Res.*,**1**, 97-105 (1968).
68. T, Frick, M.D. Francia and J.M. Wood, *Biochim. Biophys. Acta* **428**, 808-818 (1976).
69. P.Y. Law and J.M. Wood, *J. Am. Chem. Soc.*,**95**, 914-921 (1973).
70. B.C. McBride and R.S. Wolfe, *Biochemistry*,**10**, 2317-2324.
71. C.D. Taylor and R.S. Wolfe, *J. Biol. Chem.*,**249**, 4879-4885 (1974).
72. C.D. Taylor and R.S. Wolfe, *J. Biol. Chem.*,**249**, 4886-4893 (1974).
73. R.H. Abeles (these proceedings).
74. F. Challenger, *Chem. Reviews*,**36**, 315 (1945).
75. H.E. Ganther and H.S. Hsieh, "Trace Metals in Animals 2", Univ. Park Press, Baltimore, Maryland, pp. 339-353 (1974).
76. A. Cheh and J.M. Wood, unpublished results.

electron transfer mechanisms employed by metalloproteins

SCOT WHERLAND AND HARRY B. GRAY

Arthur Amos Noyes Laboratory of Chemical Physics,
California Institute of Technology, Pasadena, California 91125

(cont'd)

1. INTRODUCTION

The electron transfer reactivities of metalloproteins are of current interest for several reasons. The biological function of the simple electron carriers, long recognized for the cytochromes *(1)* and more recently elucidated for some of the blue copper proteins *(2)*, is being further investigated in terms of the mechanism of the specificity between physiological redox partners; in addition, the more complex reaction-pathway problems of the multi-electron accepting, multi-substrate oxidases are beginning to be explored. The simple one electron carriers have been singled out for study as they are not expected to have as much of a substrate-accepting role as the oxidases or enzymes catalyzing covalent changes; furthermore, there are often only two forms of the protein which must be considered (oxidized and reduced). Although structural differences between these two forms may be expected, these may be studied conveniently by standard physical and chemical techniques. The workers in the field have come from different disciplines and have provided different approaches to the problems, but many recent advances have come from the laboratories of bioinorganic chemists *(3)* and others who have studied the reactions of electron transfer proteins with small inorganic redox reagents as well as with other redox-active biomolecules.

Three general classes of metalloproteins, the c type cytochromes, the iron-sulfur proteins, and the blue, type 1 copper only, proteins, have been frequently and extensively studied. A brief summary of the available three dimensional and electronic structural information is appropriate. X-ray studies by Dickerson and coworkers *(1)* have established the essential features of the structures of both oxidized and reduced forms of horse heart cytochrome c. Further, some structural information on ferricytochrome c_{551} from *Pseudomonas aeruginosa* is also available *(1)*. In the structure of horse heart ferricytochrome c, a heme c group

(Fig. 1) is bound to the polypeptide chain through covalent sulfur linkages at Cys-17 and Cys-14, as well as by the axial iron
ligands His-18 imidazole and Met-80 sulfur.

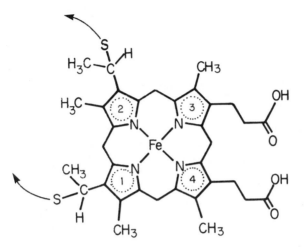

Fig. 1. *Structural formula of heme c, showing the attachment to the protein
from rings 1 and 2.*

A schematic representation of the structure in the vicinity of
the heme group is shown in Fig. 2. The heme group is buried in
the hydrophobic interior of the protein, except for one edge
(shown in boldface in Fig. 2; rings 2 and 3 in Fig. 1) which is
near the surface. In the present discussion we shall refer to
the 2,3-ring system as the "exposed heme edge". However, we
hasten to note that the exact extent of exposure of this edge to
solvent molecules and redox agents is a matter for speculation.
Examination of the models of the ferricytochrome c structure
suggests that the 2,3-edge is at least 1 Å below the protein surface and access to the region may be difficult for certain molecules. Both oxidized and reduced cytochromes c possess low-spin
electronic ground states, corresponding to the metal orbital
configurations $(t_{2g})^5$ and $(t_{2g})^6$, respectively. Outer sphere
electron transfer of a t_{2g} electron is known to be facile, as
minimal inner sphere reorganizational activation is required
(vide infra). Thus, both steric and electronic considerations
favor outer sphere electron transfer to and from the exposed heme
edge as the mechanism of choice for cytochromes c. However, adjacent inner or outer sphere attack is not precluded, but must
necessarily be accompanied by some type of conformational change
in which the heme crevice is exposed to small molecules.

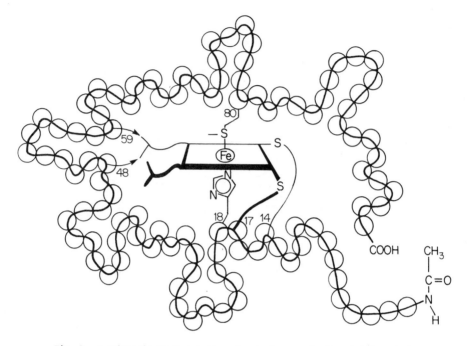

Fig. 2. A schematic representation of cytochrome c in the vicinity of the heme (see Ref. 1 for structural details).

X-ray studies of the high potential-iron sulfur protein from *Chromatium vinosum* (HiPIP) show that the cube-like $Fe_4S_4S_4^*$ cluster is completely buried (4,5) with the closest distance from a protein surface to the edge of the cluster being about 3.5 Å (6) (Fig. 3). HiPIP utilizes relatively nonbonding (e-type) orbitals (7) of the distorted tetrahedral FeS_4 centers (each Fe in the $Fe_4S_4S_4^*$ cluster is tetrahedrally coordinated to four S atoms) for electron transfer, and thus the activation requirements owing to inner sphere rearrangement are not expected to be large. As was the case for cytochrome *c*, therefore, structural considerations greatly favor outer sphere electron transfer as a pathway of choice in redox reactions involving HiPIP and other iron-sulfur proteins that contain tetrahedral FeS_4 centers.

X-ray crystal structure analysis has not been completed to date for any blue copper protein. The probability that the copper coordination environment is highly unusual, however, has long been recognized, as a result of various spectroscopic studies. A typical blue copper protein is characterized by an intense visible absorption band system (2), which peaks at about 600 nm, as well as by an extremely small $A_{\parallel}$ ESR spectral parameter.

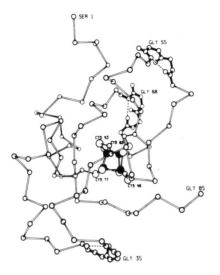

Fig. 3. Illustration of the relatively buried Fe$_4$S$_4$ cluster in HiPIP (see Ref. 4 for structural details).*

Neither of these spectral properties has been duplicated satisfactorily in low molecular weight copper(II) complexes. As square planar copper(II) centers, in particular, exhibit optical and ESR spectra that are much different from those observed for the blue proteins, most models have featured geometries based on tetrahedral or five coordination. Two explanations of the intense, 600 nm absorption have been proposed. One treats the band as arising from one or more allowed d-d transitions in a noncentrosymmetric center, and the other attributes the strong absorption to a charge transfer process, probably of the ligand-to-metal type. Spectroscopic studies of cobalt(II) derivatives of *Rhus vernicifera* stellacyanin, bean plastocyanin, and *Pseudomonas aeruginosa* azurin have established that the charge transfer interpretation is to be preferred, as intense bands analogous to the 600 nm band system are observed between 300 and 350 nm in the cobalt(II) derivatives *(8,9,10)*. Furthermore, visible and near infrared absorption, circular dichroism (CD), and magnetic circular dichroism (MCD) spectra of the cobalt(II) derivatives of stellacyanin, plastocyanin, and azurin have been interpreted *(11)* successfully in terms of the d-d transitions expected for a high-spin d^7 ion in a distorted tetrahedral binding site. In recent work, the d-d transitions have been observed for the first time in near infrared absorption, CD, and MCD studies of the native copper forms of all three proteins *(10)*. The positions of the near infrared d-d transitions are fully consistent with a slightly flattened tetrahedral structure for the blue copper centers in the native proteins.

Research aimed at the identification of the ligands comprising the flattened tetrahedral blue copper center has been particularly intense in the case of plastocyanin. Direct evidence for a sulfur ligand has come from X-ray photoelectron spectral experiments on bean plastocyanin, where a large shift of the S2p core energy of a single cysteine (Cys-85) residue in the protein upon metal incorporation has been observed *(11)*. The two histidines in spinach plastocyanin have been found in NMR titration experiments to exhibit pK values below 5 *(12)*, suggesting that they are coordinated to copper. It is reasonable to assume, therefore, that the analogous two residues in the bean protein, His-38 and His-88, are also ligands *(10)*. The fourth ligand in the proposed donor set for bean plastocyanin has been identified in extensive infrared spectral studies *(13)*. These experiments have revealed that a short section of helix in apoplastocyanin is strongly perturbed upon metal (Cu(II) or Co(II)) incorporation, thereby implicating a back-bone peptide nitrogen or oxygen as a ligand. The preference of copper for nitrogen donors, as well as evidence from charge transfer spectra, favor coordination by a deprotonated peptide nitrogen. Consideration of the bean plastocyanin sequence places the helix, and therefore the back-bone peptide nitrogen, a few residues above His-38 *(14)*.

A model for the blue copper site in bean plastocyanin based on the available spectroscopic evidence is shown in Fig. 4 *(10)*. The other blue proteins are expected to have similar (although not necessarily identical) coordination environments. The observed variation in spectroscopic parameters and reduction potentials could be accounted for by minor variations in ligand composition or coordination geometry of the blue copper center *(14)*. If the flattened tetrahedral geometry is correct for blue copper proteins, then once again we have a situation in which inner sphere reorganization associated with electron transfer (between Cu(II) and Cu(I) states) is minimized, and facile outer sphere electron transfer is to be expected. Furthermore, there is reason to believe that certain blue copper sites (especially that of azurin) *(15,16,17)* are substantially buried in hydrophobic protein interiors, and are therefore relatively inaccessible to solvent and other small molecules.

Before discussing the available kinetic data on these proteins, certain problems need to be considered. One problem that arises when comparing the reactions of a single proton with different reagents is associated with variations in the thermodynamic

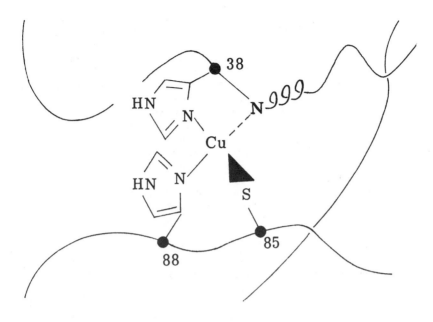

Fig. 4. Proposed model of the blue copper site in bean plastocyanin (see Ref. 10 for structural details).

driving force. As free energy/reactivity relationships are expected to exist, the question of the type of correlation between
rate constants and driving force must be considered. This correlation may be expected to be quite different for coupled systems (*e.g.*, cytochrome oxidase), which are designed to use the
excess free energy available from an oxidation-reduction cycle,
than for the simple electron carriers. The enthalpy and entropy
components of the free energy difference should also be considered. The nature of the interaction between the oxidant and
reductant must be appraised, and evidence for binding sought;
whether or not the interaction is such as to lead to measurable
binding, electrostatic and nonelectrostatic interactions between
the reactants are expected to lead to differences in rates, and
these differences might be correlated with trends in the enthalpy
and entropy of activation as well as with the overall rate constant. Further variables include steric constraints, affecting
the distance over which the electron may be transferred, which
for a protein may be dynamic and part of the reaction path (and
therefore reflected in the rate law and activation parameters),
or static but still variable with changes in pH of the medium or
salt content. Other effects of changing the medium may also be

important; the coulombic interaction between the reactants is expected to vary with the salt content of the solvent, specific activators or inhibitors for the proteins may exist, the pH and the buffer type may be influential, and the availability of bridging anions, even in outer sphere cases, is expected to have an influence on the rate which may be interpreted. The symmetry of the redox orbitals of the electron donor and acceptor and the match between them should be considered, and may be varied. The inherent protein reactivity is also a variable, and with the help of the above considerations this variable (for the same protein with different reagent as well as for different proteins) should also be considered and might be isolated.

The above problems are greatly compounded when both the oxidant and reductant are proteins; therefore, several workers have used small molecule oxidants and reductants to probe the reactivity of redox proteins. The small molecule reagents are generally more readily available in quantity than the proteins, and many changes can be made in their structures with predictable effects. By a careful selection of reagents, inner or outer sphere reactivity may be selected or at least favored, the redox potential of the reagent and thus the free energy change of the reaction may be controlled, the size and hydrophobic or hydrophilic nature of the reagent may be varied by introducing changes in the structure, the charge of the reagent, and the reactivity of the reagent (as measured by its electron self-exchange rate) may be varied. Although both inorganic and organic reagents are available, inorganic complexes are somewhat more convenient, as they have characteristic spectral changes and usually are one electron donors or acceptors (many commonly-used organic systems are two electron reagents and produce reactive free radicals on one electron transfers). After separating protein from reagent effects by judicious choices of reagents and conditions, the results of protein-protein reactions may be more critically examined.

2. THEORY

2.1 INNER AND OUTER SPHERE MECHANISMS – The distinction between inner sphere (including remote attack) and outer sphere mechanisms is a classical one in inorganic chemistry and there are several monographs and review articles available *(18,19)*, including one by Bennett which discusses bioinorganic examples up to 1973 *(3)*. A requirement for a strictly inner sphere reaction is that one of the reactants must possess an open coordination position and the other center must have a first coordination sphere ligand atom accessible for bridging. The requirements for remote attack are not so stringent in that all that must be available on the second reactant is a coordination position somewhere on a ligand. Outer sphere reactions are those that involve electron transfer in which no ligands are shared between the two redox centers. An outer sphere reaction is virtually assured if the protein metal center is buried and does not include a ligand which reaches to the protein surface. Even if there is such a ligand (like a heme), there still must be an available coordination site on it and an available position on the attacking reagent.

2.2 MARCUS THEORY - Outer sphere electron transfer reactions are among the simplest to consider theoretically because no bonds are made or broken; accordingly, several theories for treating such reactions from first principles are available. The most useful theory is due to Marcus *(20)*. The result from this treatment that will be used here is not the one for an *a priori* calculation of rate constants but rather a correlation equation for relating the cross reaction rate constant between two species to the equilibrium constant, the self-exchange rates for the two reactants, and a term which evaluates to about one for reactions with small driving force. This equation is

$$k_{12} = \sqrt{k_{11}k_{22}Kf} \tag{1}$$

where K is the equilibrium constant, k_{11} and k_{22} the two self-exchange rate constants, k_{12} the cross reaction rate constant, and

$$\ln f = (\ln K)^2/4 \ (\ln(k_{11}k_{22}/Z^2)) \tag{2}$$

The value of Z, the collision frequency of neutral molecules in solution may be estimated from the simple kinetic theory of gases, if diffusion is truly random *(21)*, and is found to be roughly constant for the reactions of interest. The theory can include corrections for adiabaticity by substituting k_{12}/p_{12}, k_{11}/p_{11}, and k_{22}/p_{22} for k_{12}, k_{11}, and k_{22}, respectively, where p is 1 for an adiabatic reaction and less than 1 for nonadiabatic reactions. Equation 1 is valid if all reactions are adiabatic ($p_{12}=p_{11}=p_{22}=1$) or if all are uniformly nonadiabatic ($p_{12}^2/p_{11}p_{22} = 1$). The Marcus relationship has been extensively applied in the inorganic literature, especially by Sutin and coworkers *(22)*, and found to be generally valid except in some cases involving cobalt amine and aquo complexes. As it will be used here, the Marcus relationship will allow compensation for differences in driving force and differences in the inherent reactivity of the reagents being used. For this purpose, the equation may be rearranged to give (for 25°)

$$k_{11} = k_{12}^2/k_{22}K = k_{12}^2/k_{22} \exp(38.94\Delta E), \qquad (3)$$

$$\text{where } \Delta E \text{ is the potential}$$
$$\text{difference in volts}$$

As used here and from this point on, the subscript 2 will refer to the small molecule reagent or the protein for which a self-exchange rate constant is assumed, and the subscript 1 will be for the protein for which a self-exchange rate constant is calculated. As presented above, the correlation is essentially one of free energies of activation, but with enough information the individual enthalpy and entropy components of the activation and equilibrium components may be treated separately.

As k_{11} values calculated using the Marcus theory equations will be used extensively in the data treatment, some more discussion of what this parameter means is warranted. First, assuming to begin with that the equation is valid, the result is only as good as the parameters which go into calculating it. The precision of the parameters will be discussed later, but it should be reiterated here that each small molecule reagent is assumed to have only one mechanism of electron transfer available, the one that is characterized by its self-exchange rate constant and activation parameters. The k_{11} calculated, then, is a quantity that characterizes the activation process (including inherent activation of the protein metal center and certain contributions from the interaction of the protein and the reagent) the protein must undergo to transfer an electron, assuming no other limiting process is involved. Thus if the redox reaction of two different reagents with the same protein leads to the same predicted k_{11} (and its activation parameters), it follows that the electron transfer mechanism is the same for the two reactions, barring fortuitous compensation. If k_{11} values vary significantly, the electron transfer mechanisms are expected to differ correspondingly.

Predicted protein self-exchange rate constants above the diffusion controlled limit may (and do) occur; diffusion control would of course set in and prevent such high rates from being attained; however, if two predicted exchange rate constants are

significantly different and both above the diffusion limit, the
conclusion is still that the two mechanisms are different. The
actual protein exchange rate, in those few cases where it has
been measured,does not "prove" or "disprove" the predictions
made from other reactions, it merely represents a mechanism that
may or may not be the same as the others that have been studied.

2.3 ELECTROSTATIC INTERACTIONS - The electrostatic interaction
between the two reacting species has been ignored in the above
presentation of the Marcus equations, as is the custom for simple
inorganic reactions. The justification for this in the simpler
systems is that reactions between reagents of similar charge type
require little if any correction and most simple inorganic reactions are between ions of similar radius and charge (often +2/+3).
For the protein reactions, however, the protein often is not of
the same charge type as the reagent, and the electrostatic interactions can no longer be ignored. In the discussion below,
several alternative analyses of the problem of electrostatic
interaction will be presented. These theories will in general
be quite simplified, and it should be possible, given the large
amount of structural and ion binding information available for
several of the proteins under consideration (especially horse
heart cytochrome *c*) and using more precise equations than those
of simple Debye-Hückel theory, to make a much more detailed analysis; however, the primary goal at this point is the development
of an approximate calculational method for charge effects, using
experimentally available data, which may be applied to proteins
for which little structural data are available.
    The most often discussed manifestation of electrostatic
interactions is the ionic strength dependence of reactions between ions. In the most common theoretical treatment, the transition state formalism is used and the ionic strength dependence
is treated as the result of the changing activity coefficients
of the reactants and the transition state with ionic strength
(23). Assuming the Debye-Hückel treatment for these activity
coefficients, the resulting equation is

$$\ln k = \ln k_0 - \frac{Z_1{}^2 \alpha \sqrt{\mu}}{1 + \beta R_1 \sqrt{\mu}} - \frac{Z_2{}^2 \alpha \sqrt{\mu}}{1 + \beta R_2 \sqrt{\mu}} + \frac{(Z_1 + Z_2)^2 \alpha \sqrt{\mu}}{1 + \beta R_{\ddagger} \sqrt{\mu}} \qquad (4)$$

where k is the rate constant at ionic strength μ, the Z's are the
charges on the reactants, the R's are the radii of the reactants
(1 and 2) and the transition state ($\ddagger$), and α and β are constants
with values 1.17 (water, 25°) and 0.329 $Å^{-1}$, respectively. If it

is assumed that the radii of the protein and the activated complex are the same, $(R_1 = R_{\ddagger})$ the equation reduces to

$$\ln k = \ln k_0 + \frac{(2Z_1Z_2 + Z_2^2)\alpha\sqrt{\mu}}{1 + \beta R_1\sqrt{\mu}} - \frac{Z_2^2\alpha\sqrt{\mu}}{1 + \beta R_2\sqrt{\mu}} \tag{5}$$

This reduces to the oft-quoted equation

$$\ln k = \ln k_0 + 2Z_1Z_2\alpha\sqrt{\mu} \tag{6}$$

only if all radii are assumed to be the same and the ionic strength is such that $1 \gg \beta R\sqrt{\mu}$.

Another approach to treating the ionic strength dependence of electron transfer reactions is to use the equations of Marcus theory and an appropriate function for the coulombic interaction *(24)*. First, the free energy change for the cross reaction, ΔG_{12}^0, can be separated into an electrostatic contribution, w^0, and a term independent of electrostatic interactions, ΔG_r^0, and thus

$$\Delta G_{12}^0 = \Delta G_r^0 + w^0 \tag{7}$$

The w^0 term in turn may be expressed as the difference between the electrostatic work to bring the reactants together, w_{12}, and that to bring the products together, w_{21},

$$\Delta G_{12}^0 = \Delta G_r^0 + w_{12} - w_{21} \tag{8}$$

Similarly, the three activation free energies, ΔG_{12}^* for the cross reaction, and ΔG_{11}^* and ΔG_{22}^* for the two electron exchange rate constants, may be expressed as the sum of work and electrostatics-independent terms

$$\Delta G_{11}^* = \Delta G_{11}^{**} + w_{11} \tag{9}$$

$$\Delta G_{22}^* = \Delta G_{22}^{**} + w_{22} \tag{10}$$

$$\Delta G_{12}^* = \Delta G_{12}^{**} + w_{12} \tag{11}$$

For the cross reaction, the part of the activation energy that is independent of electrostatic effects is expressed

$$\Delta G_{12}^{**} = (\Delta G_{11}^{**} + \Delta G_{22}^{**} + \Delta G_r^0)/2$$

$$= (\Delta G_{11}^* + \Delta G_{22}^* + \Delta G_{12}^0 - w_{12} + w_{21} - w_{11} - w_{22})/2$$

$$(12)$$

The predicted cross reaction activation free energy is then

$$\Delta G_{12}^* = \Delta G_{12}^{**} + w_{12}$$

$$= (\Delta G_{11}^* + \Delta G_{22}^* + \Delta G_{12}{}^0 + w_{12} + w_{21} - w_{11} - w_{22})/2$$

$$(13)$$

Solving for the predicted protein self-exchange activation energy results in the equation

$$\Delta G_{11}^* = 2\Delta G_{12}^* - \Delta G_{22}^* - \Delta G_{12}{}^0 - w_{21} - w_{12} + w_{11} + w_{22} \quad (14)$$

This formalism can also include the correction term f, which in this treatment is expressed through a term $1 + \alpha^{**}$

$$\Delta G_{12}^* = \tfrac{1}{2}(\Delta G_{11}^{**} + \Delta G_{12}^{**} + \Delta G_r^0 \quad (1 + \alpha^{**})) \tag{15}$$

where α^{**} is defined by Equation 16:

$$\alpha^{**} = \frac{\Delta G_r^0}{4(\Delta G_{11}^{**} + \Delta G_{22}^{**})} \tag{16}$$

Solving for ΔG_{11}^{**} gives an equation of the form

$$A(\Delta G_{11}^{**})^2 + B(\Delta G_{11}^{**}) + C = 0 \tag{17}$$

where

$$A = 4 \tag{18}$$

$$B = 8\Delta G_{22}^{**} + 4\Delta G_0{}^r - 8\Delta G_{12}^{**} \tag{19}$$

$$C = 4\Delta G_{22}{}^{**}(\Delta G_{22}{}^{**} + \Delta G_r{}^0 - 2\Delta G_{12}{}^{**}) + (\Delta G_r{}^0)^2 \tag{20}$$

which can be solved for $\Delta G_{11}{}^{**}$, using the substitutions from above for $\Delta G_{11}{}^{**}$, $\Delta G_{22}{}^{**}$, and $\Delta G_r{}^0$, and thus k_{11}. In order to calculate free energies from rate constants and potential differences (and *vice versa*), the following equations are useful:

$$\Delta G^* = RT[23.76 - \ln(R/T)]$$
$$= 592.1 \ (29.45 - \ln k) \ (T = 298^\circ K) \tag{21}$$

$$\Delta G_{12}{}^0 = -23.06 \ (\Delta E) \ \text{kcal/mole} \ (\Delta E \text{ in volts})$$
$$k = 6.21 \ (10^{12}) \ \exp \ (-\Delta G^*/0.5921) \ (T = 298^\circ K) \tag{22}$$

The terms that must be calculated in order to evaluate the above equations represent the work required to bring the two reactants from infinite separation to the separation in the activated complex. As this problem will prove difficult enough to handle, further refinements, such as protein conformation and charge distribution changes on forming the activated complex, will not be considered. The possible importance of changes in charge distribution and dipole interactions is documented *(25)* and should be considered in more detailed calculations.

The simplest model for making this calculation considers the protein to be a sphere with a totally symmetric charge distribution. The dielectric within the sphere is lower than that of the medium, but its value is not required, nor is the actual detailed charge distribution required as long as it is totally symmetric, concentric with, and located within the region of low dielectric *(26)*. The equation for the potential is *(27)*

$$V = \frac{1}{2}\left[\frac{e^{\kappa R_1}}{1 + \kappa R_1} + \frac{e^{\kappa R_2}}{1 + \kappa R_2}\right] \frac{Z_1 Z_2 e^2}{\varepsilon r} e^{-\kappa r} \tag{23}$$

where $\kappa = \alpha \mu^{\frac{1}{2}}$ A at 25° in water

$R_1 =$ radius of the protein at the distance of closest approach by the average small ion in solution

$R_2 =$ radius of reagent, defined as is R_1

$Z_1 =$ charge on the protein

$Z_2 =$ charge on the reagent

ε = dielectric constant of the medium (78.3 for water at 25°)

r = distance between the centers of the reagent and protein in the activated complex

e = charge on an electron

This equation reduces to

$$V = 2.1175 \left(\frac{e^{\kappa R_1}}{1 + \kappa R_1} + \frac{e^{\kappa R_2}}{1 + \kappa R_2} \right) \frac{Z_1 Z_2}{R_1 + R_2} \exp(-\kappa[R_1 + R_2]) \tag{24}$$

when the values of the constants are substituted. For the calculations to be done in subsequent sections, r is approximated as the sum of the radii of the protein and reagent. Fig. 5 illustrates the free energy terms according to this model.

The remaining problem that must be solved before the equations may be evaluated is the selection of the R and Z parameters; for the small molecule reagents, these values are easily determined from the molecular formula and X-ray structural data or model building. For proteins, the radius may be estimated from Equation 25 (28):

$$R^3 = \frac{3}{4\pi} \frac{M}{N} (\bar{v}_2 + \delta_1 \bar{v}_1^0) \tag{25}$$

where

R = radius

M = molecular weight

N = Avogadro's number

δ_1 = effective solvation

$\bar{v}_2$ = partial specific volume (reciprocal of the density)

v_1^0 = partial specific volume of water

The protein partial specific volume is taken as 0.73 cm³/g unless a measured value is available, the solvent partial specific volume is 1.0 cm³/g, and the effective solvation is 0.2. With these substitutions, the radius in Ångstrom units is

$$R = 0.717 \, M^{1/3} \tag{26}$$

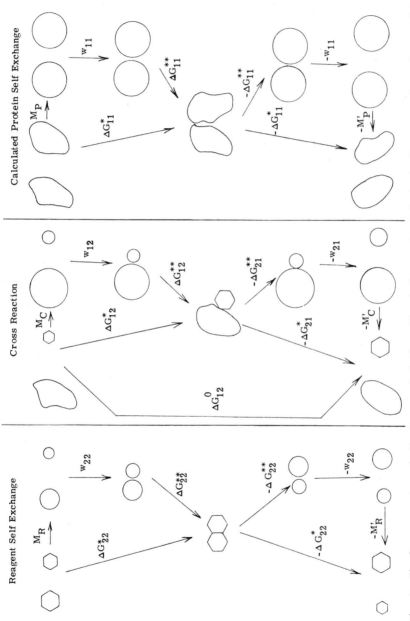

Fig. 5. The free energy terms in the Marcus calculation. The circles represent the hard sphere, smeared-out charge models used in the electrostatic calculation. The M terms represent the difference in free energy between the actual species and those assumed in the electrostatic calculation. These terms have been assumed to be zero in the calculation.

The charge on the protein may be crudely estimated from the amino acid composition. The maximum number of charged groups (at pH 7) is estimated assuming all of the glutamates and aspartates are ionized, that all the lysines are protonated, and that arginine is in its monopositive form. It is further assumed that half of the histidines are protonated; this last assumption may be somewhat of an overestimate but will be retained for simplicity in the light of the approximate nature of the whole approach. Further charged groups could include amino sugars (if they are not acetylated), the propionate side chains and deprotonated nitrogens on the heme group, the sulfides in the iron-sulfur proteins, any residues that have their pK's shifted by coordination to the metal or as a result of being located in a hydrophobic region (for the polar residues), and the metal itself. The contribution to the total charge from the metal and its ligands can be expected to be small, as is the case for well known examples. Such sites are constituted with low charges, as they are usually buried in the low dielectric of the protein interior. For example, the iron center in a heme protein is effectively neutralized by the two deprotonated nitrogens of the heme; in HiPIP with four thiolates and four sulfides, the oxidized cluster (assuming a charge of +11 from the four irons) has a charge of -1 on the oxidized and -2 on the reduced form. The copper proteins are less well characterized, but assuming a cysteine sulfur, the charge is at most +1 on the oxidized form and may be zero if a deprotonated amide is one of the ligands. The principle that the coordination site is formed so as to neutralize the charge on the metal ion may well be of general validity, at least for sites which prove to be buried, because of the high thermodynamic cost of having charged sites in a low dielectric medium.

An expression for the ionic strength dependence of the rate constant will now be derived. The free energy of activation for the cross reaction is

$$\Delta G_{12}{}^* = \tfrac{1}{2}(\Delta G_{11}{}^{**} + \Delta G_{22}{}^{**} + \Delta G_r{}^0) + w_{12} \qquad (27)$$

where the only ionic strength dependent term is w_{12}. Substituting from Equation 24 for the work terms and converting to rate form this becomes

$$\ln k = \ln k_0 - 3.576 \left[\frac{e^{-\kappa R_2}}{1 + \kappa R_1} + \frac{e^{-\kappa R_1}}{1 + \kappa R_2} \right] \left(\frac{Z_1 Z_2}{R_1 + R_2} \right) \qquad (28)$$

(where the radii of both the reagent and the protein are assumed invariant and k_0 is the rate constant at infinite ionic strength). This equation can be compared with Equations 5 and 6. The expression for the ionic strength dependence that originates from the Marcus theory treatment is preferable in terms of internal consistency when the value of the charge on the protein is sought (from ionic strength dependence data) in order to correct for the work terms in the calculation of protein self-exchange rate constants (k_{11}) from cross reaction data.

In the discussions above, the protein has been considered as a sphere of uniformly distributed charge; another model that could be considered when adequate information exists to parametrize it includes provision for an active site. In this model, the whole protein charge is divided between the site, with a given charge and radius, and the rest of the protein with the remaining charge and the full radius. Each of these spheres is treated independently with the same functions presented above.

It should be mentioned here that the general approach of treating the effects of ions using the ionic strength as the relevant variable is inaccurate away from the low ionic strength regime where the Debye-Hückel expressions are rigorously valid. Data at higher ionic strengths (above 0.1 M) have been successfully treated *(29)* when interaction parameters between all anion-cation pairs are included in the expressions, and with the use of ion association constants. These latter treatments involve many parameters and would be studies unto themselves if they were to be extended to the protein systems. They are only mentioned to emphasize the highly approximate nature of the ionic strength treatment and indicate a possible direction for future investigation.

In the discussion of electron transfer reactions, the overall reaction is often broken into three steps: (1) the formation of the precursor complex, (2) the electron transfer within this complex to form the successor complex, and (3) the dissociation of the successor complex to give the individual products. The formation constants for the precursor complexes are often too small to appear in the rate law (*i.e.*, no saturation behavior is observable), but under the assumption that the binding is purely electrostatic, these constants may be estimated from the equation *(22)*

$$K = \frac{4\pi N r^3}{3000} \ \exp(-U(r)/RT) \tag{29}$$

$$U(r) = \frac{Z_1Z_2}{\varepsilon r(1 + \kappa r)} \tag{30}$$

where r is the distance between the centers of the reactants and the rest of the symbols have already been defined. When the precursor complex is formed quickly compared to the electron transfer rate, so that the equilibrium between reactants and the precursor complex is maintained throughout the reaction, the observed rate constant may be written

$$k_{obsd} = K\, k_{et} \tag{31}$$

where k_{et} is the unimolecular electron transfer rate constant for the conversion of the precursor to the successor complex. This method of treating the electrostatic effects is required in the extreme case of a long-lived precursor complex, whereas the previously-presented work term description is proper when the precursor complex is extremely short-lived. The dividing line between the two cases is not a clear one, except in those cases where saturation is observed in the kinetics. As with the work term formulation, if the precursor and successor complex theory is used, it must be used consistently for all rate constants being considered. When this is done the Marcus cross reaction equation becomes *(20)*

$$k_{12} = \left(\frac{P_{12}P_{21}k_{11}k_{22}K_{12}f}{P_{11}P_{22}} \right)^{\frac{1}{2}} \tag{32}$$

$$\ln f = (\ln K_{12}P_{21}/P_{12})^2 \,/4\, \ln\,(k_{11}k_{22}/P_{11}P_{22}Z^2) \tag{33}$$

where P_{11}, P_{22}, and P_{12} are the stability constants for the three precursor complexes and P_{21} is the stability constant for the cross reaction successor complex (the precursor complex for the reaction in the opposite direction). Equation 32 reduces to Equation 1 when $P_{12}P_{21} \cong P_{11}P_{22}$.

2.4 HYDROPHOBIC EFFECTS - Another category of possible interactions between the reactants in an electron transfer reaction will be referred to as hydrophobic effects. For the purposes of this discussion this category will include most of the interactions that are not coulombic, including the nonpolar interactions between hydrophilic groups as well as the more important solvent interactions of nonpolar groups with water, the "freezing out" phenomenon associated with water molecules at the interface between the bulk solvent and nonpolar solutes. Hydrogen bonding is another form of interaction that may be of some importance and is also hard to analyze for or predict. Any of these interactions would affect the rate through stabilization or destabilization of the precursor and successor complexes, and would thus enter into the calculations in the same manner as discussed above. Because these interactions are less well understood than the simple coulombic ones, no attempt will be made to formulate an analogue to the work term treatment.

2.5 SPECIFIC ION EFFECTS – There are other effects that may be caused by ions (and sometimes uncharged solutes) other than the general ones discussed above. The most dramatic of these are the specific ion effects, especially specific inhibition. An example of this is the inhibition of type 2 copper containing proteins by azide. In those cases, the azide is thought to bind to the copper and thus block the binding of a substrate or modify the reactivity of the copper *(30)*. A somewhat different type of anion inhibition is shown by the interaction of azide with cytochrome *c*; in this case the azide causes a major change in the structure of the cytochrome by replacing an axial (Met-80) ligand of the iron *(31)*. Both of these types of inhibition are the result of the specific interaction of an inhibitor with the protein; a more subtle effect, and one that is usually smaller in magnitude than the previous examples, is anion assistance of outer sphere reactions. This effect has been studied by Sutin and coworkers in inorganic systems and the conclusions that have been reached are that the added anion catalyzes electron transfer by acting as a bridge *(32)*. The effectiveness of the bridge is related to the symmetry of the available bridging ligand and metal orbitals. The most effective bridges are of the same symmetry as the acceptor and donor orbitals; for example, azide is a good π bridging ligand and is especially effective between ruthenium or iron centers and reagents with π orbitals, whereas chloride is a good σ bridge and is more effective between chromium and cobalt centers and between reagents with only σ orbitals available on the ligands. The effectiveness of the bridge is also dependent on the ability of the potential bridging anion to associate with one or the other of the reactants in order to be brought into the transition state (discounting any significant number of effective ternary collisions). Even less specific is the effect of a more general binding of anions or cations to a protein surface or association with the small molecule reagents. If there is a reason to suspect such binding, a general electrostatic correction could be attempted for it by correcting the total charge on the reagent or protein for the suspected number of bound ions. Such interactions are not just speculation, as binding constants for several anions with cytochrome c have been measured *(33,34)* and the electron exchange rate for $Fe(CN)_6^{3-}/^{4-}$ is strongly affected by cations *(35)*, whereas the $Co(phen)_3^{3+}/^{2+}$ exchange is anion dependent *(36)*.

Another topic that may be considered under the general title of ion and medium effects is the effect of varying pH. Once the pH dependence is available (it should be as wide a range as possible within the capabilities of a single buffer, or separate, overlapping buffer systems may be used) attempts may be made to fit the data to one or more titration curves and pK values may be assigned. If titration data are available for the protein and the reagent, this information can first be consulted in interpreting the pH dependence of an electron exchange rate and thus it can usually be decided if the reagent or protein is involved. Usually the reagent can be eliminated as a source of a pH dependence, or at least a reagent may be found that does not have a pK in the region being analyzed if the protein pH behavior is especially important to isolate. The most probable origin of a pH dependence that originates from a pK of the small molecule reagent is electrostatic interaction, but protein proton equilibria will often involve conformational changes as well as changes in charge. Changes in conformation are especially well documented at extremely low pH values, where proteins often assume much more open structures (a good example here is again cytochrome c for which several forms have been characterized)*(1)*. Conformational changes may strongly involve the metal center, for example, the ligation of cytochrome c changes at high pH with a large decrease in the potential *(1)*, and blue proteins bleach at higher pH values *(2)*, indicating structural modification of the copper site. The particular residues responsible for a transition with a given pK are difficult to assign because of the problems associated with multiple equilibria and because the pK's of the individual side chains are often shifted by their environments. A further warning that should be made is that although a large pH dependence may be confidently interpreted, small changes of rate with pH (less than a factor of two overall) are hazardous to analyze, as there could be many reasons for such a dependence, including changes in the medium because of the changing ionic forms of the buffer and the concentration of added salt.

2.6 STERIC EFFECTS – Another variable that must be considered and may be controlled in the selection of small molecule reactants is molecular size and steric constraint. The steric

constraints and predicted effects from changes in the size of the
reagent will vary for different proteins and different proposed
mechanisms. Common examples would include a metal site that is
buried at the bottom of a pit or groove of the protein; in this
case, reagents much larger than the size of the opening should
react more slowly, and there should be a discontinuity in the
rate above a certain size reagent. If the metal site of the
protein is on the surface, then there should be no such discontinuity but other factors might be expected to change the rate
with increasing size of the reagent. Electrostatic effects would
be small, but such interactions would decrease with increasing
size, thereby leading to an increase in the rate if the reagents
are like-charged, or to a decrease if the two reagents have opposite charges. If tunnelling is suspected, the rate is expected
to be approximately proportional to e^{-r}, where r is the distance
between the donor and acceptor (which may be the metal ions or
the edges of π systems, for example)*(37)*. Strong dependences on
distance are also to be expected if changing the size of the reagent strongly affects the orbital overlap in the transition
state, thus possibly changing a case of good overlap (adiabatic)
to a case of poor overlap (possibly nonadiabatic).

2.7 SYMMETRY ASPECTS - As was hinted at before in the discussion
of bridging ligand effects, the match of symmetry between the
donor and acceptor orbitals has been found to correlate with
electron transfer rates; when the donor and acceptor orbitals
are both of the same symmetry type, the rate constant is higher
than when one is σ and the other π, all else being equal. This
generalization can be used and tested by varying the reagent as
was considered when bridging anions were discussed. Some of the
variation that arises from the symmetry properties will be compensated for in the Marcus calculation through the self-exchange
rate constant. Thus, the generally higher rate constants found
for reagents with π symmetry redox orbitals will be compensated
for by the k_{22} parameter. If the extent of orbital overlap (and
thus the adiabaticity that is assumed to parallel this overlap)
between the protein and reagent redox orbitals varies so as to
invalidate the assumption of an adiabatic or uniformly nonadiabatic reaction (see Section 2.2), then symmetry effects will
directly affect the calculated k_{11} value. As there is no way to
measure the adiabaticity, variations in k_{11}/p_{11} are all that can
really be compared; p_{11} could always be artificially chosen such
that $p_{12}^2 = p_{11}p_{22}$, and thus all of the variation placed on
changes in k_{11} (as is implicitly done when the calculations are
done without the adiabaticity factors); but from a conceptual
viewpoint the two terms, k and p, should be kept distinct and

variations in the adiabaticity of the cross reaction (relative to the adiabaticity of the reagent self-exchange reaction) should be considered in terms of p factors.

2.8 ACTIVATION PARAMETERS - Activation parameters are available for many of the electron transfer reactions to be considered. As large differences in activation parameters are sometimes found when there is little difference in rate constants, it may be expected that analysis of these parameters will show differences in reaction pathways and that these differences may be interpreted mechanistically. The Marcus theory of electron transfer reactions provides a prediction for activation parameters as well as just free energies of activation (rate constants), so this treatment will now be given *(20)*. The temperature dependence for an outer sphere reaction used in the Marcus presentation is given by

$$k_r = pZ^{-\Delta G^*/RT} \tag{34}$$

where p is the probability of electron transfer in the activated complex (transmission coefficient) and Z is the temperature dependent collision frequency between two uncharged particles in solution. Using the equations previously given for the cross reaction rate

$$k_{12} = \sqrt{k_{11}k_{22}k_{12}f} \tag{1}$$

$$\ln f = (\ln K)^2/r \ln (k_{11}k_{22}/Z^2) \tag{2}$$

the following expression for the free energy can then be derived

$$\Delta G^*_{12} = \frac{\Delta G^*_{11} + \Delta G^*_{22}}{2} + \Delta G^0_{12} (1 + \alpha) \tag{15}$$

$$\alpha = \frac{\Delta G^*_{12}}{4(\Delta G^*_{11} + \Delta G^*_{22})} \tag{16}$$

By using the expressions

$$\Delta S = -\frac{\partial \Delta G}{\partial T} \qquad \Delta H = \frac{\partial \Delta G/T}{\partial 1/T} \tag{35}$$

Equations 36 and 37 may be derived:

$$\Delta S_{12}^* = \left(\frac{\Delta S_{11}^*}{2} + \frac{\Delta S_{22}^*}{2}\right)(1 - 4\alpha^2) + \frac{\Delta S_{12}^0}{2}(1 + 2\alpha) \tag{36}$$

$$\Delta H_{12}^* = \left(\frac{\Delta H_{11}^*}{2} + \frac{\Delta H_{22}^*}{2}\right)(1 - 4\alpha^2) + \frac{\Delta H_{12}^0}{2}(1 + 2\alpha) \tag{37}$$

These activation parameters are not the same as those calculated normally from temperature dependence data, because the Eyring expression from transition state theory

$$k_r = p\frac{kT}{h} e^{-\Delta G^{\ddagger}/RT} \tag{38}$$

is different from Equation 34, which was assumed in the Marcus derivations. Therefore the following conversions are necessary to relate the two types of activation parameters (in deriving these relationships it is assumed that Z, the collision frequency, is proportional to $T^{\frac{1}{2}}$)*(21)*;

$$\Delta G^{\ddagger} = \Delta G^* - RT \ln(hZ/kT) \tag{39}$$

$$\Delta S^{\ddagger} = \Delta S^* + R \ln(hZ/kT) - \tfrac{1}{2}R \tag{40}$$

$$\Delta H^{\ddagger} = \Delta H^* - \tfrac{1}{2}RT \tag{41}$$

The terms with hZ/kT are dependent on the value selected for Z. As there is not an entirely satisfactory method for estimating Z, the correction will be left indeterminate.

The dependence of electron transfer rate on temperature in systems where a tunnelling mechanism is operating is characterized by a curved Arrhenius plot (or Eyring plot) which levels off at low temperature *(37)*.

Considering the possible importance of electrostatic (*i.e.*, coulombic) interactions, some consideration must be given to the enthalpy and entropy components of the work term contributions. The temperature dependence of the work terms (see Equation 23) is the result of the temperature dependence of κ

$$\kappa = 50.3 \text{ Å } \left(\frac{\mu}{\varepsilon T} \right)^{\frac{1}{2}} \tag{42}$$

The dielectric constant of water is somewhat temperature dependent, varying from 87.9 at 0° to 69.9 at 100°; thus, $1/\varepsilon T$ varies only from 4.17×10^{-5} at 0° to 4.82×10^{-5} at 100°, and is 4.37×10^{-5} at 20°. Assuming, then, that $\varepsilon = \varepsilon_0 + \varepsilon_1/T$, the expression for a work term

$$V = \frac{e^2}{2} \left[\frac{-\kappa R_2}{1 + \kappa R_1} + \frac{-\kappa R_1}{1 + \kappa R_2} \right] \frac{Z_1 Z_2}{\varepsilon (R_1 + R_2)} \tag{23}$$

may be solved for the entropy component using the relationship

$$\frac{\partial V}{\partial T} = -\Delta S \tag{43}$$

Carrying out the partial differentiation and substituting back for ε

$$-\Delta S = V \left\{ 25.15 \mu^{\frac{1}{2}} \varepsilon_0 \left(\frac{\varepsilon}{T} \right)^{-3/2} \left[\frac{1}{r\varepsilon} + \frac{(1 + \kappa R_1) R_2 + R_1}{(1 + \kappa R_1)^2} \right. \right.$$

$$\left. \left. + \frac{(1 + \kappa R_2) R_1 + R_2}{(1 + \kappa R_2)^2} \right] + \frac{\varepsilon_1}{r T^2 \varepsilon} \right\} \tag{44}$$

and $\Delta H = \Delta G + T \Delta S$

$$= V + T \Delta S \tag{45}$$

can be used to solve for the enthalpy component.

If the overall electron transfer rate constant is best represented as the product of an equilibrium constant for formation of a precursor complex and an intracomplex electron transfer rate constant $(k = K k_{et})$, then the enthalpy and entropy contributions to K and k_{et} may be extracted from the kinetics. If Equation 29 is used to estimate a precursor or successor complex formation constant, then under the same assumption that $\varepsilon = \varepsilon_0 + \varepsilon_1/T$,

$$K = 2.52(10^{-3})r^3 \ \exp\left[-\frac{Z_1 Z_2}{\varepsilon(1 + \kappa r)(RT)}\right] \tag{46}$$

$$\Delta G = -RT\ln k$$

$$= -RT\ln 2.52(10^{-3})r^3 + \frac{Z_1 Z_2}{\varepsilon(1 + \kappa r)r} \tag{47}$$

$$-\Delta S = \frac{\partial \Delta G}{\partial T} \tag{35}$$

$$-\Delta S = -R\ln 2.52(10^{-3})r^3 - \frac{Z_1 Z_2}{T^2 r\varepsilon(1 + \kappa)}\left[\frac{\varepsilon_1}{\varepsilon} + \frac{25.15\mu^{\frac{1}{2}}\varepsilon_0 T^{\frac{1}{2}}}{\varepsilon^{3/2}(1 + \kappa)}\right] \tag{48}$$

and $\Delta H = \Delta G + T\Delta S$.

One hazard in the interpretation of activation parameters is the possibility that in solution one of the reagents is an equilibrium mixture of several forms. This should not be a problem for the small molecule reagents but is a significant consideration for proteins. The origin of this problem may be illustrated for the case of two forms of a protein which differ by a proton and react at different rates, with the assumption that the proton transfer is fast enough with respect to the electron transfer that the two forms of the protein A and B are always in equilibrium, that there is a first order excess of the other reagent, and that the two forms of the protein are indistinguishable at the wavelength being followed.

$$A + H^+ \underset{}{\overset{K}{\rightleftharpoons}} B, K = \frac{B}{(A)(H^+)} \tag{49}$$

$$A \overset{k_1}{\rightarrow} P \tag{50}$$

$$B \overset{k_2}{\rightarrow} Q \tag{51}$$

$$k_{obsd} = \frac{k_1 + k_2 K(H^+)}{1 + K(H^+)} \tag{52}$$

As k_1, k_2, and K each have their own thermodynamic parameters, it is clear that the standard plot of $\ln(k/T)$ *vs.* $1/T$ will constitute a complex, nonlinear combination of the various contributions (this is most easily seen in the numerator where the added terms are seen to mingle under the operation of taking the logarithm). Such a plot cannot be separated into its components, except by working at pH's where only one path contributes, and by finding the temperature dependence of the equilibrium constant by doing the temperature dependence of the pH dependence. The problem of obtaining the individual enthalpies and entropies is clearly a difficult one, even in the simple case described above, and then it requires that the protein be stable at the extreme pH values where only one form (A or B) exists. The whole situation is compounded if there are overlapping proton equilibria, ion binding, and more general ionic strength dependent conformational changes. Another problem of heterogeneity involves different structural forms of the protein, resulting from partial denaturation, or from such common differences as slight variations in the sugars of glycoproteins. In general, however, these latter examples should not give clean kinetics (the different forms are not in rapid equilibrium and thus will exhibit parallel first order behavior under pseudo-first order conditions). The problem of heterogeneity of forms which are in equilibrium must be assumed at this time to be one which cannot be solved exhaustively by studying each species independently; and it must be hoped that the different redox forms follow similar enough mechanisms so that the average activation parameters actually obtained are representative. If some case of an especially clean pH dependence as described in the example is available, it would be interesting to perform the full analysis, but this would be a time-consuming task.

Once the activation parameters for a given reaction are obtained, they can be compared with others available. Within the Marcus theory formalism, the contribution to the activation parameters owing to the protein may be isolated. One of the most important questions to consider is whether the mechanisms of various reagents (small molecule or protein) reacting with a given protein are the same or different. Previously this question was considered in terms of free energies in the comparison of the calculated k_{11} values for the protein. With the additional information available in activation parameters, it may be found that rates that are similar are actually the result of coincidence (or compensation) in that the enthalpies and entropies changed drastically but the free energy remained constant.

The phenomenon of compensation (the observation of a linear relationship between the enthalpy and entropy of activation across a series of reactions) is well documented, especially for organic reactions. Under certain assumptions *(38)*, compensation is predicted when there is only one interaction between the reactants which varies within the series of reactions being considered, the remainder of the mechanism being invariant. Although the microscopic basis for compensation is not well understood *(39)*, it can be rationalized in a simple case where the difference in a given series of reactions has its origin in the binding of the two reactants. If this binding becomes stronger, the activation enthalpy will decrease, but at the same time the entropy is expected to decrease because of the more restricted orientation produced by the tighter binding. The whole situation is further complicated if the solvent is considered, as the stronger binding of the two reactants will release bound solvent (an increase in enthalpy attributable to breakage of solvent-reagent hydrogen bonds; a decrease in enthalpy for the new solvent-solvent hydrogen bonds; and an increase in entropy for the freed solvent)*(39)*. Under the assumption that the observation of a linear compensation plot indicates that only one interaction varies within the series of reactions being considered, a range of activation parameters which might otherwise be considered to indicate rather complex differences in mechanism may be more confidently considered in terms of one type of interaction. The lack of compensation similarly reinforces the conclusion that there is more than one difference among the mechanisms of the reactions being compared. In considering just the contribution from the protein activation (the calculated enthalpies and entropies derived for the protein self-exchange reaction), compensation might also be considered.

In concluding the discussion of activation parameters, it should be emphasized that comparing cross reaction activation parameters is quite hazardous without some concept of the extent to which these activation parameters are the result of the free energy change for the reaction or contributions from the different reagent self-exchange mechanisms. The best comparison, within the Marcus theory formalism, is between the activation parameters for the calculated protein self-exchange. Other comparisons should be made only with extreme caution.

2.9 PROTEIN-PROTEIN REACTIONS - Most of the above discussion has been directed at the interpretation of reactions between a protein and a small molecule reagent because these are easier to treat and there are more data available; this last part of the Theory section will discuss some of the problems in analyzing data on the reactions between two proteins. The main problem in

analyzing protein-protein reactions is that, with present knowledge, both reactants must be considered as unknowns. The great
advantage of using a small molecule reagent is that its kinetic
properties can be considered to be fixed and known, and, through
the Marcus theory analysis, these can be factored out to yield a
quantitative description of the reactivity of the protein. It
must be kept in mind that protein-protein interactions may well
be of a character significantly different from those protein-
small molecule interactions that are used as models. Good examples of interactions that should exist between proteins (especially specific physiological partners), and that are predicted
from certain calculations to be influential in reactions, are
those of the ion-ion and dipole-dipole types *(25)*. The model for
these interactions, besides including a "good fit" between the
two proteins that would line up opposite charges and opposite
dipoles in the sense of static configurations, is involved with
the induced changes in charge distribution and dipoles that can
occur when two protein surfaces interact. In this model ion-ion
(or dipole-dipole) binding may be induced between sites that were
not necessarily charged (or dipolar) in the two proteins taken
individually. Single interactions of this type might not be
large enough to give a significant rate constant differential and
thus may be missed even if a small reagent that interacts in such
a way is found, but a large reagent that can induce many such
interactions is likely to reap enormous benefits and react at a
significantly different rate. For such reasons as these, the
proteins, no matter how well understood with small molecules,
must not be taken for granted in their interactions with other
proteins.

3. DATA RELIABILITY

The most basic parameter under consideration is the electron
transfer rate constant. In the best of cases (clean pseudo-first
order kinetics giving a linear plot with a zero intercept for the
concentration dependence on the reagent in excess), the second
order rate constants under a given set of experimental conditions
should be known to better than $\pm 10\%$. The further uncertainty
involved because of possible medium dependences should always be
considered. Ionic strength and pH dependences should be determined; if they are small, it may be assumed safely that small
differences in pH or μ in, say, the various solutions used for a
concentration dependence, do not introduce significant error.
The other possible small effects (changes in buffer, changes in
salt, rigorous elimination of trace ions and product inhibition)

are usually ignored because of the time required to check for them and the difficulty in analyzing the small differences (if any) usually found on varying them when there is not a reason to specifically suspect a problem (sometimes an ion, such as perchlorate, may be suspected of denaturing the protein, or a pH dependence that fits the pK of the buffer may raise suspicions). A large dependence on any parameter (pH or a specific ion concentration) should lead to the consideration of the protein under the different conditions as separate species to be treated independently. This is easier if the relationship between the species is clearly defined by an equilibrium constant of some kind, but it is not so easy if the variation does not fit such a simple analysis, as discussed in Section 2.8. Some dependences are usually ignored in preference to taking a standard state for comparison (*e.g.*, pH 7, 0.1 *M*) and the variability is relegated to increased error limits. For the purposes of the analyses to be performed in the rest of this paper, error limits of ±10% will be used when there is no evidence that they should be larger.

A second category of problems that may be encountered when trying to assign error limits may be associated with rate laws that are more complicated than the simple second order one considered in the previous paragraph. The first problem is to accurately define the rate law and all of the parameters involved. If saturation is indicated or if the data may be fit by a parallel first order scheme, this is not difficult. However, much more data must be collected than in the simple case of a second order process. In more complicated situations, the data may be quite difficult to analyze and the various equilibrium and kinetic parameters may not be easily separable. For the later use of these parameters in a Marcus theory analysis, saturation may be handled by considering precursor complex formation and parallel paths may indicate some sort of heterogeneity of protein forms in solution, but anything more complicated may make the application of the Marcus equations inappropriate. For problems involving protein-protein reactions, it is often difficult to obtain the range of concentrations of the species required to test for binding.

Next to be considered are the reduction potentials for the proteins and reagents. For the Marcus theory calculations, the potentials are used as an indirect way of measuring the equilibrium constant between the oxidant and reductant. If this value has been measured directly for any given reaction, it is always more accurate to use the direct determination, but as such measurements are seldom performed, the measured potential values must be relied on.

For the small molecule reagents of interest, the potentials

have often been measured by direct means because the reagents react directly with electrodes at acceptable rates, but more often than not the determination has been done under conditions significantly different from the conditions of the kinetic experiments. For example, many values are available for reagents that have been used in standard inorganic electron transfer studies but these are usually measured at ionic strengths from 1 to 4 M and at 0.1 M or higher acid concentration to match the conditions of the experiments then under consideration. For some classic cases, careful medium dependences have been performed (the best example here is the $Fe(CN)_6^{3-/4-}$ couple); these studies show strong dependences on the identity and concentration of ions. Although the lower charged ions are not expected to be so medium dependent in their electrochemistry, the possibility in general remains unexplored. For these reasons, error limits for most reagent potentials are set at ± 20 mV, or ± 10 mV in the best of cases.

The situation for the protein potentials is even less encouraging. The potentials cannot be measured directly because the proteins do not react quickly at electrodes, and therefore either titrations must be done or electrode techniques using mediators must be used. Most of the data in the literature have been obtained by titration with $Fe(CN)_6^{3-/4-}$, because this is a simple technique. There are problems with this technique, however; these include the need to know the $Fe(CN)_6^{3-/4-}$ potential under the experimental conditions, and the problem of the interaction between the reagent and the protein, which is especially severe with such a highly charged reagent. Besides the problem of the method used, many available potentials for metalloproteins are determined in only one medium, and thus the same problem of an unknown medium dependence is involved here as was involved with the reagent potentials; one advantage here is that the differences between the medium of interest and the medium for which the measurement was done is less in the case of the protein data, as the proteins often are not stable under a wide range of conditions. The best data available have been obtained with a mediator technique (the mediator being kept at a low enough concentration that binding problems are minimized). With this technique data can be taken with an optically transparent electrode, thereby allowing direct monitoring of the concentrations of the oxidized and reduced protein species. Because of the problem of the poor techniques used, most of the protein potentials will be given error limits of ± 20 mV.

The activation parameters determined from Eyring plots are seldom quoted with error limits; these limits are obtained from least squares fits only if the data are carefully weighted and the propagation of error is properly accounted for in the weighting; less care than might be proper is sometimes given to considering whether the plots are linear, and the range of temperatures that can be covered is usually so small (0 to 30°) that only extreme curvature would be apparent anyway. For these reasons, interpretations of activation enthalpy differences of less than two kcal/mol and entropy differences of less than five cal/mol-deg will not be attempted.

The individual components of the free energy changes for a reaction, ΔH^0 and ΔS^0, are evaluated most conveniently by the same method as is the equilibrium constant, *i.e.*, by taking the sum of the thermodynamic changes for the oxidation and reduction of the reductant and oxidant, respectively. These data are especially scanty; as they are usually obtained by determining the temperature dependence of the reduction potential, all of the problems associated with determining potentials apply here also. When the data have been determined for the small molecule reagents, they are usually for the state of infinite dilution and medium dependences of enthalpy and entropy changes are even rarer than medium dependences of potentials. For the determination of potential parameters, the ferricyanide titration technique requires a knowledge of the enthalpy and entropy changes for the $Fe(CN)_6^{3-/4-}$ couple instead of just its potential, and the influence of solvation of these highly charged ions is likely to result in compensation behavior. Thus the knowledge that the potential of the couple is insensitive to a certain medium change does not insure that the entropy and enthalpy components are similarly invariant. The conclusion here, as for the determination of the potentials, is that a great deal of careful work is necessary with techniques such as the optically transparent electrode approach described previously. No attempt will be made to give general error limits here as the few examples available will be treated individually as needed.

The next parameter to be discussed which is of interest in some of the calculations that follow is the measured electron exchange rate between the two oxidation states of a small molecule reagent or a protein. One of the first problems in measuring the rate of such a reaction is finding a property to monitor, because the reactants and products are chemically identical. The most common approach used for the slower reactions of certain metal ion complexes is isotopic labelling, where one oxidation state of the reactants is originally a radioactive isotope; aliquots of the reaction mixture are taken periodically and the two oxidation states separated, and counted for label in each. This method is clearly limited to rather slow reactions, but great ingenuity has been used in fast separation techniques that allow rates as fast as that of the $Fe(CN)_6^{3-/4-}$ exchange to be measured by this method. For faster reactions, NMR and ESR line broading measurements may be used, if there is some magnetic resonance signal that is shifted by the change in oxidation state. There is a gulf between these two methods where reactions are too fast to be measured by labelling (given the separation methods that must be used) and too slow to be measured by magnetic resonance; some of the attempts to measure exchange rates have failed for this reason, whereas the difficulty of the isotopic labelling method has inhibited the collection of data by that method. Other methods for obtaining estimates of the true exchange rate involve making some sort of small change in one of the reactants that makes the reaction easier to follow, but presumably does not change the activation requirements and thus the rate appreciably. Such changes include making one of the reactants optically active (*e.g.*, a substitution inert metal ion complex), or changing the ligand just enough to perturb the ultraviolet spectrum slightly. Such approaches as these are only as good as the assumption that the kinetics have not been changed by the modification, as the data usually become relatively easy to acquire with good accuracy. Even for those systems that have been studied, the now familiar problem of medium effects is again involved. Especially for the highly charged complexes, strong dependences on cations and anions have been observed. In general, when the available data are suspect because of the approach that had to be used (*e.g.*, changing the ligand slightly) or if medium dependences are expected but not studied, a range of a factor of two to three is not an unwarranted estimate of the possible error of the self-exchange rate constant.

The radii used for the reagents are evaluated from crystal structure data when possible, by taking the distance from the metal ion in the complex to a counter ion in the crystal, and subtracting the ionic radius of the counter ion. For $Fe(EDTA)^{2-/-}$ a larger value is required, as the complex is not actually spherical and the counter ion is especially close. As the radii of the protein and reagent are not accurately known (about ± 0.5 Å for the metal ion complexes), and as it is difficult to determine the average sizes of the other ions in solution, the small reagents have been given radii 0.2 or 0.3 Å larger than actually calculated and no correction has been made for the loss of solvating ions in going to the transition state. This procedure simplifies the calculations and does not decrease their accuracy, as the parameters are poorly defined in the first place.

Given the estimates of the reliability of the parameter values for the Marcus calculation, an overestimate of the error in k_{11} is

$$k_{11} = \frac{(1.1\ k_{12})^2}{1/3\ k_{22}\ ^{39.94(\Delta E - 0.03)}} \quad to \quad \frac{(0.9\ k_{12})^2}{3\ k_{22}\ ^{38.94(\Delta E + 0.03)}}$$

$$= 12\ to\ 0.08\ \frac{k_{12}^2}{k_{22}\ ^{38.94(\Delta E)}} \tag{53}$$

Thus the k_{11} values are determined to at least an order of magnitude.

4.   PROPERTIES OF THE REACTANTS

4.1   SMALL MOLECULE REAGENTS - This section presents a catalogue
of the properties of the small molecule reagents and the proteins
of interest. For the small molecules, properties such as reduction potential, electron self-exchange rate, size, stability, and
some discussion of reactivity characteristics (such as inner or
outer sphere predilections) are given. For the proteins, the
properties presented are molecular weight, size, isoelectric
point, amino acid sequence, function, and the self-exchange rate,
if it has been measured. Some indication of the reliability of
the various parameters will be given.
     Table 1 gives the available data *(40-58)* on the small mole-
cule reagents of interest and some other information will be presented briefly now.
     The three reagents chromium(II), dithionite, and hydroquinone
are generally considered to use inner sphere mechanisms. Chro-
mium(II) has been used for much of the inorganic electron transfer research available, especially the extensive work with
various cobalt(III) pentammine derivatives that has defined
bridging ligand effects *(18,19)*. One problem in the use of
Cr(II) for studying metalloprotein electron transfer rates is the
instability of the ion above about pH 4; as a result, this reagent has had only limited use. Both hydroquinone and dithionite
can transfer one or two electrons. Dithionite can be a one electron reagent by producing a sulfur(IV) species and SO_2^-, the
latter of which is in equilibrium with dithionite itself *(59)*:

$$S_2O_4^{-2} \xrightleftharpoons{-e^-} S(IV) + SO_2^- \tag{54}$$

$$2SO_2^- \rightleftharpoons S_2O_4^{-2} \tag{55}$$

TABLE 1: PROPERTIES OF THE SMALL MOLECULE REAGENTS

Reagent	E(mV, 25°)	μ(M)	Salt	k_{22}(M^{-1}s^{-1})	$\Delta H^\ddagger$(kcal/mole)	$\Delta S^\ddagger$(eu)	pH	Buffer	μ(M)	Salt	R(Å)	Ref.
$Fe(EDTA)^{-/2-}$	120[a]	0.1	KCl	$3(10^4)$[b]	4.0	-25	4.5-6.5	A,L[c]	0.1		4	40-43
$Co(phen)_3^{3+/2+}$	370	0.1	P/Na₂SO₄	45[d]	5.1	-34			0.1	KNO₃	7	36,44-46
$Co(5\text{-}Cl\text{-}phen)_3^{3+/2+}$	430	0.05	NaCl	$1.3(10^{-2})$[e]				P	0.1[e]	NaCl	8.5[f]	47
$Co(5,6\text{-}(Me)_2\text{-}phen)_3^{3+/2+}$	420	0.05	NaCl	$5.0(10^{-2})$[e]				P	0.1[e]	NaCl	8[f]	47
$Co(4,7\text{-}(Me)_2\text{-}phen)_3^{3+/2+}$	340	0.05	NaCl	$7.7(10^{-1})$[e]				P	0.1[e]	NaCl	7[f]	47
$Co(4,7\text{-}(\phi\text{-}SO_3^-)_2\text{-}phen)_3^{3-/4-}$	330	0.05	NaCl	$7.0(10^3)$[e]				P	0.1[e]	NaCl	11.5[f,g]	47,48
	330	0.05	NaCl	$1.4(10^3)$[e]				P	0.1[e]	NaCl	7[g]	47,48
	330	0.05	NaCl	$3.5(10^2)$[e]				P	0.1[e]	NaCl	7[g]	47,48
$Co(bipy)_3^{3+/2+}$	305	0.1	P/Na₂SO₄	18[d]	7.7	-27			0.1	KNO₃	7[o]	44,45,49
$Co(terpy)_3^{3+/2+}$	270	0.1	P/Na₂SO₄	$1.94(10^3)$[h]	10±1	-11±3	~2	CF₃CO₂D (D₂O)	0.1		7[o]	44,49,50
$Ru(NH_3)_6^{3+/2+}$	51	0.1	NaBF₄	$(8\pm1)(10^2)$[i]					0.013		3	51-53
$Fe(CN)_6^{3-/4-}$	416	0.1	K₂SO₄	$2.2(10^3)$[i]					0.02	KX[j]	4.5	35,54-56
	409	0.1	Na₂SO₄	$7.3(10^3)$					0.05	KX	4.5	35,54-56
	409	0.1	Na₂HPO₄	$1.5(10^4)$					0.1	KX	4.5	35,54-56
	424	0.1	KCl	$2.6(10^4)$					0.2	KX	4.5	35,54-56
	431	0.26	K₂SO₄	$3.3(10^4)$					0.3	KX	4.5	35,54-56
	423	0.26	Na₂SO₄	$1.8(10^3)$	3.6	-23.7			3.2	$Fe(CN)_6^{3-/4-}$	4.5	35,54-56
	423	0.26	Na₂HPO₄								4.5	35,54-56
	442	0.26	KCl								4.5	35,54-56
	355±1[k]	~0		25	8.56	-23.4			→0		4.5	35,54-57
$Fe(CN)_5N_3^{3-/4-}$	240[l]	0.05	KNO₃	$5.0(10^3)$[m]			5.5-7.0		0.05	KNO₃	4.5[n]	58
$Fe(CN)_5NH_3^{2-/3-}$	330[l]	0.05	KNO₃	$1.8(10^5)$[m]			5.5-7.0		0.05	KNO₃	4.5[n]	58
$Fe(CN)_5P\phi_3^{2-/3-}$	540[l]	0.05	KNO₃	$7.0(10^3)$[m]			5.5-7.0		0.05	KNO₃	4.5[n]	58
$Fe(CN)_4bipy^{-/2-}$	550[l]	0.05	KNO₃	$9.5(10^6)$[m]			5.5-7.0		0.05	KNO₃	4.5[n]	58

TABLE 1 FOOTNOTES

^{a}*pH 4-6, 20°.*

^{b}*For the cross reaction with Fe(CyDTA)$^-$.*

^{c}*The buffers are A (acetate), T (tris), P (phosphate), G (glycine), C (cacodylate) and L (lutidine).*

^{d}*By optical rotation; 25° value from activation parameters.*

^{e}*Calculated from the cross reaction with Co(terpy)$_2$$^{3+/2+}$, corrected from 0.5 to 0.1 M NaCl.*

^{f}*Radii calculated from covalent radii and standard bond lengths.*

^{g}*See footnote 48 for a description of the three models.*

^{h}*Calculated from the cross reaction with Co(phen)$_3$$^{3+}$ at μ = 0.05 M (Ref. 49).*

^{i}*Calculated by extrapolation of the equation of Ref. 35 and assuming k_{298}= 4(k_{273}), approximately pH independent.*

^{j}*Where X is any anion.*

^{k}*Extrapolated to zero ionic strength. $\Delta H°$ = -26.7 ± 0.3 kcal/mole, $\Delta S°$ = -62 ± 1 eu (Ref. 57).*

^{l}*pH 5.5 to 7.0.*

^{m}*Recalculated, including electrostatic corrections, from the data of Ref. 58 for the cross reaction with Fe(CN)$_6$$^{3-/4-}$, assuming a potential of 409 mV and an exchange rate of 7.3(10^3) $M^{-1}s^{-1}$ for Fe(CN)$_6$$^{3-/4-}$.*

^{n}*Assumed equal to Fe(CN)$_6$$^{3-/4-}$.*

^{o}*Assumed equal to Co(phen)$_3$$^{3+}$.*

Hydroquinone goes to the semiquinone after a one electron transfer and two moles of semiquinone disproportionate to give benzoquinone and hydroquinone. A further important property of hydroquinone is that it has a pK of 9.85 (0.65 M, 25°)(60). A value of 10 at 0.04 M has also been given (61).

The properties of those reagents that are required to be outer sphere, or at least usually observed to be so, are treated next. A water molecule is probably coordinated to Fe(EDTA)$^-$ in aqueous solution; the pK of this coordinated water is 7.58 (62). It is possible that the pK of 9.1 found (63) for Fe(EDTA)$^{2-}$ also refers to a coordinated water. The Fe(EDTA)$^-$ complex forms an oxobridged dimer; a value of 340 M^{-1}(26°, 1.0 M NaClO , 0.05 M EDTA buffer, pH 9) has been reported (64) for the equilibrium

$$K = \frac{[Fe(EDTA)_2O^{-4}]}{[Fe(EDTA)OH^{-2}]^2}.$$

The estimates available for the electron exchange rate are from cross reactions of the iron complexes of EDTA and CyDTA. As well as water or hydroxide, Fe(EDTA)⁻ can bind anions in the seventh coordination position; however, this binding is expected to be considerably weaker for the Fe(II) complex than for Fe(III). Because of this labile position, inner sphere pathways must be considered a possibility. If inner sphere mechanisms seem likely, it may be preferable to use Fe(HEDTA)⁻ as a test, as there are two open coordination sites (the seventh and the one left when one acetate of EDTA is replaced by the alcohol of HEDTA) and binding constants for exogenous ligands are somewhat higher.

The tris complexes of cobalt(III) with phenanthroline and its derivatives are an informative series of reagents. The original study of the $Co(phen)_3^{3+/2+}$ exchange was rather imprecise but did highlight the problem of anion catalysis *(36)*. Later studies with an optically active Co(III) complex *(45)* have provided more precise rate constants and good activation parameters. The potentials of the derivatives are given in Table 1. Their self-exchange rate constants have been calculated from cross reaction data with $Co(terpy)_2^{3+}$, and these values are also set out in Table 1.

One of the most common reagents used in electron transfer studies of metalloproteins is $Fe(CN)_6^{3-/4-}$. This is the result of the commercial availability of the reagent, its stability, convenient potential, and large spectral change on changing oxidation state. The properties of the reagent itself have also been studied extensively, probably for the same reason that biochemists have used it; this is fortunate because this system appears to have more significant complications from medium dependences than any other reagent under consideration. The highest pK of $Fe(CN)_6^{3-/4-}$ is about 4 *(65)*. The potential is quite dependent on ionic strength and the type of ion, as is the self-exchange rate; these and other problems of binding of the reagent to proteins probably originate in the solvation properties of such a small, negative ion. Some care is given in the Marcus calculations to choosing the proper potential and exchange rate for a given medium, but because the media of interest are seldom exactly like those used in the physical measurements, some extrapolation is usually necessary and thus the errors in the parameters chosen, especially the self-exchange rate, are larger than might be preferred. Derivatives of the iron hexacyanides where one cyanide is replaced by another ligand have also been studied and some exchange rates are available *(58)*.

The last outer sphere reagent to be discussed is $Ru(NH_3)_6^{2+}$. This reagent is less stable than any of the others discussed except chromous; it is light sensitive and is not stable at neutral

TABLE 2: *PROTEIN PROPERTIES*

	E(mV,25°)	pH	Buffer[a]	μ(M)	Salt	MW(10⁻³)	R(Å)[b]	Lysines
Cytochrome *c* (horse heart)	261[g]	7	T,P,C	0.01-0.23		12.5	16.6	19
Cytochrome *c₅₅₁* (*P. aeruginosa*)	260[j]	7	P	0.2	KCl	8.1	14.4	8
Cytochrome *f* (parsley)	340	7	P	0.1	NaCl	34.0	23.2	24
Cytochrome *c₅₅₂* (*E. gracilis*)	370	6-8	TAGP	0.1		10.7	15.8	4
Cytochrome *c* (*C. krusei*)	264[k]	7		0.01		12.0	16.4	11
Cytochrome *c₃* (*R. rubrum*)	320	7	TAGP	0.1	NaCl	12.8	16.6	17
HiPIP (*Chromatium*)	350	7				10.1	15.5	5
Rubredoxin (*Clostridium*)	-57	7	T	0.1		6.0	13.0	4
Azurin (*P. aeruginosa*)	328 304	6.4 7	P	0.05		13.9	17.2	11
Plastocyanin (bean, *P. vulgarus*)	350 360[p] 370[q]	6.6	P	0.1		10.7	15.8	5
Stellacyanin (*R. vernicifera*)	184	7.1	P	0.3		20	19.5	11

TABLE 2 FOOTNOTES:

[a]*The buffers are as given in Table 1.*

[b]*Calculated from Equation 26.*

[c]*The total is in parentheses, the actual number used (outside parentheses) differs if some pK's are known to be shifted.*

[d]*If the number is in parentheses, only the total of glutamines and glutamates or asparagines and aspartates is known.*

[e]*Oxidized/reduced.*

[f]*Calculated as described in the text; oxidized/reduced.*

[g]ΔH^o = -16.8 ± 0.5 *kcal/mole*, ΔS^o = -36 ± 1.5 *eu.*

[h]*Deprotonated heme nitrogens contribute -2.*

[i]*The two propionate side chains of the heme contribute -2, and for horse heart cytochrome c the N terminus is acetylated; thus there is an additional -1 contribution from the C terminal carboxylate for this protein.*

Arginines	Histidines[c]	Glutamates[d]	Aspartates[d]	Metal Site[e]	Other	Charge[f]	pI	Ref.
2	1(3)	9	3	$1/0^h$	-3^i	7.5/6.5	10.5	1,66
1	0(1)	5	5	$1/0^h$	-2^i	-2/-3	4.7	1,67
11	1(3)	(35)	(37)	$1/0^h$	-2^i		4.7	1,68
1	0(1)	7	5	$1/0^h$	-2^i	-8/-9	5.5	1,69,70
4	2(4)	7	4	$1/0^h$	-2^i	4/3	> 7	1,66
0	0(2)	9	6	$1/0^h$	-2^i	1/0	6.2	71,72
2	1	4	5	$-1/-2^l$		-2.5/-3.5	3.7	73
0	0	(11)	(6)	$-1/-2^m$			< 7.3	74
1	$2(4)^n$	4	11	$0/-1^o$		-1/-2	5.2	75,76,77
0	$0(2)^n$	9	5	$0/-1^o$		-9/-10	$< 6^h$	66,78-80
4	$2(4)^n$	2(4)	5(21)	$0/-1^o$	1^s	11/10	9.9	81-83

^{j}From $K = 1$ with horse heart cytochrome c (Ref. 67).

$^k \Delta H^o = -17.8$ kcal/mole, $\Delta S^o = -39$ eu.

^{l}Includes a contribution of -4 from thiolates, -8 for four sulfides and +11 for the four irons in the oxidized form.

^{m}Four thiolates are coordinated to the single iron.

^{n}Assuming two histidines are coordinated.

^{o}Assuming a deprotonated amide and a thiolate as ligands.

^{p}Parsley plastocyanin.

^{q}Spinach plastocyanin.

^{r}Less than 4.2 of the spinach protein (Ref. 78).

^{s}See text.

pH; furthermore, there are often acid dependences of the rates of ruthenium complexes.

4.2 PROTEINS - The various properties *(1,66-83)* of the proteins are collected in Table 2. The errors given in the table are those claimed by the authors and are smaller than those settled on in an earlier section. The radii were calculated from Equation 26, the charges were calculated from the amino acid sequences as described in Section 2.3, the isoelectric points are experimental except in those cases where only an upper limit is given; in those cases the behavior of the protein on ion exchange resins is used. Some comments are in order for the charge calculation; the metal site charge calculation has already been discussed, and the assumption that both propionates of the heme of a cytochrome are charged is from the result for cytochrome *c* (horse heart). There are not adequate data to make a safe estimate of the charge of stellacyanin, but as there is quite a bit of data available on this protein, an effort was made to make an educated guess in this case, and the assumptions made deserve some comment. The main problem involves the fact that stellacyanin is 20% carbohydrate by weight and the number and type of sugar residues, the number of sugar chains, and the linkage of the sugars to the peptide are unknowns. The problem is further aggravated because the amino acid data available do not distinguish between glutamine and glutamate, or between asparagine and aspartate, and there is the added complication that sugars could be bound to these residues. In charge counting, the linkage of sugars to hydroxyl moieties has no effect, but the more typical asparagine linkages need to be considered. An important distinction that must be made for the amino sugars is whether they have free amines and are charged at neutral pH or are acetylated and thus neutral. The following guesses are made:

(1) Ten of the asparagine/aspartates are covalently linked to sugars (this makes the average chain about four sugar residues long, assuming 40 sugar residues of molecular weight 200 make up the 8,000 dalton sugar contribution); (2) of the remaining 15 aspartate, asparagine, glutamate, and glutamine residues, about half, or 7, are in the acid form; (3) assuming that all but one of the amino sugars are acetylated, and that they compose a total of 4,000 daltons (half of the sugar contribution has been determined to be hexoseamine), there are 17 acetylated positions. This still leaves a charge of 11 on the oxidized protein and clearly more information would be helpful.

Protein electron self-exchange rate constants have been left out of Table 2 because so few are known. The only measured numbers are for cytochrome *c* (both horse heart and *Candida krusei*), determined by NMR line broadening *(84)*. The measured rate constants for the horse heart protein are 1×10^3 $M^{-1}s^{-1}$ (pH 7, tris, 0.1 M, $40°$), with an Arrhenius activation energy of 13 ± 1 kcal/mole, and 1×10^4 $M^{-1}s^{-1}$, with an activation energy of 7 ± 1 kcal/mole under the same conditions except at 1.0 M ionic strength. For the *Candida krusei* protein under the same pair of conditions the rate constants are 1×10^2 and 1×10^3 $M^{-1}s^{-1}$. The rate has also been measured at pH 10 for horse heart cytochrome c and the rate constant is 2×10^4 $M^{-1}s^{-1}$, independent of ionic strength. Similar measurements have been attempted for bean plastocyanin, but the only conclusion that could be reached was that the rate constant was much less than 2×10^4 $M^{-1}s^{-1}$ (pH 7, phosphate, 0.1 M, $50°$) *(85)*.

5. INNER SPHERE PROTEIN-SMALL MOLECULE REACTIONS

There are few well supported examples of inner sphere mechanisms, but those that are available are deserving of comment because of the information they give concerning the protein activation required to take advantage of such a pathway. For the case of the Cr(II) reduction of cytochrome c *(86)*, the data have been interpreted as involving crevice opening and adjacent attack at the iron at low pH in the presence of a high concentration of chloride (1 M). The saturation rate at high Cr(II) concentrations is about 60 s^{-1}. Even in this system, however, at higher pH or in the presence of good π bridging ions (such as azide or thiocyanate), another mechanism is favored that most likely involves remote attack at the heme edge. The data for the reduction of cytochrome c by dithionite (pH 6.5, 1.0 M) have also been interpreted as involving remote and adjacent pathways *(59)*, with the latter pathway again being indicated by saturation of the rate constant to a value (30 s^{-1}) near that for the crevice opening process. This issue is not entirely settled, however, as another possible interpretation involves reduction by dithionite and the free radical SO_2^- by parallel paths *(87,88)*. Inner sphere penetration of a metal center in a protein by a reductant has also been proposed as a pathway in the hydroquinone reduction of tree laccase *(89)*. In this case the type 2 copper was suggested as the site of attack by the hydroquinone monoanion. Unfortunately, the oxidation state of the type 2 copper cannot be monitored by optical absorption methods, and the reduction of the separate type 3 and type 1 sites must be followed. The results show that these two sites are not reduced at quite the same rate or with quite the same activation parameters. Possible mechanisms include prior reduction of the type 2 center followed

by electron transfer from the type 2 copper (reduced) to another
site, or electron transfer from a coordinated hydroquinone
directly to one of the sites. For the fungal form of the enzyme,
fast quench ESR analysis has shown that the type 2 copper is
transiently reduced and reoxidized during the reduction of the
other two sites *(90)*, but caution must be exercised in comparing
the two systems, as their redox mechanisms may well be different
(2).

6. OUTER SPHERE PROTEIN-SMALL MOLECULE REACTIONS

The cross reaction equations of Marcus theory will be employed in many of the calculations to be discussed in this section. Some general justification for the use of these equations,
besides the argument that the reactions must be outer sphere, can
be given. The prediction of a uniform k_{11} value, from given k_{12},
k_{22}, and K values, is not expected because of the presumed mechanistic options for each protein. However, the general pattern of
large differences in driving force being paralleled by a difference in rate in the same direction, and large differences in k_{22}
leading to a faster cross reaction rate constant corresponding to
the reagent with the much larger k_{22}, may be expected. A dramatic difference between the $Co(phen)_3^{3+}$ and the analogous $Co(EDTA)^-$
oxidations of ferrocytochrome *c* may be cited, where the latter
reagent has a self-exchange rate about 10^8 lower *(91)* than that
(45) of the former. The k_{12} for the $Co(EDTA)^-$ oxidation of
ferrocytochrome *c* is at least 10^6 smaller than that for the
corresponding $Co(phen)_3^{3+}$ reaction *(92)*.

The available data for electron transfer reactions between
metalloproteins and inorganic complexes are collected in Table 3
(93-110), and the calculated self-exchange rate constants using
the parameters from Tables 1, 2, and 3 are given in Table 5.

Electrostatic effects are reflected in the ionic strength
dependences of the cross reaction rate constants. As there are
several ionic strength dependence studies available, it is of
interest to fit the available data to the three equations (Equations 5, 6, 28) given in the Theory section. The results of the
fits are set out in Table 6 *(111)*, along with the isoelectric
points, and the charges taken from the sequences (Table 2).
Equation 6, which predicts a linear dependence of log(k) on $\mu^{\frac{1}{2}}$,
is included because of its common use, but it should not be considered as an alternative, as Equation 5 gives the ionic strength
dependence predicted from the standard activity coefficient
theory. In these calculations, only the protein charge has been
considered as a variable, and the rest of the parameters have

TABLE 3: PROTEIN-SMALL MOLECULE CROSS REACTION RATE DATA

Protein	Reagent	$k(M^{-1}s^{-1})$	T	$\Delta H^{\ddagger}$(kcal/mole)	$\Delta S^{\ddagger}$(eu)	pH	Buffer[a]	μ(M)	Salt	Ref.
Cytochrome c (Horse heart)	$Fe(EDTA)^{2-}$	$2.57(10^4)$	25	6.0	-18	7.0	P[b]	0.10	NaCl	93
	$Ru(NH_3)_6^{2+}$	$3.8(10^4)^c$	25	2.9^c	-28^c	3.3-7.0	T	0.10	HCl	94
	$Co(phen)_3^{3+}$	$1.50(10^3)$	25	11.3	-6.2	7.0	P	0.10	NaCl	95
	$Co(5-Cl-phen)_3^{3+}$	$1.26(10^2)$	25	11.3	-6	7.0	P	0.10	NaCl	47
	$Co(5,6-(Me)_2-phen)_3^{3+}$	$2.66(10^2)$	25	9.6	-16	7.0	P	0.10	NaCl	47
	$Co(4,7-(Me)_2-phen)_3^{3+}$	$2.76(10^1)$	25	14.6	-3	7.0	P	0.10	NaCl	47
	$Co(4,7-(\phi-SO_3)_2-phen)_3^{3-}$	$2.87(10^4)$	25	12.8	0	7.0	P	0.10	NaCl	47
	$Fe(CN)_6^{4-,d}$	$2.4(10^4)$	25			7.0	P,T,C	0.10		96
	$Fe(CN)_6^{4-}$	$2.6(10^4)$	22			7.0	P,EDTA	0.18	Na_2SO_4	97
	$Fe(CN)_6^{3-}$	$8.1(10^6)$	22			7.0	P,EDTA	0.18	Na_2SO_4	97
	$Fe(CN)_6^{3-}$	$1.2(10^7)$	25			7.2	P,EDTA	0.10	NaCl	59
		$6.5(10^6)^e$	25			7.0	P	0.10		96
		$6.7(10^6)$	25	0	-27	7.0	P	0.18^f	KCl	67
		$8.0(10^6)$	25	1.1	-23	7.0	P	0.10	KCl	98
	$Fe(CN)_4bipy^-$	$1.6(10^8)$	25			7.0	P	0.10	KCl	98
	$Fe(CN)_5P\phi_3^{2-}$	$3.0(10^7)$	25	1.2	-20	7.0	P	0.10	KCl	98
	$Fe(CN)_5NH_3^{2-}$	$2.5(10^6)$	25	2.4	-21	7.0	P	0.10	KCl	98
	$Fe(CN)_5N_3^{3-}$	$9.0(10^5)$	25	2.9	-21	7.0	P	0.10	KCl	98
Cytochrome $c551$ (*Pseudomonas aeruginosa*)	$Fe(EDTA)^{2-}$	$4.2(10^3)$	25	3.2	-30	7.0	P	0.10		99
	$Co(phen)_3^{3+}$	$5.3(10^4)$	25	12.3	4	7.0	P	0.10		100
	$Co(5-Cl-phen)_3^{3+}$	$4.42(10^3)$	25	9.4	-10	7.0	P	0.10	NaCl	47
	$Co(5,6-(Me)_2-phen)_3^{3+}$	$2.70(10^3)$	25	13.6	7	7.0	P	0.10	NaCl	47

TABLE 3 (Continued)

Protein	Reagent	$k(M^{-1}s^{-1})$	T	$\Delta H^{\ddagger}$(kcal/mole)	$\Delta S^{\ddagger}$(eu)	pH	Buffer[a]	μ(M)	Salt	Ref.
Cytochrome c_{551} (*Pseudomonas aeruginosa*)	$Co(4,7\text{-}(Me)_2\text{-phen})_3^{3-}$	$3.17(10^3)$	25	10.9	-6	7.0	P	0.10	NaCl	47
	$Co(4,7\text{-}(\phi\text{-}SO_3)_2\text{-phen})_3^{3-}$	$2.75(10^4)$	25	13.8	8	7.0	P	0.10	NaCl	47
	$Fe(CN)_6^{3-}$	$8(10^4)$	20			7.0	P,EDTA	0.10		101
Cytochrome c (*Candida krusei*)	$Co(phen)_3^{3+}$	$2.7(10^3)$	25			7.2	P	0.10	NaCl	95
	$Fe(CN)_6^{3-}$	$2.1(10^7)$	25			7.2	P	0.10	NaCl	59
Cytochrome c (*Euglena gracilis*)	$Fe(CN)_6^{4-}$	$7.8(10^3)^{g}$	23	5.2	-21	7.0	TAGP			102
	$Fe(CN)_6^{3-}$	$4.0(10^4)$	23	0	-37	7.0	TAGP			102
Cytochrome f (Parsley)	$Fe(CN)_6^{3-}$	$8.0(10^4)$	25			7.0	P	0.10	NaCl	68
Cytochrome c_2 (*Rhodospirillum rubrum*)	$Fe(CN)_6^{4-}$	$6.6(10^4)^{h}$	20	11.4	1.8	7.0	P	0.10	NaCl	104
	$Fe(CN)_6^{3-}$	$2.2(10^6)$	20	5.2	-10.2	7.0	P	0.10	NaCl	104
HiPIP (*Chromatium vinosum*)	$Fe(EDTA)^{2-}$	$1.7(10^3)$	25	0.4	-41	7.0	P	0.1	NaCl	105
	$Co(phen)_3^{3+}$	$2.8(10^3)$	26	14	4	7.0	P	0.1	NaCl	105
	$Fe(CN)_6^{4-}$	$1.8(10^2)$	25			7.0	T	0.10		3
	$Fe(CN)_6^{4-}$	$1.49(10^2)$	20	4.2	-35	7.3	T	0.008		102
	$Fe(CN)_6^{3-}$	$2.4(10^3)$	25	-0.2	-44	7.0	P	0.10	NaCl	105
	$Fe(CN)_6^{3-}$	$4.2(10^3)$	25			7.0	T	0.10		3
	$Fe(CN)_6^{3-}$	$1.15(10^3)$	20	0	-45	7.3	T	0.008		102

TABLE 3 (Continued)

Protein	Reagent	$k(M^{-1}s^{-1})$	T	$\Delta H^{\ddagger}(kcal/mole)$	$\Delta S^{\ddagger}(eu)$	pH	Buffer[a]	$\mu(M)$	Salt	Ref.
HiPIP *(Chromatium vinosum)*	$Ru(NH_3)_6^{2+}$	$3.1(10^5)$	25			7.0	T	0.10		3
	$Co(bipy)_3^{3+}$	$4.5(10^2)$	25			7.0	T	0.10		107
Rubridoxin *(Clostridium)*	$Ru(NH_3)_6^{2+}$	$9.5(10^4)$	36	~ 1.4	~ -31	6.3–7.0	T	0.10	HCl	108
Azurin *(Pseudomonas aeruginosa)*	$Fe(EDTA)^{2-}$	$1.3(10^3)$	25	2.0	-37	7.0^e	P	0.20	$(NH_4)_2SO_4$	109
	$Co(phen)_3^{3+}$	$3.2(10^3)$	25	14.3	7	7.0	P	0.20	NaCl	100
	$Co(5-Cl-phen)_3^{3+}$	$4.21(10^2)$	25	8.0	-17	7.0	P	0.20	NaCl	100
	$Co(5,6-(Me)_2-phen)_3^{3+}$	$1.54(10^3)$	25	11.6	-5	7.0	P	0.20	NaCl	100
	$Co(4,7-(Me)_2-phen)_3^{3+}$	$8.41(10^1)$	25	9.9	-17	7.0	P	0.20	NaCl	100
	$Fe(CN)_6^{3-}$	$1.2(10^4)$	20			7.0	P	0.10		101
	$Fe(CN)_6^{3-}$	$2.7(10^4)$	25	-4.1	-52	7.0	P,EDTA	0.22		106
	$Fe(CN)_6^{4-}$	$3.4(10^2)$	25	5.9	-27	7.0	P,EDTA	0.22		106
Plastocyanin *(Bean)*	$Fe(EDTA)^{2-}$	$8.2(10^4)$	25	2.1	-29	7.0^e	P	0.20	$(NH_4)_2SO_4$	109
	$Co(phen)_3^{3+}$	$4.9(10^3)$	25	14.0	5	7.0	P	0.10	$(NH_4)_2SO_4$	100
	$Co(5-Cl-phen)_3^{3+}$	$6.96(10^2)$	25	11.5	-6	7.0	P	0.10	$(NH_4)_2SO_4$	100
	$Co(5,6-(Me)_2-phen)_3^{3+}$	$7.97(10^2)$	25	13.6	1	7.0	P	0.10	$(NH_4)_2SO_4$	100
	$Co(4,7-(\phi-SO_3)_2-phen)_3^{3-}$	$2.59(10^1)$	25	7.8	-26	7.0	P	0.10	$(NH_4)_2SO_4$	100
(Spinach)	$Fe(CN)_6^{4-}$	$1.9(10^4)$	25	8.4	-11	6.0	A	0.20		110
(Parsley)	$Fe(CN)_6^{4-}$	$2.0(10^4)$	20	8.8	-9.1	6.0	A	0.20	NaCl	110
	$Fe(CN)_6^{3-}$	$7(10^4)$	25			7.0	P	0.10		68
Stellacyanin *(Rhus vernicifera)*	$Fe(EDTA)^{2-}$	$4.3(10^5)$	25	3.0	-21	7.0^e	P	0.50	$(NH_4)_2SO_4$	109
	$Co(phen)_3^{3+}$	$1.8(10^5)$	25	6	-13	7.0	P	0.10	$(NH_4)_2SO_4$	110
	$Co(5,6-(Me)_2-phen)_3^{3+}$	$1.85(10^4)$	25	9.5	-7	7.0	P	0.10	$(NH_4)_2SO_4$	110
	$Co(4,7-(\phi-SO_3)_2-phen)_3^{3-}$	$2.31(10^6)$	25	5.9	-10	7.0	P	0.10	$(NH_4)_2SO_4$	110

been assumed to have the values given in Tables 1 and 2. This is justified as the charge and radius of each of the reagents are well known, and the radius of the protein may reasonably be estimated from Equation 26, as shown by the comparison to the dimensions of cytochrome c (25 x 30 x 35 Å *(1)*, compared to a predicted diameter for the equivalent sphere of 33.2 Å). Another approach that has been considered involves the "active site" concept. In this model, there is a small region of the protein that is assumed to have a certain charge and radius and to contribute to the electrostatic interactions in an independent way. If the rest of the protein is considered also, the problem becomes one of considering two spheres, one within the other, which are tangential at the point of attack by the reagent. The charge and radius of the small "site" are set and the radius of the protein is used for the larger sphere. The charge on the larger sphere is the difference between the total charge (as determined from the sequence) and that assigned to the site. Reasonable guesses for the site can only be made for proteins for which X-ray data are available; therefore, the example calculation given in Table 4 is for cytochrome c. In principle, this approach could also be used for fitting the ionic strength data, but the number of parameters is larger and the results would be difficult to interpret; therefore, the site model will only be used in some example calculations of work terms *(vide infra)*.

Using the work term theory presented in Section 2.3, the calculated protein self-exchange rate constants may be corrected for electrostatic (coulombic) effects. In order to make these corrections, a charge and radius are needed for the protein; for this purpose the charges calculated from the sequences and the radii given in Table 2 will be used. The resulting $k_{11}{}^{corr}$ values are presented in Table 5. The Marcus theory fit to the ionic strength dependences could be used for estimating the charge in order to maintain internal consistency. As the charges calculated in this way (except for $Fe(CN)_6{}^{3-/4-}$) are close to the sequence charges, and, as ionic strength dependence data are not available for many reactions, these calculations are not presented.

Reagent	Model	$w_{12}{}^a$	$w_{21}{}^a$	$w_{11}{}^a$	$w_{22}{}^a$	$\Delta G_{12}{}^{\star corr}{}^a$	$k_{11}{}^{corr}$ $(M^{-1}s^{-1})$
$Fe(EDTA)^{2-}$	1^b	-0.567	-0.246	0.406	0.493	16.36	6.2
	$2^{c,d}$	-0.471	-0.236	0.251	0.493	16.10	9.7
	3^e	-0.790	-0.395	0.712	0.493	17.05	1.9
$Co(phen)_3{}^{3+}$	1	0.490	0.377	0.406	0.507	13.55	$7.1(10^2)$
	2	0.436	0.291	0.251	0.507	18.64	$6.2(10^2)$
	3	0.614	0.409	0.712	0.507	13.70	$5.6(10^2)$
$Ru(NH_3)_6{}^{2+}$	1	0.655	0.851	0.406	3.402	15.82	$1.6(10^1)$
	2	0.563	0.844	0.251	3.402	15.77	$1.7(10^1)$
	3	1.037	1.556	0.712	3.402	15.06	$5.6(10^1)$
$Fe(CN)_6{}^{3-f}$	1	-0.433	-0.667	0.406	1.752	11.88	$1.2(10^4)$
	2	-0.461	-0.614	0.251	1.752	11.71	$1.6(10^4)$
	3	-0.862	-1.149	0.712	1.752	13.10	$1.5(10^3)$

TABLE 4: SITE MODEL FOR CYTOCHROME c SELF-EXCHANGE RATE CONSTANT CALCULATION

[a] *All energies in kcal/mole; see footnotes b,c,d, and f of Table 5 for a description of the conditions for which the calculations are made.*

[b] *The same as Table 5 (a radius of 16.6 Å and charges of 7.5 and 6.5.*

[c] *The site parameters are derived from measurement of a model of oxidized tuna cytochrome c and the assumption that the attack site is near the point the heme edge comes nearest the surface of the protein; model 3 is considered as a single lysine; model 2 is a larger site including several charged groups.*

[d] *A radius of 8 Å and a charge of 2.*

[e] *A radius of 2 Å and a charge of 1.*

[f] *Ref. 67.*

TABLE 5: CALCULATED PROTEIN SELF-EXCHANGE RATE CONSTANTS, INCLUDING ELECTROSTATIC CORRECTIONS

Protein	Reagent	$w_{12}^{a,b}$	$w_{21}^{a,b}$	$w_{11}^{a,c}$	$w_{22}^{a,d}$	$\Delta G_{11}^{*a,e}$	$k_{11}(M^{-1}s^{-1})^e$	$\Delta G_{11}^{*corr,f}$	$k_{11}^{corr}(M^{-1}s^{-1})^f$
Cytochrome c (Horse heart)	Fe(EDTA)$^{2-}$	-0.567	-0.246	0.406	0.493	14.63	1.2(10^2)	16.36	6.2
	Ru(NH$_3$)$_6$$^{2+}$	0.655	0.851	0.406	3.401	13.51	7.6(10^2)	15.82	1.6(10^1)
	Co(phen)$_3$$^{3+}$	0.490	0.377	0.406	0.507	13.51	7.6(10^2)	13.55	7.1(10^2)
	Co(5-Cl-phen)$_3$$^{3+}$	0.404	0.311	0.406	0.328	12.94	2.0(10^3)	12.95	2.0(10^3)
	Co(5,6-(Me)$_2$-phen)$_3$$^{3+}$	0.431	0.331	0.406	0.377	12.63	3.4(10^3)	12.65	3.3(10^3)
	Co(4,7-(Me)$_2$-phen)$_3$$^{3+}$	0.490	0.377	0.406	0.507	15.17	4.6(10^1)	15.22	4.3(10^1)
	Co(4,7-(ϕ-SO$_3$)$_2$phen)$_3$$^{3-g}$	-0.282	-0.434	0.406	0.304	12.11	8.1(10^3)	13.53	7.4(10^2)
		-0.490	-0.754	0.406	1.014	11.78	1.4(10^4)	14.44	1.6(10^2)
		-0.731	-1.031	0.406	0.956	10.34	1.6(10^5)	13.45	8.4(10^2)
	Fe(CN)$_6$$^{4-h}$	-1.056	-0.686	0.406	2.408	7.19	3.3(10^7)	11.73	1.6(10^4)
	Fe(CN)$_6$$^{3-h,i}$	-0.686	-1.056	0.406	2.408	8.18	6.2(10^6)	12.71	2.9(10^3)
	Fe(CN)$_6$$^{3-j,k}$	-0.433	-0.667	0.406	1.752	8.634	2.9(10^6)	11.88	1.2(10^4)
	Fe(CN)$_5$N$_3$$^{3-}$	-0.686	-1.056	0.406	2.408	5.78	3.6(10^8)	10.34	1.6(10^5)
	Fe(CN)$_5$NH$_3$$^{2-}$	-0.458	-0.792	0.406	1.204	8.74	2.4(10^6)	11.58	2.0(10^4)
	Fe(CN)$_5$P$\phi_3$$^{2-}$	-0.458	-0.792	0.406	1.204	8.24	5.6(10^6)	11.08	4.7(10^4)
	Fe(CN)$_4$bipy$^-$	-0.229	-0.528	0.406	0.401	10.67	9.3(10^4)	12.20	7.0(10^3)
Cytochrome c_{551} (*Pseudomonas aeruginosa*)	Fe(EDTA)$^{2-}$	0.194	0.146	0.079	0.493	16.80	3.0	17.03	2.0
	Co(phen)$_3$$^{3+}$	-0.287	-0.128	0.079	0.507	9.28	9.7(10^5)	10.29	1.8(10^5)
	Co(5-Cl-phen)$_3$$^{3+}$	-0.236	-0.105	0.079	0.328	8.71	2.5(10^6)	9.47	7.0(10^5)
	Co(5,6-(Me)$_2$-phen)$_3$$^{3+}$	-0.252	-0.112	0.079	0.377	9.88	2.5(10^5)	10.71	8.7(10^4)
	Co(4,7-(Me)$_2$-phen)$_3$$^{3+}$	-0.287	-0.128	0.079	0.507	9.55	6.1(10^5)	10.56	1.1(10^5)
	Co(4,7-(ϕ-SO$_3$)$_2$-phen)$_3$$^{3-}$	0.164	0.146	0.079	0.304	12.16	7.4(10^3)	12.24	6.6(10^3)
		0.287	0.255	0.079	1.014	11.21	3.7(10^4)	11.76	1.5(10^4)
		0.434	0.353	0.079	0.956	10.40	1.5(10^5)	10.64	9.7(10^4)
	Fe(CN)$_6$$^{3-i}$	0.406	0.361	0.079	2.408	13.42	8.9(10^2)	15.14	4.9(10^1)

TABLE 5 (Continued)

Protein	Reagent	$w_{12}^{a,b}$	$w_{21}^{a,b}$	$w_{11}^{a,c}$	$w_{22}^{a,d}$	$\Delta G_{11}^{*a,e}$	$k_{11}(M^{-1}s^{-1})^e$	$\Delta G_{11}^{*}corr^{a,f}$	$k_{11}corr(M^{-1}s^{-1})^f$
Cytochrome c (*Candida krusei*)	Co(phen)$_3^{3+}$	0.231	0.205	0.104	0.507	12.73	2.9(10^3)	12.90	2.2(10^3)
	Fe(CN)$_6^{3-}$	-0.324	-0.576	0.104	2.408	6.78	6.7(10^7)	10.17	2.2(10^5)
Cytochrome c_{552} (*Euglena gracilis*)	Fe(CN)$_6^{4-}$	1.230	1.038	0.705	2.408	11.00	5.4(10^4)	11.85	1.3(10^4)
	Fe(CN)$_6^{3-}$	1.038	1.230	0.705	2.408	12.06	8.9(10^3)	12.91	2.1(10^3)
Cytochrome c_2 (*Rhodospirillum rubrum*)	Fe(CN)$_6^{4-}$	-0.141	0	0	2.408	7.49	2.0(10^7)	10.03	2.7(10^5)
	Fe(CN)$_6^{3-}$	0	-0.141	0	2.408	8.19	6.2(10^6)	10.73	8.4(10^4)
HiPIP (*Chromatium vinosum*)	Fe(EDTA)$^{2-}$	0.213	0.149	0.091	0.493	19.81	1.9(10^{-2})	20.02	1.3(10^{-2})
	Co(phen)$_3^{3+}$	-0.296	-0.141	0.091	0.507	10.75	8.1(10^4)	11.79	1.4(10^4)
	Fe(CN)$_6^{4-}$	0.398	0.417	0.091	2.408	15.22	4.2(10^1)	16.91	2.5
	Fe(CN)$_6^{3-}$	0.417	0.397	0.091	2.408	15.62	2.2(10^1)	17.30	1.3
	Ru(NH$_3$)$_6^{2+}$	-0.247	-0.519	0.019	3.401	12.72	2.9(10^3)	16.96	2.3
	Co(bipy)$_3^{3+}$	-0.056	-0.170	0.021	0.507	12.94	2.0(10^3)	13.70	5.6(10^2)
Azurin[1] (*Pseudomonas aeruginosa*)	Fe(EDTA)$^{2-}$	0.041	0.041	0.015	0.493	19.66	2.4(10^{-2})	20.09	1.2(10^{-2})
	Co(phen)$_3^{3+}$	-0.074	-0.025	0.015	0.507	11.09	4.5(10^4)	11.71	1.6(10^4)
	Co(5-Cl-phen)$_3^{3+}$	-0.058	-0.019	0.015	0.328	10.02	2.8(10^5)	10.44	1.4(10^5)
	Co(5,6-(Me)$_2$-phen)$_3^{3+}$	-0.062	-0.021	0.015	0.377	9.06	1.4(10^6)	9.53	6.3(10^5)
	Co(4,7-(Me)$_2$-phen)$_3^{3+}$	-0.074	-0.025	0.015	0.507	12.31	5.8(10^3)	12.93	2.0(10^3)
	Fe(CN)$_6^{3-}$	0.198	0.132	0.015	2.408	14.15	2.6(10^2)	16.24	7.6
	Fe(CN)$_6^{3-m}$	0.104	0.070	0.015	1.752	14.07	3.0(10^2)	15.66	2.0(10^1)
	Fe(CN)$_6^{4-m}$	0.070	0.104	0.015	1.752	13.71	5.5(10^2)	15.30	3.7(10^1)
Plastocyanin (Bean)	Fe(EDTA)$^{2-}$	0.743	0.413	0.882	0.493	15.18	4.6(10^1)	15.35	3.4(10^1)
	Co(phen)$_3^{3+}$	-0.820	-0.492	0.882	0.507	10.09	2.5(10^5)	12.79	2.6(10^3)
	Co(5-Cl-phen)$_3^{3+}$	-0.677	-0.406	0.882	0.328	8.93	1.7(10^6)	11.23	3.6(10^4)
	Co(5,6-(Me)$_2$-phen)$_3^{3+}$	-0.721	-0.432	0.882	0.377	9.34	8.7(10^5)	11.76	1.5(10^4)

TABLE 5 (Continued)

Protein	Reagent	$w_{12}^{a,b}$	$w_{21}^{a,b}$	$w_{11}^{a,c}$	$w_{22}^{a,d}$	$\Delta G_{11}^{*a,e}$	$k_{11}(M^{-1}s^{-1})^e$	$\Delta G_{11}^{*corr\,a,f}$	$k_{11}^{corr}(M^{-1}s^{-1})^f$
Plastocyanin (Bean)	Co(4,7-(ϕ-SO$_3$)$_2$-phen)$_3$$^{3-}$	0.471	0.565	0.882	0.304	18.36	$2.1(10^{-1})$	18.51	$1.6(10^{-1})$
		0.820	0.984	0.882	1.014	17.41	1.1	17.50	$9.1(10^{-1})$
	Fe(CN)$_6$$^{4-\,k}$	1.229	1.352	0.882	0.956	16.59	4.2	15.84	$1.5(10^{1})$
Plastocyanin (Spinach)	Fe(CN)$_6$$^{4-\,n}$	0.804	0.670	0.882	1.752	9.83	$3.8(10^{5})$	11.00	$5.3(10^{4})$
					1.752	10.25	$1.9(10^{5})$	11.42	$2.6(10^{4})$
Plastocyanin (Parsley)	Fe(CN)$_6$$^{3-}$	1.153	1.384	0.882	2.408	11.40	$2.7(10^{4})$	12.15	$7.6(10^{3})$
Stellacyanin^o (*Rhus vernicifera*)	Fe(EDTA)$^{2-}$	0	0	0	0.493	9.63	$5.4(10^{5})$	10.12	$2.3(10^{5})$
	Co(phen)$_3$$^{3+}$	0	0	0	0.507	9.46	$7.1(10^{5})$	9.97	$3.0(10^{5})$
	Co(5,6-(Me)$_2$-phen)$_3$$^{3+}$	0	0	0	0.377	9.21	$1.1(10^{6})$	9.59	$5.8(10^{5})$
	Co(4,7-ϕ-SO$_3$)$_2$-phen$_3$$^{3-}$	0	0	0	0.304	8.56	$3.3(10^{6})$	8.86	$2.0(10^{6})$
		0	0	0	1.014	7.61	$1.6(10^{7})$	8.62	$3.0(10^{6})$
		0	0	0	0.965	6.78	$6.6(10^{7})$	7.74	$1.3(10^{7})$

a) All energies are in kcal/mole.
b) Work terms are calculated using the conditions given in Table 3 and assuming protein charges calculated from the sequences (Table 2).
c) Work term calculated for 0.1 M ionic strength, pH 7, and assuming protein charges calculated from the sequences (Table 2).
d) Work term calculated for the condition of the measured self-exchange rate constant given in Table 1.
e) Calculated without consideration of electrostatic interaction.
f) Calculated for 0.1 M ionic strength, pH 7, including compensation for electrostatic interactions assuming protein charges calculated from the sequences (Table 2).
g) The three different values are for three models for the electrostatic interaction; see footnote 48.
h) From the data of Ref. 96.

i) Using $k_{22}(Fe(CN)_6^{3-/4-})$ of $1.5(10^4)$ $M^{-1}s^{-1}$ and a potential of 425 mV from the 0.1 M ionic strength (KCl) data.
j) From the data of Ref. 67.
k) Using $k_{22}(Fe(CN)_6^{3-/4-})$ of $2.6(10^4)$ $M^{-1}s^{-1}$ and a potential of 433 mV from the 0.2 M ionic strength data.
l) A potential of 330 mV is used for azurin, as given in Ref. 75.
m) A potential difference of 120 mV is used (Ref. 106) and a $Fe(CN)_6^{3-/4-}$ exchange rate of $2.6(10^4)$ $M^{-1}s^{-1}$ for 0.2 M ionic strength.
n) Using $k_{22}(Fe(CN)_6^{3-/4-})$ of $2.6(10^4)$ $M^{-1}s^{-1}$ and a potential of 435 mV.
o) Assuming $Z = 0$ from the ionic strength dependence with Co(phen)$_3$$^{3+}$; see Table 6.

Another approach to treating the problem of electrostatic interaction is to do the calculations with the formula that includes the equilibrium constants for the formation of the precursor and successor complexes (Equation 29). This is the preferred method when pre-equilibrium binding is shown by saturation behavior in the plot of k_{obsd} *vs.* reagent concentration. There is, however, a problem. Although some of the binding constants may be estimated from the kinetics, others are not so available (this is generally true for the exchange reactions, where both oxidation states are usually positive or negative, and therefore not expected to show discernible binding based on electrostatics), and must be estimated using Equation 29. An important pair of protein cross reactions which may be treated in this manner involve the iron hexacyanides with horse heart cytochrome c, where saturation behavior is observed (as are exceptionally high rates). The cytochrome c self-exchange has been calculated using the equilibrium constant approach (Table 7). The result may be compared to predictions based on treatment of the same data with work terms (Table 5).

Shown in Fig. 6 is a plot of the log of the ratio between the calculated protein self-exchange rate constant with a variety of reagents and with $Fe(EDTA)^{2-}$ *vs.* log $k_{11}{}^{corr}$ based on $Fe(EDTA)^{2-}$ alone. The exchange rate constants used in making up the figure are the best calculated values, using the work and f terms. Ignoring, for now, the abscissa, the range of ordinate values can be considered. If a protein acts like a simple inorganic reagent, then the self-exchange rate constant should be independent of the cross reaction it is calculated from, and the calculated $k_{11}{}^{corr}$ values should cluster about a single point for each protein. If, on the other hand, the $k_{11}{}^{corr}$ values calculated from the different reagents do not agree, as is usually the case, there can be several reasons. The contributions to $\Delta G_{11}{}^{**}$ include: (1) the inherent activation of the protein metal center, which should to a good approximation be invariant, (2) the non-electrostatic (*i.e.*, noncoulombic) interactions between the protein and reagent involved in attaining the activated complex configuration, (3) breakdown of the assumption that the adiabaticity factors may be ignored, and (4) deviations from the model used for making the electrostatic calculations (these include contributions to the M terms in Fig. 5: non-uniform charge distribution, induced changes in charge distribution, deviations from Debye-Hückel behavior, and any specific medium effects). Several of these factors are related; for example, contributions (2) and (3) are linked because orbital overlap will usually be improved if the reagent can more closely approach the protein metal center; but, as the metal center is at least to some extent protected by the peptide chain, approach by the reagent will involve interactions that are likely to include noncoulombic ones. Whatever the precise origin of the variation, the wider the range of calculated k_{11}^{corr} values for a single protein from a given set of reagents, the greater is the variety of mechanistic pathways employed by it. These are lowest free energy routes adopted in response to the varying demands (steric, charge, hydrophobicity, and orbital overlap) of the reagent. The calculated k_{11}^{corr} values clearly show that some proteins exhibit a greater reactivity spread than others, ranging from more than six orders of magnitude for azurin to much less than one for stellacyanin. The proteins on the left side of Fig. 6 are interpreted as having more available mechanisms (a greater sensitivity to the differences between the reagents), whereas stellacyanin, at the extreme right, apparently employs a single mechanism in its electron transfer reactions.

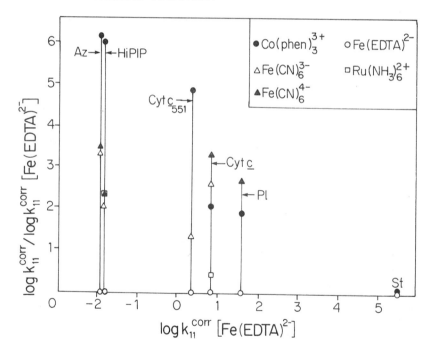

Fig. 6. Accessibility of protein redox centers to small molecule reagents based on a plot of $\log k_{11}^{corr}$ [reagent]/k_{11}^{corr} [Fe(EDTA)$^{2-}$] vs. $\log k_{11}^{corr}$ [Fe(EDTA)$^{2-}$].

It is of interest to consider what protein properties might parallel the variation in the k_{11}^{corr} spread observed; the most obvious general property is the degree to which the metal site (or the periphery of its conjugated ligand systems) is buried within the peptide. HiPIP is the best available example of a buried site, with the distance of closest approach of the iron-sulfur cluster to the surface estimated *(6)* as at least 3.5 Å; azurin is another example, as its copper center has been estimated from solution studies to be similarly isolated from the solvent *(15-17)*. One heme edge of cytochrome *c* is known from crystal structure studies *(1)* to be positioned near (2 Å or less) the surface of the protein at one point, and thus is more available for electron transfer than the azurin and HiPIP centers. Less solution data are available for plastocyanin and stellacyanin, but based on the aforementioned reactivity correlations, their copper sites, which are assumed similar in coordination environment to that of azurin, must be more available to reagents in solution than any of the other proteins considered. The property that leads to the range of calculated k_{11}^{corr} values can be called the kinetic availability or accessibility of the metal center. This property is not expected to correlate precisely with X-ray structural evidence, as the availability of a site to external reagents intimately involves protein-reagent interactions as well as the dynamics of conformational changes of the protein molecule in solution. It is comforting, however, that in most cases the static physical picture of buried *vs.* exposed sites does relate to the kinetic patterns observed.

We shall now consider why $Fe(EDTA)^{2-}$ cross reactions give such a large range of calculated k_{11}^{corr} values, why this range seems to parallel the kinetic availability of the protein metal center, and why the other reagents cause the protein to react so differently from the mechanism with $Fe(EDTA)^{2-}$ in most cases. First, $Fe(EDTA)^{2-}$ is far from spherically symmetrical, with one face that is hydrophobic, being mostly methylene hydrogens, and the opposite face being hydrophilic, owing to the carbonyl oxygens and a coordinated water. Only the hydrophilic side has any π symmetry ligand orbitals, but only the other side is likely to be able to penetrate the hydrophobic residues protecting the metal centers of most proteins. Thus, $Fe(EDTA)^{2-}$ would have to pay a high enthalpic cost to force its π orbitals in close enough to overlap significantly with the orbitals at the redox center of the protein. The alternative of penetration with the hydrophobic side leading is not attractive, as this side is insulating with respect to π electron transfer. Thus, $Fe(EDTA)^{2-}$ is expected to be quite sensitive to the kinetic accessibility of the protein metal site, as seems to be borne out by the data. For this reason the k_{11}corr values based on Fe(EDTA)$^{2-}$ were selected for the abscissa of Fig. 6.

The reagent that usually predicts the highest k_{11}corr is Co(phen)$_3^{3+}$. In this case binding due to electrostatic interactions should not be significant and the work term calculations should make up for the general coulombic interactions; however, the hydrophobic nature of the phenanthroline ligands may be expected to encourage penetration of the protein surface. The blade-like nature of the phenanthroline rings should also encourage penetration as compared to a complex that is more nearly spherical and has the same radius as Co(phen)$_3^{3+}$ (about 7 Å). The matching of the π symmetry of the phenanthroline orbitals with the π symmetry of conjugated ligand systems and the redox orbital of the protein metal center should make the cobalt system more reactive than Fe(EDTA)$^{2-}$. One problem that remains in the Co(phen)$_3^{3+}$ system is that the planar, bidentate phen groups are held in rigid geometric position with respect to each other; thus the two phenanthrolines not involved in the π overlap will still affect the orientation of the pseudo-bridging ligand to the extent that they have preferred interactions with the peptide.

The reactions of Fe(CN)$_6^{3-/4-}$ are often quite fast, beating out Co(phen)$_3^{3+}$ in predicted k_{11}corr in several cases. The main problem in interpreting the cross reactions of the iron hexacyanides is the possibility (and sometimes the observation) of binding. The work term treatment used to obtain the k_{11}corr values shown in Fig. 6 is applicable in the extreme of no precursor complex formation. The precursor complex formulas should be used when such complexation is observed. One problem, as has been discussed before, is that the estimation of precursor complex formation constants is difficult and usually still necessary for the exchange reactions (and this method apparently does not work well in the Fe(CN)$_6^{3-/4-}$ case (see Table 7) or at least predicts a rather small k_{11}). On symmetry grounds, Fe(CN)$_6^{3-/4-}$ should be quite reactive with proteins, because the electron transfer path is through ligands that have a cylindrically symmetrical π orbital set and that will not have such precise requirements for orientation as does phenanthroline; however, the cyanide ligands do not have as great an advantage in promoting hydrophobic interactions as do the phen groups in the Co(phen)$_3^{3+}$ reactions.

The two reactions given for Ru(NH$_3$)$_6^{2+}$ are insufficient information for much discussion of this reagent, but in both cases the reactivity shown is between that of Fe(EDTA)$^{2-}$ and Co(phen)$_3^{3+}$. As the ammine ligands are hydrophilic, the ruthenium complex is expected to behave more like Fe(EDTA)$^{2-}$ in this regard, and the ligands do not have π orbitals to facilitate electron transfer into and out of the ruthenium center. However, the t_{2g} orbitals of the second row transition elements extend much farther from the nucleus than those of the first transition series, and in this sense the overall small size of $Ru(NH_3)_6{}^{2+}$ could be an advantage for effective π electron transfer.

The second large set of related data involves the oxidation of several of the proteins by tris complexes of various phenanthroline derivatives of Co(III). These data are presented in Fig. 7. For convenience, the five derivatives will be given numbers, as follows: **1** is the parent complex, $Co(phen)_3{}^{3+}$; **2** is the complex with 5-Cl-phen; **3** is 5,6-Me$_2$ substituted; **4** is 4,7-Me$_2$ substituted; and **5** has phenylsulfonates in the four and seven positions of the ring, 4,7-(ϕ-SO$_3$)$_2$-phen.

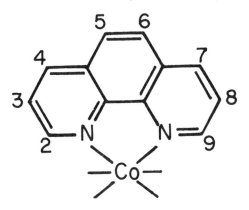

Complex **2** differs from the parent compound primarily because of the inductive effect of the electron-withdrawing chlorine; the perturbation such an effect might have on the potential is compensated for in the Marcus calculation, but the result of different binding capabilities of the Cl compared to an H will show up in the $k_{11}{}^{corr}$ value. The chlorine substituent also blocks the 5-position. The self-exchange rate predicted for **2** from the cross reaction with $Co(terpy)_2{}^{2+}$ (see Table 1) is $1.3(10^{-2})M^{-1}s^{-1}$, or about 3,000-fold below that for **1** and similar to that for **3**, which is also blocked in the 5-position. The inductive effect and any interaction such as a hydrogen bond to the Cl are not expected to be influential in the reaction with $Co(terpy)_2{}^{2+}$, so this result is quite reasonable given the previous discussion. Complex **3** has the distal edge of each of the phenanthroline rings blocked, but the 4 and 7 positions are left unhindered; this last point is best seen in space filling models of the complex. The methyl substituents are electron donating with respect to hydrogen, and, as with **2**, the redox potential of **3** is higher than that of the parent ion. Substitution at the 4 and 7 positions also produces electronic effects on Co(III), as complexes **4** and **5** have potentials that are 30 and 40 mV, respectively, below that of the parent complex (and roughly 90 mV below **2** and **3**). Sterically, substitution in the 4 and 7 positions blocks both approaches to the phenanthroline rings in the complex (again, this is best seen by examining space filling models); thus, steric effects are expected to be more pronounced in **4** than in **3**. Little difference is observed between the rate constants for the cross reactions of Co(terpy)$_2{}^{2+}$ with **3** and **4**, but, because of the potential difference, the calculated k_{22} value for **4** is 0.77 $M^{-1}s^{-1}$, or about 50-fold less than **1**. The fact that the calculated $k_{11}{}^{corr}$ for **4** is greater than that for **3** or **2** is puzzling, and could mean that hydrophobic interactions are more important than steric hindrance in this case. Reagent **5** is the only negatively-charged member of the set and is also by far the largest in size, if the radius is taken to include the sulfonate group (various estimates are given in Table 1). The phenyl group may or may not extend the conjugation of the system out to the sulfonate, depending on whether its rings can become coplanar (or approximately coplanar) with the phenanthroline; examination of space filling models makes it clear that the interaction between the phenyl hydrogens *alpha* to the point of attachment to the phenanthroline and the 3,5 or 4,6 hydrogens should strongly encourage the phenyl rings to be well out of the ligand plane. However, attaining approximate coplanarity in a transition state cannot be ruled out, and it should be noted further that in **5** there are large hydrophobic channels between the phenyls of each ligand that lead to a phenanthroline edge. The cross reaction with Co(terpy)$_2{}^{2+}$ predicts a $k_{11}{}^{corr}$ of 3.5 x 10^2 to 7 x 10^3 $M^{-1}s^{-1}$, depending on the way the work terms are calculated. Anywhere in this range the rate constant is much greater than that for any of the other derivatives; although something special about the interaction between **5** and Co(terpy)$_2{}^{2+}$ is a possibility, it is not likely because the estimate of the $k_{11}{}^{corr}$ for **5** from the cross reaction with Ru(NH$_3$)$_5$py^{2+} is also relatively high. Despite the general agreement between the ruthenium and cobalt reagent cross reactions with **5**, erratic or at least different behavior may be expected for the proteins whose redox centers cannot penetrate to the phenanthroline edge as easily as smaller reactants.

Based on the discussion of the previous paragraph and the model of kinetic accessibility developed earlier, the results for the derivatives of Co(phen)$_3{}^{3+}$ shown in Fig. 7 may now be considered. Complexes **1**, **2**, and **3** predict larger $k_{11}{}^{corr}$ values than

complexes **4** and **5**, except with stellacyanin, where **5** predicts a
$k_{11}{}^{corr}$ about an order of magnitude larger than the value predicted from **1** and **3**. The differences among reagents **1**, **2**, and **3**
are probably too small to interpret, but it is interesting that
2 or **3** (or both) always does better than **1**. Thus one of these
derivatives always leads to some advantage in protein-reagent
interaction.

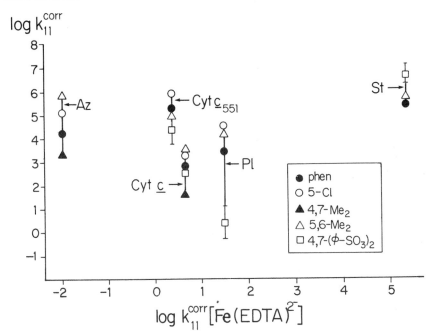

*Fig. 7. Comparison of the $k_{11}{}^{corr}$ values based on reactions involving the
ring-substituted $Co(phen)_3{}^{3+}$ reagents.*

The observation that stellacyanin is generally less sensitive
than the other proteins to the differences among the cobalt complexes is consistent with the previous conclusion that its copper
site is kinetically accessible and thus behaves more like a
simple reagent. Considering the other four proteins, **4** or **5**
always predicts a $k_{11}{}^{corr}$ at least an order of magnitude smaller
than the others. The result is consistent with the general
steric arguments given in the preceding section, as substitution
at the 4,7 positions gives more overall steric hindrance than
substitution at the 5,6 positions; thus the proteins (except
stellacyanin) are uniformly more sensitive to steric hindrance

than is $Co(terpy)_2^{2+}$. The most extreme difference in reactivity involves plastocyanin with **5** as compared with all of the other reagents. If the plastocyanin result is accepted *(112)*, then this response to the difference between **5** and **1**, **2**, or **3** is totally out of character with the differences seen with the other proteins (in no other case is the gap between the $k_{11}corr$ from one to the next nearest reagent greater than an order of magnitude). Apparently the redox centers in plastocyanin and **5** cannot easily come together, either because of unfavorable interactions of the protein with the charged sulfonates or an inability to penetrate the hydrophobic channels.

Cytochrome c is uniformly about 2 orders of magnitude less reactive with the Co(III) reagents than is cytochrome c_{551}. The results indicate that the electron transfer mechanism of cytochrome c_{551} involves a site that can accept a hydrophobic reagent such as a phenanthroline complex in such a way as to give it good kinetic accessibility to the porphyrin (or the iron). Similarly, plastocyanin is about one order of magnitude less reactive than azurin. Overall, therefore, the protein reactivity order that may be inferred from cobalt phenanthroline derivatives is much different from that indicated by the cross reactions with $Fe(EDTA)^{2-}$. Although azurin is quite inaccessible to solvent and reagents such as $Fe(EDTA)^{2-}$, the hydrophobic cobalt phenanthroline complexes have little difficulty in gaining access to its blue copper center. The high reactivity of stellacyanin is still consistent, as its site is exposed to all reagents.

Owing to their stability and commercial availability, the iron hexacyanides are the reagents that have been used most widely to study the electron transfer reactivity of metalloproteins. In retrospect, the choice was a poor one, as these small, highly-charged ions often show complicated kinetic behavior because of electrostatic binding, and both the exchange rate and the redox potential of $Fe(CN)_6^{3-/4-}$ are strongly medium dependent. Some discussion of the available data can reasonably be made, provided that we keep in mind the problems associated with binding, even in those cases where it has not been proven. Table 5 and Fig. 8 present the available data for the reactions of ferri- or ferrocyanide with ten proteins. Because correction for driving force is made in the Marcus calculation, the calculated value of the protein self-exchange rate should be independent of the direction (reduction or oxidation of the protein) in which the reaction is run.

For those proteins for which data are available for both ferricyanide oxidation and ferrocyanide reduction, the $k_{11}corr$ values are in sizable disagreement when the ionic strength dependence charges are used. When the sequence charges are

used, there is reasonable agreement between the ferri- and ferrocyanide results (except for the *Euglena gracilis* cytochrome *c*
data) and the proteins may be ranked according to their $k_{11}{}^{corr}$
values, as was previously done for Fe(EDTA)$^{2-}$ as reagent. The
order with Fe(CN)$_6{}^{3-}$ (Fe(CN)$_6{}^{4-}$ gives a similar order) is *Candida*
cytochrome *c* > *R. rubrum* cytochrome c_2 > plastocyanin > horse
heart cytochrome *c* $\cong$ *Euglena* cytochrome *c* > cytochrome c_{551} >
azurin > HiPIP. This order is similar to that based on Fe(EDTA)$^{2-}$
except that azurin and HiPIP have switched places. The ordering
from Fe(EDTA)$^{2-}$ is probably more to be trusted because of the
binding problem, but the Fe(CN)$_6{}^{3-/4-}$ results do place the first
three cytochromes in the high reactivity category, and suggest
some interesting future experiments involving the more trust-
worthy reagents.

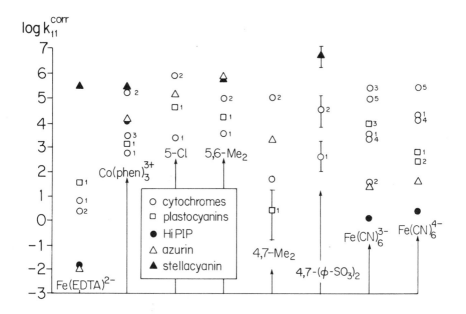

*Fig. 8. Relative protein electron transfer reactivities with various inorganic
reagents. The cytochromes are (1) horse heart, (2) Pseudomonas aeruginosa,
(3) Candida krusei, (4) Euglena gracilis, and (5) Rhodospirillum rubrum. The
plastocyanins are (1) bean, (2) spinach, and (3) parsley.*

The use of the Marcus theory equations including binding constants might make more sense of the $Fe(CN)_6^{3-/4-}$ data, but the binding constants are not well enough defined and the one example treated (see Table 7) gives such an unexpectedly low k_{11}^{corr} that the method must be considered suspect. This may well be the result of the binding involved not conforming to the equation (Equation 29) used for estimating the unmeasured precursor and successor binding constants, because of the ionic strength used or because nonelectrostatic influences are important.

Related to the analysis of the $Fe(CN)_6^{3-/4-}$ reactions is a study of the oxidation of horse heart cytochrome c with various derivatives of ferricyanide as listed in Table 5. The comparison of the k_{11}^{corr} values shows that cytochrome c is most reactive with $Fe(CN)_5N_3^{3-}$, followed by $Fe(CN)_5P\phi_3^{2-}$, then $Fe(CN)_5NH_3^{2-}$ and $Fe(CN)_6^{3-}$ itself, with $Fe(CN)_4bipy^-$ the poorest reagent. The ordering would indicate that the azide must make a better bridge than cyanide. Triphenylphosphine is second best at facilitating the reaction; it could act as a bridge, or it could assist the reaction through favorable protein-$P\phi_3$ hydrophobic interactions, or both. The next two reagents do about as well and are quite similar in shape and ligand composition; they differ mainly in charge (which should have been compensated for anyway). Thus the ammonia does not help the reaction, but the loss of one cyanide does not hurt either. The fact that the bipyridine complex is significantly less reactive is hard to understand and no explanation will be offered. The range of kinetic effects is smaller than would be preferred for solid conclusions, especially as there are large differences in potentials and self-exchange rates for the various reagents. It does seem clear, however, that small complexes containing linear ligands possessing π systems are especially reactive, and hydrophobic interactions are likely to be important.

The analysis of ionic strength (general salt) dependences will now be discussed. Table 6 gives the results of fitting the available ionic strength dependence data to the three equations from Section 2.3. Equation 6 will be ignored for the reasons discussed before. First, comparing the fits to the Marcus theory equation (Equation 28) with the charges derived from the sequence data, it can be seen that the values of Z_1 are consistently similar to the calculated charge except in a few cases. The most general exception to the success of the Marcus theory fits is from the $Fe(CN)_6^{3-/4-}$ data, which give some large values of the protein charge (*e.g.*, a value of 14.6 for the cross reaction between $Fe(CN)_6^{4-}$ and horse heart cytochrome c and the charges derived from the data for the cytochrome c_2 reactions). The other major inconsistency is the zero charge calculated for

TABLE 6: IONIC STRENGTH DEPENDENCE FITS[a]

Protein	Reagent	Equation 6			Equation 5			Equation 28			Ref.
		Z_1	k_0	S.E.[b]	Z_1	k_0	S.E.[b]	Z_1	k_0	S.E.[b]	
Cytochrome c (7.5/6.5, 10.5)[c] (Horse heart)	$Fe(EDTA)^{2-}$	1.69	$2.68(10^5)$	$7.8(10^2)$	5.49	$7.96(10^5)$	$1.2(10^3)$	8.10	$8.58(10^3)$	$1.9(10^3)$	93
	$Co(phen)_3^{3+}$	0.42	$6.00(10^2)$	$1.2(10^1)$	5.73	$3.08(10^1)$	$1.8(10^1)$	4.71	$2.89(10^3)$	$2.0(10^1)$	95
	$Fe(CN)_6^{3-}$	0.64	$6.94(10^7)$	$1.2(10^6)$	2.01	$2.25(10^8)$	$5.1(10^5)$	6.61	$3.81(10^6)$	$5.8(10^5)$	67
	$Fe(CN)_6^{4-}$	0.49	$2.93(10^5)$	$1.1(10^6)$	1.01	$1.90(10^6)$	$9.1(10^2)$	14.6	$5.74(10^3)$	$7.6(10^2)$	67
	Cytochrome c[d] (Horse heart)	1.55	$2.84(10^2)$	$4.8(10^4)$	5.75	$1.9(10^{-1})$	$5.4(10^2)$	10.3	$9.51(10^3)$	$1.1(10^3)$	84
Cytochrome c_{551} (-2/-3, 4.7) (*Pseudomonas aeruginosa*)	$Co(phen)_3^{3+}$	0.55	$2.32(10^5)$	$8.5(10^3)$	-2.0	$6.52(10^5)$	$7.3(10^3)$	-4.33	$3.47(10^4)$	$7.2(10^3)$	100
Cytochrome c_{552} (-8/-9, 5.5) (*Euglena gracilis*)	$Fe(CN)_6^{4-}$	-0.18	$1.73(10^3)$	$1.90(10^2)$	-5.9	$5.63(10^1)$	$5.6(10^2)$	-1.30	$4.78(10^3)$	$2.5(10^3)$	69
	$Fe(CN)_6^{3-}$	-0.42	$8.60(10^3)$	$3.1(10^3)$	-7.0	$3.73(10^2)$	$2.6(10^3)$	-2.98	$4.94(10^4)$	$2.5(10^3)$	69
Cytochrome c_2 (1/0, 6.2) (*Rhodospirillum rubrum*)	$Fe(CN)_6^{4-e}$	0.73	$1.06(10^6)$	$6.8(10^4)$	4.46	$9.10(10^7)$	$3.9(10^3)$	13.96	$2.54(10^3)$	$7.3(10^4)$	104
	$Fe(CN)_6^{3-f}$	0.39	$7.61(10^6)$	$3.1(10^3)$	0.90	$1.90(10^7)$	$2.6(10^3)$	7.26	$6.96(10^5)$	$2.9(10^5)$	104
	$Fe(CN)_6^{4-g}$	0.08	$4.80(10^2)$	9.4	-4.8	$1.72(10^1)$	9.6	1.54	$2.47(10^2)$	9.0	104
Azurin (-1/-2, 5.2) (*Pseudomonas aeruginosa*)	$Fe(EDTA)^{2-}$	-0.15	$1.12(10^3)$	$1.0(10^1)$	-3.55	$3.96(10^2)$	$3.4(10^1)$	-1.10	$1.68(10^3)$	$1.4(10^1)$	111
	$Co(phen)_3^{3+}$	-0.16	$3.18(10^3)$	$3.1(10^1)$	1.04	$2.01(10^3)$	$3.1(10^1)$	-1.89	$1.72(10^3)$	$2.6(10^1)$	100
Plastocyanin (-9/-10, <6) (Bean)	$Fe(EDTA)^{2-}$	-0.55	$2.05(10^4)$	$3.9(10^3)$	-8.88	$5.62(10^2)$	$7.9(10^2)$	-7.20	$1.39(10^5)$	$1.2(10^3)$	111
Stellacyanin (10, 9.9) (*Rhus vernicifera*)	$Co(phen)_3^{3+}$	0.00	$1.92(10^5)$	$1.4(10^3)$	3.49	$3.51(10^4)$	$2.2(10^3)$	0.00	$1.89(10^5)$	$1.4(10^3)$	100
HiPIP (-2.5/-3.5, 3.7) (*Chromatium vinosum*)	$Fe(CN)_6^{3-}$	-0.46	$8.52(10^2)$	$2.6(10^1)$	-5.2	$9.3(10^1)$	$1.5(10^2)$	-2.25	$4.1(10^3)$	$1.3(10^2)$	102
	$Fe(CN)_6^{4-}$	0.00	$1.51(10^2)$	5.0	-3.2	$2.6(10^1)$	9.8	-0.01	$1.50(10^2)$	5.0	102

TABLE 6 FOOTNOTES

[a]*Least squares procedure.*

[b]*The standard error, defined as*
$(\Sigma(k_{obs}-k_{fit})^2)/N$.

[c]*These numbers are the charges calculated from the sequence (oxidized/reduced) and the isoelectric point.*

[d]*Solving for the charge assuming both oxidation states have the same charge.*

[e]*This is k_{12} of Ref. 104.*

[f]*This is k_{43} of Ref. 104.*

[g]*This is k_{21} of Ref. 104.*

stellacyanin from the cross reaction with $Co(phen)_3{}^{3+}$. The result that stellacyanin appears uncharged to $Co(phen)_3{}^{3+}$ is surprising based on the isoelectric point of the protein, but, as the sequence data are so sketchy and the carbohydrate contributions are not defined, it is not clear that there is a contradiction.

The inconsistencies based on the iron hexacyanide data probably have their origin in the previously-mentioned problems of the medium dependence of many of the properties of this reagent. Foremost among these problems in the present case is the medium dependence of the exchange rate and the ion pairing between ferri- or ferrocyanide and most cations. Because of ion association, the effective $Fe(CN)_6{}^{3-/4-}$ self-exchange rate varies with the concentration of potassium ion, for example, and this dependence is quite large; at $0.1°$, the self-exchange rate varies from 1.75×10^2 $M^{-1}s^{-1}$ at 0.01 M to 1.5×10^4 $M^{-1}s^{-1}$ at 0.1 M K^+ (extrapolated value). As this variation is attributable to a high degree of ion association, the ionic strength dependence equations that are used are invalid and the actual reagents in solution in many cases would be better considered as the K^+-$Fe(CN)_6{}^{3-/4-}$ pairs. In this regard, it should be emphasized that in the excellent experimental study of the $Fe(CN)_6{}^{3-/4-}$ exchange rate, electrostatic interaction theory was not used to predict the nature of the dependences on various cations; instead, a rate law was presented that included reaction paths for $Fe(CN)_6{}^{3-/4-}$ with one or two cations, and different rate constants and the binding constants (or functions thereof) were assigned for each path. In the previous calculations with ferri- or ferrocyanide, the reactants have been assumed to be the free ions (without bound cations), and the work terms have been calculated for these species, but the exchange rates have been chosen to reflect the presence of the cations in the reaction medium. The result of this treatment is to attribute to the medium some of the dependence that is actually electrostatic and belongs in a work term. This trade-off is not as much a problem for the terms that describe the $Fe(CN)_6{}^{3-/4-}$ exchange as it is

for the work terms of the cross reaction. Thus, a better approach would be to include all of the different paths that might
be involved, but such an approach could quickly get out of hand
and would require a complete re-analysis of the $Fe(CN)_6{}^{3-}/^{4-}$
exchange rate data, as well as accurate ion pair binding constants under all conditions of interest, which are not available.
In some of the cases, the ferricyanide data give quite a reasonable value for Z_1, whereas ferrocyanide leads to a much different
value; this difference is probably attributable to the difference
in ion association constants between ferrocyanide (K^+ binding
constant about 31 M^{-1} at 0.1 M ionic strength)*(113,114)* and
ferricyanide (K^+ binding constant 7 M^{-1} under the same conditions
(113,114).

Equation 5, from transition state theory, is somewhat more
erratic in its predictions of ionic strength dependences than is
the Marcus theory equation, as many of the fits produce a change
of sign of the slope in the k *vs.* μ plot. Furthermore, in the
case of azurin, Equation 5 leads to a protein charge that is
opposite in sign to that predicted from the Marcus theory equation and the sequence calculation. One apparent improvement in
the parameters from Equation 5 is that the ferrocyanide data do
not give the large values for Z_1 derived from the Marcus treatment.

The fits from any of these equations are not so successful
that they should be considered really accurate models for the
electrostatic interactions between proteins and small molecules,
as can be seen from the plots of the theoretical curves and experimental points in Figs. 2 and 3 in Rosenberg *et al (111)*. A
more dramatic example is the difference observed in changing
from acetate to phosphate buffer in the ionic strength dependence
of the reaction between cytochrome *c* (horse heart) and
$Ru(NH_3)_5py^{3+}$ *(44)*; although only a slightly higher charge is
expected on cytochrome *c* at the lower pH, the plots show a much
larger effect than would be predicted from such a small change.
Clearly, binding of the buffer ions to the protein is involved
and only further careful study of these effects can unravel all
of the contributions to this complex problem.

The dependence on pH will next be discussed. Because of all
the possible ways that changing hydrogen ion concentration can
lead to variation in rate constants, the origin of pH dependences can be quite difficult to establish. The approach to be
taken in this analysis is to try first to factor out all of the
contributions that have previously been considered, and see if a
significant pH dependence is left. The preferred method, then,
is to do the full Marcus calculation with potentials and work
terms calculated for each pH, thereby accounting for all proton

binding to the reagent ($Fe(CN)_6^{4-}$ with a pK of 3.17 at 0.1 M or 3.85 at 0.01 M ionic strength is again the greatest problem)*(65)* and the protein. Unfortunately, the information necessary for the full calculation is not often available. The pH dependence of redox potentials is seldom determined, and titration data for determining the charge on the protein are also rare.

An example of a set of pH dependences that are considered too small to warrant detailed analysis involves the reduction of several blue proteins by $Fe(EDTA)^{2-}$ *(111)*. The small differences observed in this case could well involve such elusive factors as the influence of changes in the form of the buffer or in ion association. Some of the more complex and interesting pH dependences that have been measured are for the $Fe(CN)_6^{3-/4-}$ reactions with cytochrome c from *Euglena gracilis (69)* and HiPIP from *Chromatium vinosum (102)*. In these two studies from Cusanovich's laboratory, the pH dependences both of the ferricyanide and ferrocyanide reactions and of the potentials of the proteins were determined. The two systems show an intriguingly similar pattern in the pH dependences, and the HiPIP results will be discussed in some detail. Fig. 9 shows the measured pH dependences for oxidation and reduction of HiPIP by iron hexacyanides.

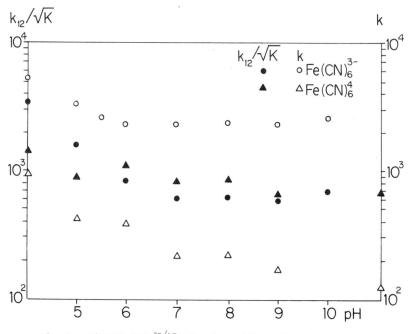

Fig. 9. HiPIP-Fe(CN)$_6^{3-/4-}$ pH analyses (data taken from Ref. 102).

The filled-in symbols represent the same data corrected for the change in driving force, calculated from the pH dependence of the potential of HiPIP, and assuming the ferri-ferrocyanide potential is 425 mV (there are data *(65)* that support the contention that the potential should be constant down to pH 6). The corrections are made, therefore, by dividing the observed rate constant by the square root of the equilibrium constant for the reaction in the direction measured. This correction accounts for most of the difference between the ferricyanide and ferrocyanide reactions, but there is still a variance below pH 6. Although the pH dependence of the $Fe(CN)_6^{3-/4-}$ couple is not known in higher ionic strength media, if the potential increases with pH in the medium used for this experiment, as it does in dilute solution, then ferrocyanide and ferricyanide $k_{12}/K^{\frac{1}{2}}$ values could be brought into agreement.

The general increase in $k_{12}/K^{\frac{1}{2}}$ below pH 7 must still be explained (Fig. 9). For a negative protein that becomes protonated and thus less negative as the pH is taken below 7, the electrostatic work terms w_{12} and w_{21} will become less positive (since the reagent is negative), and the rate will increase. The same result may be predicted from the protonation of $Fe(CN)_6^{4-}$. The self-exchange work terms, w_{11} and w_{22}, will become less positive, and, as they are subtracted, the rate will increase owing to this contribution. The w_{11} terms are usually small because of the large radius of the protein, so they are not very important, especially for a protein with as low a charge as HiPIP. Although the w_{22} term could be rather large and would change by 25% in the extreme case of complete protonation of $Fe(CN)_6^{3-/4-}$, not even half-protonation is expected at the pH values considered in the HiPIP study. Thus the cross reaction work terms can be expected to dominate. A similar general trend of increasing rate with decreasing pH is to be expected of the protein ever becomes positive, as then the magnitude of the work terms will increase with decreasing pH.

The pH dependences for the reactions of the iron hexacyanides with cytochrome c_{552} from *Euglena gracilis* show similar patterns to the one just discussed (Fig. 10); when corrections are made for the equilibrium constant, there is general agreement between data for reduction and oxidation except below pH 6, where there is divergence (the $Fe(CN)_6^{3-}$ results again predict a higher value of $k_{12}/K^{\frac{1}{2}}$). Also, there is a general increase in rate for the whole pH range measured, and the dependence becomes steeper in the lower pH region, where protonation of the reagent is expected to contribute.

For the activation parameters for the reactions of small molecules with proteins, the most informative interpretation

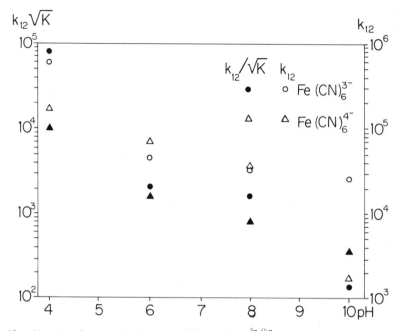

Fig. 10. Cytochrome c (Euglena gracilis)-Fe(CN)$_6^{3-/4-}$ pH analyses (data taken from Ref. 69). Filled symbols, left vertical axis; open symbols, right vertical axis.

requires their conversion to $\Delta H_{11}^{\ddagger}$ and $\Delta S_{11}^{\ddagger}$, as was discussed in Section 2.8. Unfortunately, the lack of free energy data for the equilibria between the reagents and the proteins (or, the equivalent, the enthalpy and entropy of reduction from the temperature dependence of the potentials) makes these calculations impossible at this time. The existence of some activation parameters that show near zero $\Delta H_{12}^{\ddagger}$ and very negative $\Delta S_{12}^{\ddagger}$ (see especially the reactions of HiPIP) is intriguing, and could mean that the involved reactions employ long-distance electron transfer mechanisms that are highly "forbidden". The clear involvement of the standard free energy terms can be seen in the uniformly high $\Delta H_{12}^{\ddagger}$ for Co(phen)$_3^{3+}$ reactions, whereas the Fe(CN)$_6^{3-}$ reactions often show low $\Delta H_{12}^{\ddagger}$ and negative $\Delta S_{12}^{\ddagger}$ values. The latter trend correlates with the quite negative ΔH° and ΔS° for the Fe(CN)$_6^{3-/4-}$ couple itself (see Table 1). Further analysis of the activation parameters with the aid of the equilibrium thermodynamic parameters should prove enlightening, but with the data presently available generalizations are difficult to make and any interpretation is hazardous.

7. PROTEIN-PROTEIN REACTIONS

The available data on protein-protein electron transfer reactions are set out in Table 8. In Table 9 calculated and observed protein-protein k_{12} values are compared; the calculated rate constants are based on the largest predicted k_{11}^{corr} value for each protein. This approach has been taken as any deviation from "predicted" behavior is as likely to be a function of complementary interactions between the two proteins as it is to be a difference in the activation process of only one of them. The first clear observation is that most of the reactions (all except three with cytochrome c (horse heart)) are faster than predicted (even with the bias that a large predicted k_{11}^{corr} is used), and the ratio $k_{12}^{obsd}/k_{12}^{corr}$ varies greatly. The former observation would seem to indicate that most proteins share some properties that allow them to interact better with each other than with small molecules; such interactions may well be hydrophobic in nature and on the average outweigh specific coulombic contacts. Unfortunately, most of the reactions between physiological partners cannot be compared to calculated values as the information for making the calculation is not available. One possible exception is the reaction between cytochrome c_{551} and azurin. Taking the data of Rosen and Pecht *(77)*, which is considered to be the most reliable, the ratios are 210 for the oxidation of azurin and 68 for the reduction. A similar pattern of disagreement between oxidation and reduction is found in the results of another study *(101)*, but in this case the ratios are smaller. The fact that the ratios do not agree tends to indicate that the potential differences being used are improper (however, the potentials used in the first case give the equilibrium constant as measured by Rosen and Pecht). Although it is tempting to suggest that the large value of the ratio indicates that specific interaction between physiological partners is involved, such a conclusion is shaky at best. In this regard, it should be noted that the ratio for cytochrome c_{552} and HiPIP is similar to that for azurin-cytochrome c_{551}, and the former two proteins are clearly not physiological partners.

8. SUMMARY AND CONCLUSIONS

In this paper a method has been presented for sorting out several of the contributions to the energetics of protein-small molecule electron transfer reactions. This method involves the formalism of the Marcus theory of outer sphere electron transfer, which allows isolation of the following components of the

activation energy: (1) the activation of the reagent metal
center; (2) the contribution to the activation energy from the
free energy change for the reaction; and (3) the electrostatic
(coulombic) contributions to the reagent self-exchange, the
protein self-exchange, and the cross reaction, assuming a model
of uniformly-charged hard spheres. The calculated activation
energy that results from this treatment refers to the protein
self-exchange reaction were the protein to react with itself in
the same manner it reacts with the reagent. Important contributions to this activation energy are: (1) the nonelectrostatic
(*i.e.*, noncoulombic) interaction between the protein and reagent
required to position the reagent for most favorable electron
transfer; (2) the contributions from the varying adiabaticity
(overlap between reagent and protein redox orbitals) of the reactions; (3) the inherent activation of the metal center, resulting from the need to attain, for example, bond lengths between
the protein metal center and its ligands that are intermediate
between those of the oxidized and reduced state; and (4) electrostatic interactions, both those covered by the model used for
making the work term calculations, and those not considered, including induced changes in charge distribution and shape of the
protein and reagent.

The result of treating protein-small molecule electron transfer data in the manner outlined above is a collection of calculated protein self-exchange rate constants. Large variations in
the k_{11}^{corr} values are observed for most proteins considered.
It is concluded that the two most important contributions to this
large variation are the extent of orbital overlap in the transition state and the nonelectrostatic interaction between the protein and reagent, and together these influences define the kinetic accessibility of the protein redox center to a given reagent.
The contributions from orbital overlap and nonelectrostatic
interaction are closely related and coupled in their influence.
As most protein metal sites are, to some extent at least, protected from reactants in the medium by the peptide, any near-
approach to the metal center will require interaction with the
blocking residues; as electron transfer will be much more efficient if there is good orbital overlap between the redox orbitals
of the protein and reagent, close approach is expected to greatly
increase the rate of electron transfer. A compromise thus must
be reached between the benefit gained from the reagent penetrat-
ing close to the protein metal site, and the thermodynamic cost
of such penetration. Analysis of the data indicates that hydrophobic reagents penetrate more easily and ligands with π symmetry
orbitals available facilitate overlap with the redox centers of
the proteins. The kinetic accessibility based on $Fe(EDTA)^{2-}$

increases according to azurin $\cong$ HiPIP $\langle$ cytochrome c_{551} $\langle$ cytochrome c (horse heart) $\langle$ plastocyanin $\langle$ stellacyanin (see Fig. 6), which parallels the extent to which the metal center is buried in the peptide for those cases where structural data are available.

Protein-protein reactions are in general more difficult to treat. However, most measured protein-protein rate constants are larger than would be expected based on even the largest k_{11}^{corr} values. Interestingly, we have found no clear cut evidence for specificity between natural partners from the limited amount of evidence available. The protein-protein reactions apparently benefit from numerous interactions that are individually small but as a whole can be much more significant than those between a small molecule and a protein. Clearly, this area needs much more careful experimentation and analysis.

9. ACKNOWLEDGEMENTS

We have benefited greatly from helpful discussions with all our coworkers. We cite especially the contributions of Diane Cummins, Cathy Coyle, Napapon Sailásuta, Jim McArdle, Bob Holwerda, Pam Peerce, Barry Dohner, Steve Schichman, Bob Rosenberg, and Bob Scott. We gratefully acknowledge the support of the National Science Foundation. This is Contribution No. 5377 from the Arthur Amos Noyes Laboratory.

TABLE 7: CYTOCHROME c SELF-EXCHANGE RATE CONSTANT CALCULATION
WITH COMPLEX FORMATION $(Fe(CN)_6^{3-/4-})$

P_{21} $Fe(CN)_6^{4-}$ - cytochrome c(III)

     Experimental   400 M^{-1}

     Calculated[a]   370 M^{-1}

P_{12} $Fe(CN)_6^{3-}$ - cytochrome c(II)

     Experimental   400 M^{-1}

     Calculated[a]   260 M^{-1}

P_{11} cytochrome c(II) - cytochrome c(III)

     Calculated[a]   8.7 M^{-1}

P_{22} $Fe(CN)_6^{4-}$ - $Fe(CN)_6^{3-}$

     Calculated[a]   0.013 M^{-1}

K = 618 (oxidation)[b]

k_{12}[c] $(M^{-1}s^{-1})$ $6.5(10^6)$ (oxidation), $2.5(10^4)$ (reduction)

k_{22}[d] $1.5(10^4)$ $M^{-1}s^{-1}$

k_{11}[e] $(M^{-1}s^{-1})$ 3.2 (oxidation) 6.5 (reduction) (1.7 times
higher if calculated values are used for P_{12}
and P_{21})

[a] Using Equations 29 and 30, and taking the radius of $Fe(CN)_6^{3-/4-}$ as 4.5 Å, the radius of cytochrome c (6.5/7.5) as 16.6 Å, and 0.1 M ionic strength.

[b] Using potentials of 425 mV $(Fe(CN)_6^{3-/4-})$ and 260 mV (cytochrome c).

[c] Ref. 96.

[d] Table 1.

[e] Equation 32.

TABLE 8: PROTEIN-PROTEIN REACTION DATA

Oxidant	Reductant	$k(M^{-1}s^{-1})$	T	$\Delta H^{\ddagger}$(kcal/mole)	$\Delta S^{\ddagger}$(eu)	pH	Buffer[a]	μ(M)	Salt	Ref.
Plastocyanin[b]	Cytochrome c[c]	$1.0(10^6)$	25			7.0	P	0.1	NaCl	68
Plastocyanin	Cytochrome c_{551}[d]	$7.5(10^5)$	25			7.0	P	0.1	NaCl	68
Plastocyanin	Cytochrome f[e]	$3.6(10^7)$	25			7.0	P	0.1	NaCl	68
Azurin	Cytochrome f	$1.0(10^6)$	25			7.0	P	0.1	NaCl	68
Cytochrome c	Azurin	$1.1(10^3)$	20	13	0	7.0	P,EDTA	0.1	NaCl	103
Cytochrome c	Azurin	$\sim 1(10^3)$	20			7.0	P,EDTA	0.1		101
Cytochrome c_{551}	Azurin	$6.1(10^6)$[f]	25	13.5	-10.7	7.0	P	0.05		77
Azurin	Cytochrome c_{551}	$8.1(10^6)$	25	10.6	8.6	7.0	P	0.05		77
Cytochrome c_{551}	Azurin	$3(10^6)$	20	16	26	7.0	P	0.1		101
Azurin	Cytochrome c_{551}	$1.4(10^6)$	20	9	25	7.0	P	0.1		101
Cytochrome c	Cytochrome c_{551}	$8(10^4)$[g]	25	12	4	7.2	P,EDTA	0.1	NaCl	103
Cytochrome c	Cytochrome c_{551}	$1.6(10^4)$[h]	4.5	11	1	7.0	P	0.2	KCl	67
Cytochrome c_{551}	Cytochrome c	$1.6(10^4)$	4.5			7.0	P	0.2	KCl	67
Cytochrome c_{552}	HiPIP[i]	$5.8(10^6)$[j]	24			7.0	P	0.12	NaCl	69
HiPIP	Cytochrome c_{552}	$3.0(10^6)$	24			7.0	P	0.12	NaCl	69
Cytochrome c_2	HiPIP	$3.1(10^4)$[k]	20			7.0	P	0.12	NaCl	102
HiPIP	Cytochrome c_2	$1.5(10^5)$[k]	20			7.0	P	0.12	NaCl	102

[a] Buffer abbreviations are as given in Table 1.

[b] Plastocyanin is from parsley.

[c] Cytochrome c from horse heart unless otherwise indicated.

[d] Azurin and cytochrome c_{551} are from Pseudomonas (whether from Pseudomonas aeruginosa or fluorescens is not clear).

[e] Cytochrome f is from parsley.

[f] Two forms of azurin, one not kinetically active with cytochrome c_{551}, are indicated.

[g] Ionic strength dependence has been performed.

[h] Ionic strength independent 0.03 to 0.3 M; $k = 4(10^4)$ $M^{-1}s^{-1}$ at 25° from $E_A = 12$ kcal/mole.

[i] HiPIP is from Chromatium vinosum.

[j] Temperature jump data indicate that this reaction is not simply bimolecular.

[k] Ionic strength independent.

TABLE 9:   CALCULATED PROTEIN–PROTEIN RATE CONSTANTS, INCLUDING ELECTROSTATIC CORRECTIONS

Oxidant	Reductant	w_{12}[a,b,c]	w_{21}[a,b,c]	w_{11}[a,b,d]	w_{22}[a,b,d]	ΔG_{12}^{*} corr[b,e]	k_{12}^{corr} (M⁻¹s⁻¹)[e]	k_{12}^{obsd} (M⁻¹s⁻¹)[f]	$k_{12}^{obsd}/k_{12}^{corr}$[f]
Plastocyanin[g,h]	Cytochrome c[i]	-0.528	-0.677	0.882	0.406	10.48	1.3(10⁵)	1.0(10⁶)	7.7
	Cytochrome c551[i]	0.307	0.227	0.882	0.079	9.89	3.5(10⁵)	7.5(10⁵)	2.1
Cytochrome c[i]	Azurin[i]	-0.118	-0.051	0.406	0.015	13.13	1.5(10³)	1.1(10³)	0.73
Cytochrome c551[i]	Azurin[i,j]	0.078	0.058	0.079	0.015	11.52	2.2(10⁴)	6.1(10⁶)[j]	2.8(10²)
Azurin[i,j]	Cytochrome c551[i]	0.058	0.078	0.015	0.079	10.51	1.2(10⁵)	8.1(10⁶)[j]	6.8(10¹)
Cytochrome c551[i]	Azurin[i,k]	0.040	0.030	0.079	0.015	11.78	1.4(10⁴)	3(10⁶)[k]	2.1(10²)
Azurin	Cytochrome c551[i]	0.030	0.040	0.015	0.079	10.21	2.0(10⁵)	1.4(10⁶)[k]	7.0
Cytochrome c[i]	Cytochrome c551[i]	-0.235	-0.136	0.406	0.079	11.48	2.3(10⁴)	8.0(10⁴)[l]	3.5
Cytochrome c[i]	Cytochrome c551[i]	-0.096	-0.056	0.406	0.079	11.59	2.0(10⁴)	1.6(10⁴)[m]	0.80
Cytochrome c551[i]	Cytochrome c[i]	-0.056	-0.096	0.079	0.406	11.59	2.0(10⁴)	1.6(10⁴)[m]	0.80
Cytochrome c552[n]	HiPIP[i]	0.229	0.184	0.705	0.091	11.54	2.1(10⁴)	5.8(10⁶)	2.8(10²)
HiPIP[i]	Cytochrome c552[n]	0.184	0.229	0.091	0.705	12.01	9.7(10³)	3.0(10⁶)	3.1(10²)

[a] The subscript 1 refers to the oxidant and 2 to the reductant.

[b] All energies in kcal/mole.

[c] Work terms are calculated using the conditions given in Table 4 and assuming protein charges calculated from the sequences (Table 2).

[d] Work terms are calculated for the conditions appropriate to the calculated self-exchange rate constant (μ0.1 M, ph 7).

[e] Calculated for the cross reaction condition of Table 8.

[f] From Table 8.

[g] Conventions for protein names are as given in footnotes b to e, Table 8.

[h] The self-exchange rate constant is calculated from the cross reaction with Fe(CN)₆³⁻.

[i] The self-exchange rate constant is calculated from the cross reaction with Co(phen)₃³⁺.

[j] From the cross reaction and potential (azurin = 304 mV) data of Ref. 77.

[k] From the cross reaction data of Ref. 101 and using E = (azurin) of 330 mV.

[l] From the cross reaction data of Ref. 103.

[m] From the cross reaction data of Ref. 67.

[n] The self-exchange rate constant is calculated from the cross reactions with Fe(CN)₆³⁻′⁴⁻.

SUPPLEMENTAL DATA TO TABLE 6:
IONIC STRENGTH DEPENDENCE DATA

Protein	Reagent	$k(M^{-1}s^{-1})^{b,c}$	$\mu(M)$	Ref.
Cytochrome c^{a} (Horse heart)	Fe(EDTA)$^{2-}$ ^{a}	8.21(10^{4})	0.02	93
		5.55	0.04	
		5.44	0.04	
		3.88	0.06	
		4.24	0.06	
		2.73	0.08	
		3.03	0.08	
		2.06	0.10	
		2.13	0.10	
	Co(phen)$_3$$^{3+}$ ^{a}	1.27(10^{3})	0.06	95
		1.18	0.06	
		1.51	0.10	
		1.53	0.10	
		1.83	0.14	
		1.75	0.14	
		2.05	0.17	
		2.02	0.17	
		2.22	0.20	
		2.15	0.20	
	Fe(CN)$_6$$^{4-}$ ^{a}	> 18(10^{4})	0.035	67
		3.2	0.242	
		1.9	0.321	
		1.7	0.400	
Cytochrome c	Fe(CN)$_6$$^{3-}$ ^{a}	4.20(10^{7})	0.028	67
		2.89	0.039	
		1.66	0.090	
		1.05	0.153	
		0.660	0.202	
		0.598	0.280	
		0.604	0.313	
		0.482	0.423	
	Cytochrome c^{a} (Horse heart)	1.24(10^{3})	0.068	84
		2.48	0.148	
		4.70	0.231	
		6.11	0.305	
		8.90	0.384	
		14.6	0.469	

Protein	Reagent	$\mu(M)$	$k(M^{-1}s^{-1})^{b,c}$	Ref.
Cytochrome c_{551}	Co(phen)$_3$$^{3+}$ ^{a}	0.02	2.13(10^{5})	100
		0.02	1.13	
		0.06	7.60	
		0.06	7.40	
		0.10	6.22	
		0.10	6.09	
		0.15	5.28	
		0.15	5.25	
		0.20	4.56	
		0.20	4.46	
Cytochrome c (*Euglena gracilis*)	Fe(CN)$_6$$^{4-}$ ^{d}	0.027	1.54(10^{3})	69
		0.033	2.22	
		0.055	2.84	
		0.082	3.41	
		0.132	3.63	
		0.520	5.25	
	Fe(CN)$_6$$^{3-}$	0.027	0.713(10^{4})	69
		0.033	1.32	
		0.055	2.03	
		0.082	2.59	
		0.132	3.31	
		0.520	5.75	
Cytochrome c_2 (*Rhodospirillum rubrum*)	Fe(CN)$_6$$^{4-}$ ^{d,e} (k_{12})	0.035	7.0(10^{4})	104
		0.085	13	
		0.135	6.6	
		0.235	1.5	
		0.535	0.63	
		1.035	0.15	
	Fe(CN)$_6$$^{3-}$ ^{d,e} (k_{43})	0.035	6.0(10^{6})	104
		0.085	3.6	
		0.135	2.2	
		0.235	1.5	
		0.535	1.1	
		1.035	0.5	
	Fe(CN)$_6$$^{3-}$ ^{d,e} (k_{21})	0.035	4.5(10^{2})	104
		0.085	4.0	
		0.135	3.5	

SUPPLEMENTAL DATA (Continued)

Protein	Reagent	μ(M)	k(M^{-1}s^{-1})[b,c]	Ref.
Cytochrome c_2 (*Rhodospirillum rubrum*)	Fe(CN)$_6$$^{3-}$ (k_{21})[d,e]	0.235	3.0(10^2)	104
		0.535	2.6	
		1.035	2.4	
HiPIP (*Chromatium vinosum*)	Fe(CN)$_6$$^{4-}$	0.008	1.49(10^2)	102
		0.033	1.38	
		0.058	1.73	
		0.108	1.52	
		0.158	1.36	
		0.208	1.54	
	Fe(CN)$_6$$^{3-}$[d]	0.008	1.15(10^3)	102
		0.033	1.47	
		0.058	1.95	
		0.108	2.53	
		0.158	3.01	
		0.208	3.82	
Azurin	Fe(EDTA)$^{2-}$[f]	0.010	1.16(10^3)	111
		0.010	1.13	
		0.035	1.31	
		0.035	1.33	
		0.075	1.39	
		0.075	1.40	
		0.130	1.48	
		0.130	1.46	
		0.200	1.51	
		0.200	1.51	
		0.350	1.70	
		0.350	1.69	
	Co(phen)$_3$$^{3+}$[a]	0.05	2.61(10^3)	100
		0.05	2.43	
		0.08	2.20	
		0.08	2.26	
		0.12	2.11	
		0.12	2.08	
		0.16	2.11	
		0.16	1.83	
		0.20	2.02	
		0.20	1.89	

Protein	Reagent	μ(M)	k(M^{-1}s^{-1})[b,c]	Ref.
Plastocyanin (Bean)	Fe(EDTA)$^{2-}$[g]	0.05	3.11(10^4)	111
		0.10	4.95	
		0.20	7.59	
		0.30	8.91	
		0.50	11.3	
Stellacyanin	Co(phen)$_3$$^{3+}$[a]	0.052	1.89(10^5)	100
		0.194	1.89	
		0.296	1.93	
		0.390	1.92	
		0.502	1.84	

[a] pH 7, phosphate, 25°.

[b] When two values are given at an ionic strength, they reflect two reagent concentrations.

[c] For a given protein-reagent pair all rate constants carry the same multiplicative factor.

[d] pH 7, phosphate, 20° (NaCl).

[e] Rate constants are identified as given by the authors for the mechanism involving complex formation.

[f] pH 6.8, phosphate, 25°.

[g] pH 6.9, phosphate, 25°.

REFERENCES

1. R.E. Dickerson and R. Timkovich, "The Enzymes", P.D. Boyer, Ed., Academic Press, New York, 3rd Edition, Vol. 11, Chap. 7 (1976).
2. J.A. Fee, *Structure and Bonding,* **23**, 1 (1975).
3. L.E. Bennett, *Prog. Inorg. Chem.,* **18**, 1 (1973).
4. C.W. Carter, Jr., J. Kraut, S.T. Freer, N.H. Xuong, R.A. Alden and R.G. Bartsch, *J. Biol. Chem.,* **249**, 4212 (1974).
5. C.W. Carter, Jr., J. Kraut, S.T. Freer and R.A. Alden, *J. Biol. Chem.,* **249**, 6339 (1974).
6. L.E. Bennett, personal communication.
7. R.H. Holm, B.A. Averill, T. Herskovitz, R.B. Frankel, H.B. Gray, O. Siiman and F.J. Grunthaner, *J. Am. Chem. Soc.,* **96**, 2644 (1974).
8. D.R. McMillin, R.A. Holwerda and H.B. Gray, *Proc. Nat. Acad. Sci. (U.S.A.),* **71**, 1339 (1974).
9. D.R. McMillin, R.C. Rosenberg and H.B. Gray, *Proc. Nat. Acad. Sci. (U.S.A.),* **71**, 4760 (1974).
10. E.I. Solomon, J.W. Hare and H.B. Gray, *Proc. Nat. Acad. Sci. (U.S.A.),* **73**, 1389 (1976).
11. E.I. Solomon, P.J. Clendening, H.B. Gray and F.J. Grunthaner, *J. Am. Chem. Soc.,* **97**, 3878 (1975).
12. J.K. Markley, E.L. Ulrich, S.P. Berg and D.W. Krogmann, *Biochemistry,* **14**, 4428 (1975).
13. J.W. Hare, E.I. Solomon and H.B. Gray, *J. Am. Chem. Soc.,* **98**, 3205 (1976).
14. J.W. Hare, Ph.D. Thesis, California Institute of Technology, 1976.
15. N. Boden, M.C. Holmes and P.F. Knowles, *Biochem. Biophys. Res. Comm.,* **57**, 845 (1974).
16. S.H. Koenig and R.D. Brown, *Ann.N.Y. Acad. Sci.,* **222**, 752 (1973).
17. G.H. Rist, J.S. Hyde and T. Vänngård, *Proc. Nat. Acad. Sci. (U.S.A.),* **67**, 79 (1970).
18. F. Basolo and R.G. Pearson, "Mechanisms of Inorganic Reactions", 2nd Ed., John Wiley, New York (1967).
19. R.G. Wilkins, "The Study of Kinetics and Mechanism of Reactions of Transition Metal Complexes", Allyn and Bacon, Boston (1974).
20. R.A. Marcus and N. Sutin, *Inorg. Chem.,* **14**, 213 (1975).
21. The equation is $Z = pr^2(8\pi kT/\mu)^{1/2}$ where p is the probability of reaction during an encounter (including "cage" effects), r is the sum of reactant radii, k is the Boltzmann constant and μ is the reduced mass $(mm_2/(m+m_2))$; Z is insensitive to changes in radius and mass since the mass is proportional to about $4/3\pi r^3$ and thus Z is proportional to $r^{1/2}$. Z is about

twice as large for two proteins of molecular weight 10,000
and radius 15 Å as it is for two small molecules of molecular
weight 200 and radius 4 Å. In the Marcus treatment, electrostatic influences are relegated to the activation energy
term. R.M. Noyes, *Prog. Reac. Kinetics*, **1**, 129 (1961).

22. N. Sutin, "Inorganic Biochemistry", G.L. Eichhorn, Ed.,
Elsevier, New York, Vol. 2, p. 611.

23. S. Glasstone, K.J. Laidler and H. Eyring, "Theory of Rate
Processes", McGraw Hill, New York (1940).

24. A. Haim and N. Sutin, *Inorg. Chem.*, **15**, 476 (1976).

25. J.G. Kirkwood and J.B. Shumaker, *Proc. Nat. Acad. Sci.
(U.S.A.)*, **38**, 855, 863 (1952).

26. C. Tanford, "Physical Chemistry of Macromolecules", John
Wiley, New York, p. 466 (1961).

27. R.A. Alberty and G.G. Hammes, *J. Chem. Phys.*, **62**, 154 (1958).

28. Ref. 26, p. 467.

29. A.D. Pethybridge and J.E. Prus, *Prog. Inorg. Chem.*, **17**, 327
(1972).

30. R.A. Holwerda, S. Wherland and H.B. Gray, *Ann. Rev. Bioph.
Bioeng.*, **5**, 363 (1976).

31. A. Shejter, I. Aviram, R. Margalit and T. Goldkorn, *Ann. New
York Acad. Sci.*, **244**, 51 (1975).

32. T.J. Przystas and N. Sutin, *J. Am. Chem. Soc.*, **95**, 5545
(1973).

33. R. Margalit and A. Shejter, *Eur. J. Biochem.*, **32**, 500 (1973).

34. E. Stellwagen and R.D. Cass, *J. Biol. Chem.*, **250**, 2095 (1975).

35. R.J. Campion, C.F. Deck, Jr. and A.C. Wahl, *Inorg. Chem.*, **6**,
672 (1967).

36. B.R. Baker, F. Basolo and H.M. Neumann, *J. Phys. Chem.*, **63**,
371 (1959).

37. J.J. Hopfield, *Proc. Nat. Acad. Sci. (U.S.A.)*, **71**, 3640 (1974).

38. J.E. Leffler and E. Grunwald, "Rates and Equilibria of
Organic Reactions", Wiley, New York, Chap. 9 (1963).

39. R. Lumry and S. Rajender, *Biopolymers*, **9**, 1125 (1970).

40. G. Schwarzenbach and J. Heller, *Helv. Chim. Acta*, **34**, 576
(1951).

41. O.K. Borggaard, *Acta Chem. Scand.*, **26**, 393 (1972).

42. R.G. Wilkins and R.E. Yelin, *Inorg. Chem.*, **7**, 2667 (1968).

43. M.D. Lind and J.L. Hoard, *Inorg. Chem.*, **3**, 34 (1964).

44. D. Cummins, unpublished results.

45. H.M. Neumann, quoted in R. Farina and R.G. Wilkins, *Inorg.
Chem.*, **7**, 514 (1968).

46. G.P. Khare and R. Eisenberg, *Inorg. Chem.*, **9**, 2211 (1970).

47. J.V. McArdle, Ph.D. Thesis, California Institute of Technology, 1976.

48. The treatment of the reagents as spheres with the charge centered on the metal ion is least likely to be correct for $Co(4,7-(\phi-SO_3)_2-phen)_3{}^{3-}$, as in this case two-thirds of the charge is on the sulfonate groups at the periphery of the ligands. Furthermore, estimation of the radius of this complex is complicated by the large, open channels that exist between the ligands. Because of this latter problem, two of the three cases considered in all calculations differ in the radius chosen, using two extremes of 11.5 (estimated distance from the cobalt to the sulfonate oxygens) and 7 Å (the radius used for the unsubstituted parent complex). The third model is designed to compensate for the location of the charged groups; in this approach the cobalt(III) center and all but two of the sulfonates are taken as a unit of radius 7 Å and charge of -1 (-2 for Co(II)); the remaining two sulfonates are treated as individual centers of radius 3 Å and -1 charge. This is designed to overestimate the charge contributions, as two of the sulfonates are given heavy weight through their small radii, and the "trailing" sulfonates are grouped with the nearer cobalt and this composite center is given a small radius.

49. R. Farina and R.G. Wilkins, *Inorg. Chem.*, **7**, 514 (1968).
50. D. Cummins and H.B. Gray, *J. Am. Chem. Soc.*, submitted.
51. H.S. Lim, D.J. Barclay and F.C. Anson, *Inorg. Chem.*, **11**, 1460 (1972).
52. T.J. Meyer and H. Taube, *Inorg. Chem.*, **7**, 2369 (1968).
53. H.C. Stynes and J.A. Ibers, *Inorg. Chem.*, **10**, 2304 (1971).
54. I.M. Kolthoff and W.J. Tomsicek, *J. Phys. Chem.*, **39**, 945 (1935).
55. B.I. Swanson and R.R. Ryan, *Inorg. Chem.*, **12**, 283 (1973).
56. D.B. Brown and D.F. Shriver, *Inorg. Chem.*, **8**, 37 (1969).
57. G.I. Hanania, D.H. Irvine and P. George, *J. Phys. Chem.*, **71**, 2022 (1967).
58. R. Stasiw and R.G. Wilkins, *Inorg. Chem.*, **8**, 156 (1969).
59. C. Creutz and N. Sutin, *Proc. Nat. Acad. Sci. (U.S.A.)*, **70**, 1701 (1973).
60. J. Baxendale and H.R. Hardy, *Trans. Farad. Soc.*, **49**, 1140 (1953).
61. J. Baxendale, H.R. Hardy and C.H. Sutcliffe, *Trans. Farad. Soc.*, **47**, 963 (1951).
62. R. Skochdople and S. Chaberek, *J. Inorg. Nucl. Chem.*, **11**, 222 (1959).
63. R.L. Gustafson and A.E. Martell, *J. Phys. Chem.*, **67**, 576 (1963).
64. H.J. Schugar, A.T. Hubbard, F.C. Anson and H.B. Gray, *J. Am. Chem. Soc.*, **91**, 71 (1969).

65. J. Jordan and G.J. Ewing, *Inorg. Chem.*,**1**, 587 (1962).
66. R. Margalit and A. Shejter, *Eur. J. Biochem.*,**32**, 492 (1973).
67. R.A. Morton, J. Overnell and H.A. Harbury, *J. Biol. Chem.*, **245**, 4653 (1970).
68. P.M. Wood, *Biochim. Biophys. Acta*,**357**, 370 (1974).
69. F.E. Wood and M.A. Cusanovich, *Arch. Bioch. Bioph.*,**168**, 333 (1975).
70. G.W. Pettigrew, *Biochem. J.*,**139**, 449 (1974).
71. K. Sletten, K. Dus, H. DeKlerk and M.D. Kamen, *J. Biol. Chem.*, **243**, 5492 (1968).
72. K. Dus, K. Sletten and M.D. Kamen, *J. Biol. Chem.*,**243**, 5507 (1968).
73. K. Dus, H. DeKlerk, K. Sletten and R.G. Bartsch, *Biochim. Biophys. Acta*,**140**, 291 (1967).
74. W. Lovenberg and B.E. Sobel, *Proc. Nat. Acad. Sci. (U.S.A.)*, **54**, 193 (1967).
75. T. Yamanaka, "Biochemistry of Copper", J. Peisach, P. Aisen and W.E. Blumberg, Eds., Academic Press, New York, pp. 275-292 (1966).
76. R.P. Ambler and L.H. Brown, *Biochem. J.*,**104**, 784 (1967).
77. P. Rosen and I. Pecht, *Biochemistry*,**15**, 775 (1976).
78. J.A.M. Ramshaw, R.H. Brown, M.D. Scawen and D. Boulter, *Biochim. Biophys. Acta*,**303**, 269 (1973).
79. N. Sailasuta, F.C. Anson and H.B. Gray, to be submitted for publication.
80. P.R. Milne, J.R.E. Wells and R.P. Ambler, *Biochem. J.*,**143**, 691 (1974).
81. B. Reinhammar, *Biochim. Biophys. Acta*,**153**, 299 (1968).
82. B. Reinhammar, *Biochim. Biophys. Acta*,**205**, 35 (1970).
83. J. Peisach, W.G. Levine and W.E. Blumberg, *J. Biol. Chem.*, **242**, 2847 (1967).
84. (a) R.K. Gupta, *Biochim. Biophys. Acta*,**292**, 291 (1973); (b) R.K. Gupta, S.H. Koenig and A.G. Redfield, *J. Magn. Res.*, **7**, 66 (1972).
85. J.K. Beattie, D.J. Fensom, H.C. Freeman, E. Woodcock, H.A.O. Hill and A.M. Stokes, *Biochim. Biophys. Acta*,**405**, 109 (1975).
86. J.K. Yandell, D.P. Fay and N. Sutin, *J. Am. Chem. Soc.*, **95**, 1131 (1973).
87. W.G. Miller and M.A. Cusanovich, *Bioph. Struc. Mech.*,**1**, 97 (1975).
88. D.O. Lambeth and G. Palmer, *J. Biol. Chem.*,**248**, 6095 (1973).
89. R.A. Holwerda and H.B. Gray, *J. Am. Chem. Soc.*,**96**, 6008 (1974).
90. R. Aasa, R. Branden J. Deinum, B.G. Malmstrom, B. Reinhammar and T. Vänngård, *FEBS Lett.*,**61**, 115 (1976).

91. Y.A. Im and D.H. Busch, *J. Am. Chem. Soc.*, **83**, 3357 (1961).

92. C.L. Coyle, unpublished results.

93. H.L. Hodges, R.A. Holwerda and H.B. Gray, *J. Am. Chem. Soc.*, **96**, 3132 (1974).

94. R.X. Ewall and L.E. Bennett, *J. Am. Chem. Soc.*, **96**, 940 (1974).

95. J.V. McArdle, H.B. Gray, C. Creutz and N. Sutin, *J. Am. Chem. Soc.*, **96**, 5737 (1974).

96. R.M. Zabinski, K. Tatti and G.H. Czerlinski, *J. Biol. Chem.*, **249**, 6125 (1974).

97. K.G. Brandt, P.C. Parks, G.H. Czerlinski and G.P. Hess, *J. Biol. Chem.*, **241**, 4180 (1966).

98. J.C. Cassatt and C.P. Marini, *Biochemistry*, **13**, 5323 (1974).

99. C.L. Coyle and H.B. Gray, *Biochem. Biophys. Res.Commun.*, in press.

100. (a) J.V. McArdle, C.L. Coyle, H.B. Gray, G.S. Yoneda and R.A. Holwerda, *J. Am. Chem. Soc.*, submitted for publication; (b) J.V. McArdle, K. Yocom and H.B. Gray, *J. Am. Chem. Soc.*, submitted for publication.

101. E. Antonini, A. Finazzi-Agro, P. Guerrieri, G. Rotilio and B. Mondovi, *J. Biol. Chem.*, **245**, 4847 (1970).

102. I.A. Mizraki, F.E. Wood and M.A. Cusanovich, *Biochemistry*, **15**, 343 (1976).

103. C. Greenwood, A. Finazzi-Agro, P. Guerrieri, L. Avigliano, B. Mondovi and E. Antonini, *Eur. J. Biochem.*, **23**, 321 (1971).

104. F.E. Wood and M. Cusanovich, *Bioinorg. Chem.*, **4**, 337 (1975).

105. J. Rawlings, S. Wherland and H.B. Gray, *J. Am. Chem. Soc.*, **98**, 2177 (1976).

106. M. Goldberg and I. Pecht, *Biochemistry*, **15**, 4197 (1976).

107. L.E. Bennett, "The Iron-Sulfur Proteins", Vol. 3, W. Lovenberg, Ed., Academic, New York, Chap. 9 (1976).

108. C.A. Jacks, L.E. Bennett, W.N. Raymond and W. Lovenberg, *Proc. Nat. Acad. Sci. (U.S.A.)*, **71**, 1118 (1974).

109. S. Wherland, R.A. Holwerda, R.C. Rosenberg and H.B. Gray, *J. Am. Chem. Soc.*, **97**, 5260 (1975).

110. D. Fensom, quoted in R.A. Holwerda and H.B. Gray, *J. Am. Chem. Soc.*, **97**, 6036 (1975).

111. R.C. Rosenberg, S. Wherland, R.A. Holwerda and H.B. Gray, *J. Am. Chem. Soc.*, **98**, 6364 (1976).

112. This reaction might be somewhat suspect, as it was run against the driving force for the reduction (by 20 mV), but the 100- to 1,000-fold excess of oxidant used gives minimum 98 percent completion. Therefore, coupled with the result that the concentration dependence on **5** was found to be linear with a zero intercept, the driving force problem is

not expected to be too serious (the same may not be said for the cross reaction run at a 10-fold excess, and the reaction of **5** with azurin is apparently complicated by this and other problems).

113. R.W. Chlebek and M.W. Lister, *Can. J. Chem.*, **44**, 437 (1966).
114. W.A. Eaton, P. George and G.I.H. Hanania, *J. Phys. Chem.*, **71**, 2016 (1967).

oxidation-reduction properties of cytochromes and peroxidases

R.J.P. WILLIAMS, G.R. MOORE AND P.E. WRIGHT

Inorganic Chemistry Laboratory,
Oxford University, England

1. INTRODUCTION

This article will describe the work of some members of the Oxford Enzyme Group on redox reactions in metalloproteins. It concentrates upon three types of proteins, cytochrome-*c* (from various species), peroxidases (both from various species and iso-enzymes from the same species) and cytochromes P-450. The work is due to Dr. P.E. Wright and Dr. G.R. Moore with considerable help from several senior and junior people: Dr. I. Campbell, Dr. C.M. Dobson, Mrs. P. Burns, Mrs. S. Graham and Dr. G. Mazza (Marseille). The work has been assisted by collaboration with Professor J. Ricard (Marseille), Dr. K. Welinder (Copenhagen) and Dr. H.A.O. Hill (Oxford). We have also benefitted from sabbatical visits by Professor B. Dunford (Alberta, Canada) and Professor J. Harrod (Toronto, Canada).

First, the three series of proteins will be described individually. However the action of cytochromes P-450 requires a combination of the actions of a protein very like cytochrome-*c*, *i.e.*, a one-electron acceptor/donor, plus a peroxidase, *i.e.*, a second multi-electron acceptor/donor but one which can react with molecular oxygen, and its overall reaction is a multi-electron process. Thus, at the end of the description of the individual proteins a discussion of protein/protein interaction in multi-protein combinations will be developed so that the full power of the cytochromes P-450 can be appreciated.

2. CYTOCHROMES-*c*

2.1 INTRODUCTION - Cytochromes-*c* are usually monomeric proteins associated with membranes which are involved in simple one electron transfer to and from heme iron, Fig. 1. They have no known small substrate except the electron and therefore they can function in only two ways; as connecting links between other redox proteins allowing electrons to pass over a considerable distance while keeping the reductants and oxidants apart, and/or as temporary electron stores. Interest in them centres upon how the electron moves through the proteins. For many years there has been good reason to suppose that in the crystalline form the protein in one oxidation state might have a different conformation from that in the other oxidation state *(1)*. In solution there has been suggestive evidence that there is a conformation change on change of oxidation state too, but it has been difficult to describe this change in any detail. The situation is confused further by the facts that the early description of the conformational states in the crystal of the first cytochrome-*c*

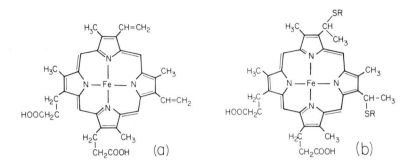

Fig. 1. (a)   Protoporphyrin IX - the most common porphyrin in biology.
(b)   The thio-ether links from protein to protoporphyrin found in
the cytochromes-c.

to be examined were not totally correct and fully revised structures are only just available (June, 1976), and that the protein
in solution is now known to have some mobile side-chains. Thus a
first necessity is a full description of the structures.

2.2 GENERAL STRUCTURE - It has been shown both by NMR and crystallography that in cytochrome-*c* the low-spin iron in both oxidation states has six ligands: four porphyrin pyrrole nitrogens,
one thio-ether and one histidine. It is generally agreed that
one of the coordinated ligands in the iron(III) state changes at
pH > 9, when the methionine thio-ether is replaced by a lysine
nitrogen keeping the iron low-spin, and at pH < 3 when the histidine nitrogen is replaced by water and the iron(III) goes high-
spin. It would appear that these reactions do not greatly alter
the fold of the protein, suggesting that the axial ligands and
the pocket in which the iron sits are relatively readily adjustable within a protein which is highly structured. Surprising as
it may seem, this limited unfolding may be a feature of other
heme proteins such as hemoglobin, peroxidase, cytochrome-*c*' and
cytochrome P-450, where in all cases the heme pocket appears to
be a relatively labile region within a rather fixed protein frame.
The importance of this observation will be stressed in what follows and it permits the substitution of the iron porphyrin by
other metal porphyrins. There is no suggestion that the function
of cytochrome-*c* is dependent upon change of the metal ligands,
however, but limited mobility within a heme pocket may be very
important in removing various kinetic barriers to reaction.
    Our initial aims in studying this protein have been to deter-
mine structural features in solution enabling discussion of electron transfer paths. Our studies use high resolution proton NMR

spectroscopy and have developed directly from those of MacDonald and Phillips, Redfield and Gupta, and Shulman and Wüthrich, and are greatly assisted by the crystal structure data. Like ourselves, all these workers set out to assign the proton NMR spectrum as a first objective but this was really beyond the capacity of the early NMR equipment. The design of the Bruker 270 MHz Fourier Transform NMR spectrometer and its development at Oxford have changed this situation. It is now possible to carry out all the NMR assignment methods on a protein of 25,000 daltons, which could be carried out on a small molecule. Together with various modes of NMR difference spectroscopy *(2)* to isolate particular parts of the total spectrum, *i.e.*, to obtain a given class of sub-spectra, these methods have allowed us to assign considerable parts of the NMR spectrum of several proteins including cyto-chrome-*c*, and to go forward to a partial structure *(2)*.

2.3 METHODS OF ASSIGNMENT OF NMR SPECTRA - Our methods of study-ing the NMR spectrum of a protein have been published in many dispersed papers *(2)*. Campbell and Dobson will shortly provide a general description of the methods.

Our assignment procedures are as follows. Firstly, it is possible to make assignments to amino acid type by determining the characteristic coupling patterns and coupling constants of as many resonances of each residue as possible. The next stage is assignment of resonances to a particular amino acid in the se-quence. Sequence comparisons simplify this task. We stress also the use of paramagnetic probes in conjunction with outline struc-tures from X-ray studies of crystals *(2)*. Intrinsic and extrin-sic probes, both paramagnetic and diamagnetic, can be used to perturb the spectrum (by shifting resonances and changing their relaxation times) and enable *spatial* assignment of resonances to specific amino acids. The heme itself in its various oxidation and spin-states provides an invaluable intrinsic probe in cyto-chrome-*c* and other heme proteins. However, we can also use a variety of low molecular weight extrinsic paramagnetic probes with cytochrome-*c*, locating their binding sites while assigning the structure. The location of the reagent binding site will also allow us to discuss electron transfer paths when our struc-tural studies are sufficiently complete. Notice, too, that assignment to specific amino-acids in a protein using structure-dependent perturbation automatically involves a piece-by-piece assembly of the structure of the protein (see Campbell, Dobson and Williams)*(2)*. Cross-saturation methods allow assignment of one oxidation state in terms of another.

se properties are common to some five or six proteins
e have looked at in detail. Further details of the struc-
ɲ now be tackled using isomorphous replacement.

TAL-REPLACEMENT IN CYTOCHROMES-c - Recently cytochromes-*c*
fferent metal ions in the centre of the porphyrin have
epared *(3)*. This development permits the use of the
ɔf isomorphous metal replacement in cytochrome-*c*, as has
ɛd previously in other heme proteins and by Vallee and
ɔ, and Malmstrom and coworkers in a long series of studies
ɔus metalloenzymes *(4)*. We note that the use of this
in heme proteins was attempted very early by biochemists
ɛ not aware of the physical-chemical studies which are
ɛd by such replacement. Table 3 lists some of the heme-
that can be used.

TABLE 3: SOME METALLO-PORPHYRIN PROBES[a]		
Metal Ion	Spin State	Use in NMR
Fe(II)	4/2	Shift and Relaxation Probe
	zero	Diamagnetic Blank
Fe(III)	5/2	Shift and Relaxation Probe
	1/2	Shift Probe (largely)
Mn(II)	5/2	Relaxation Probe
Co(II)	1/2	Shift Probe
Ln(III)	zero to 7/2	All types of Probe
Cu(II)	1/2	Relaxation Probe
VO(II)	1/2	Relaxation Probe
Co(III)	zero	Diamagnetic Blank

[a]*It is feasible that a much greater variety of metals can
be incorporated but this has not yet been attempted.*

cular interest in cytochrome-*c* is the use of the diamag-
(III) porphyrin in place of heme for it can serve as a
tic reference state for the Fe(III) low-spin state of the
allowing by difference a direct determination of the
ɛtic shifts of the NMR lines in the Fe(III) state, which
ɛntial data in structure determination. It is also possi-
ompare diamagnetic Fe(II) and Co(III) proteins looking
ɛnce of a conformation change due to a change of oxidation
Again cobalt(II) porphyrin is an excellent shift reagent
as strictly axial symmetry and thus the methods used in

TABLE 1: AROMATIC AMINO-ACIDS VISIBLE IN NMR SPECTRUM OF HORSE-HEART CYTOCHROME-c AT 30°C, pH = 7			
Residue	No. in Protein	No. seen by NMR reduced	oxidised
Tryptophan	1	1	1
Phenylalanine	4	4[a]	2
Tyrosine	4	2	2
Histidine	3	3	2

[a]*All protons of the four phenylalanines are only seen
clearly at 360°K. One of the two tyrosines is seen
clearly only below 310°K and above 360°K.*

2.4 SOLUTION STRUCTURE AND DYNAMICS - A remarkable finding after
fairly extensive assignment of the spectrum of cytochrome-*c* was
that at 30°C only about one half of the expected number of resonances of the aromatic residues could be resolved (Table 1).
Of those that can be seen the magnetic equivalence of the 2' and
6', and the 3' and 5' protons of tyrosines and phenylalanines
shows that a proportion of the aromatic residues are flipping at
rates > 10^4 times per sec while others, those that can not be
seen readily, must be much less mobile and probably cover a range
of mobilities with flipping times of 10^3 or less per second. Our
assignments, which are virtually complete for the aromatic amino-
acids, place the mobile amino-acids as in Table 2. *Thus the protein is mobile to different extents in different regions.* These
observations, which were prompted by the study of lysozyme *(2)*,
have led us to change our attitude to the nature of the conformation of a protein in solution and we have been able to describe a
variety of dynamic features of these molecules (lysozyme and
cytochrome-*c*) which were previously unsuspected through the real
and apparent success of X-ray crystal structure studies *(2)*.

There is a very considerable difference in the NMR spectrum
of the Fe(III) and Fe(II) forms of cytochrome-*c* and separate
assignment of the different states was needed, again aided by
reference to the outline crystal structures. The assignments
were confirmed in many cases using cross-saturation techniques.
The NMR study shows a previously unrecognised difference between
the oxidation states, for while the oxidised protein exists in a
dynamic equilibrium mixture of spin states at temperatures above
40°C (see Fig. 2), we have failed to find any evidence for spin-
state equilibria in the reduced form. Although the mobilities of
the aromatic residues in the two oxidation states are very

TABLE 2: MOBILITIES OF AROMATIC AMINO-ACIDS IN CYTOCHROME-c

Aromatic Amino Acid	Mobility Fe(III) State	Mobility Fe(II) State
Tryptophan 59	Not Mobile	Not Mobile
Phenylalanine 10	Mobile	Mobile
Phenylalanine 36	Mobile	Mobile
Phenylalanine 46	?	Only Mobile
Phenylalanine 82	?	Above 350° K
Tyrosine 97	Mobile	Mobile
Tyrosine 48	?	Only Mobile Above 350° K
Tyrosine 74	Not Observed	Not Observed
Tyrosine 67		

NOTES: (a) *An assignment is only complete when all protons are assigned and therefore the assignment of some of the above groups is not yet certain.*
(b) *The amino-acids are placed in relation to certain channels in the protein structure in Fig. 4.*
(c) *Mobile implies a flip rate of greater than 10^4 per sec at 300° K and immobile a flip rate of less than 10^3 per sec at the same temperature.*

similar, there is a marked increase in overall stability to denaturation in the reduced form. This form is stable at 90°C and over a wide pH range, 2-12, at room temperature, while the oxidised form is only stable to about 50°C and in the pH range 3.5 to 9.0. Thus the change of unit charge on the iron atom has brought about a remarkable change in the cooperativity of the protein structure.

We have examined the NMR spectra of cytochromes-c from horse heart, bonito, yeast, pigeon, rabbit, cow, and chicken. All the cytochromes have the same fold in solution. We state this with confidence as the NMR spectrum of a protein is a very good diagnostic fingerprint of its conformation and this is especially true of paramagnetic proteins (see under peroxidases). The most important use of the different species has been therefore in assisting assignment. It should be observed that there is really no functional difference between the different cytochromes-c from the different eucaryotic species. However, we do observe that changes in the sequence alter the mobility of aromatic residues to some extent. Thus in horse heart cytochrome-c the relative mobilities of tyr 48 and phe 46 are different from the mobilities of tyr 48 and tyr 46 of tuna cytochrome-c.

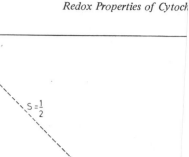

*Fig. 2. The spin-states of iron(III) in diffe...
increasing ligand field strength. Note that...
the high-spin state is stable. The intermedia...
rarely of greatest stability. The ligands, p...
groups, place iron in biological systems clos...
points. A similar diagram can be drawn for t...
of iron(II).*

A major problem which has to be overco...
proteins is the correct alignment of the g...
Fe(III) state, for once this is achieved s...
easier. We have proceeded through the ass...
rin ring resonances of the heme-group rely...
internal probes. For example, some of the...
be assigned to their appropriate pyrrole r...
observations of the effects of externally...
$[Cr(CN)_6]^{3-}$. The full details of the assi...
lished later. It suffices to say here tha...
outline knowledge of the disposition of r...
tic groups in the structure from crystallo...
limited knowledge of the Fe(III) heme g-t...
lowing conclusions:

1. The fold of the protein as seen i...
   as that seen in solution.
2. The positions of many side chains...
   due to internal motion.
3. The exterior of the protein can r...
   shape as motion of surface groups...

the study of molecular complexes of metal porphyrins by Hill and Williams *(5)* will become possible. Relaxation studies can be used if Mn(II) high-spin or, probably better, Cr(III) high-spin (d^3) porphyrins can be inserted, but note that high-spin Fe(III) is both a relaxation and shift probe. If the heme protein does not contain a six-coordinated low-spin central iron atom to which access is blocked, then shift and relaxation probes may be developed in an alternative way by using different ligands to the heme. This procedure can not be used in cytochrome-*c* without affecting its native structure and we leave its discussion until later (see peroxidases). The last method we call heme-tickling.

In collaboration with Professor Chien, who has prepared several metallocytochromes-c_3, we have now examined the NMR spectrum of cobalt(III) cytochromes-*c* in detail. The isomorphous replacement is very precise. In fact a comparison of the diamagnetic Co(III) and the diamagnetic Fe(II) enzymes shows that the perturbations due to the intrinsic ring current probes within the two molecules, including the effect of the metal-porphyrin, are almost precisely the same (Table 4).

TABLE 4: CONFORMATIONAL PARAMETERS FOR TRYPTOPHAN						
Cytochrome *c*	Position of Resonance (ppm)					
	d	*d*	*t*	*t*	*s*	shifts
Fe(II)	7.07	7.60	6.68	5.74	(6.99)	ring current
Co(III)	7.07	7.59	6.62	5.77	(7.06)	ring current
Fe(III)	7.57	7.37	6.54	6.31	(6.86)	ring current + pseudo-contact
Fe(III)-Co(III)	+0.50	-0.22	-0.08	+0.54	(-0.20)	pseudo-contact
d = doublet, t = triplet, s = singlet						

Thus within the limits of NMR studies Co(III) and Fe(II) enzymes show no conformational changes in the heme pocket or elsewhere, although there is probably a 5% change in bond lengths to the ligands. Apart from the structural implication - that there is little or no conformation change in the heme region of cytochrome-*c* on change of oxidation state from Fe(II) to Co(III), and presumable therefore to Fe(III), there is the mechanistic implication that the electron-transfer changes only the metal-ligand bond distances by a small amount. (Note: the discussion applies to low temperatures only, *i.e.*, < 50°C). This result is in direct contradiction with deductions based on general physical properties and on the early crystal structure of the two oxidation states. There may be small conformation changes elsewhere in cytochrome *c* on oxidation/reduction, but the changes in solution are indeed very small internally. (Very recently we have been informed by Prof. Dickerson, who was unaware of our NMR results, that the refined accurate crystal structure analyses of Tuna cytochrome *c* in the two oxidation states shows no really detectable conformation change (June, 1976)).

2.6   TEMPERATURE AND pH CHANGES - Monitoring the NMR spectrum of cytochrome-*c* (Fe(III) form) with changing temperature shows that there are different states of Fe(III) cytochrome-*c* in fast exchange. We find, for example, that some of the resonances with large paramagnetic shifts do not follow the expected $1/T$ dependence, and indeed some of them show an inverted dependence on temperature which implies that the magnetic field experienced by these protons, *i.e.*, the local susceptibility, increases with temperature. The simplest explanation is that there is a fast spin-equilibrium

$$\begin{array}{ccc} \text{low spin state} & \longrightarrow & \text{higher spin state} \\ (\text{low } T^\circ C) & \rightleftharpoons & (\text{high } T^\circ C) \end{array}$$

These two states must have somewhat different geometries with the average bond distances, Fe to ligand atoms, somewhat longer in the high-spin than in the low-spin state (Fig. 3). As the above equilibration is fast on the NMR time scale, ligand motions of at least 0.1 Å (and perhaps as much as five times this) must occur at rates exceeding 10^4 times per second. The same temperature effects are not seen in the low-spin Fe(II) state, but note that the low-spin Fe(II)-ligand bond lengths are expected to be somewhat longer than those of low-spin Fe(III) complexes.

On changing the pH of the Fe(III) form of cytochrome-*c* from 4 to 3 there is again a continuous change in the NMR spectrum which shows that the dissociation of the ligand(s) from the heme

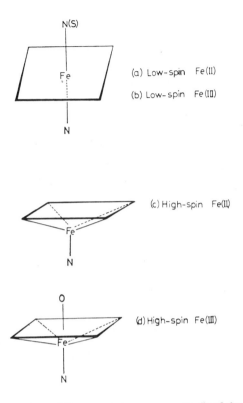

Fig. 3. *Some of the geometric arrangements found for iron in porphyrin complexes and in different spin-states.*

which produces the observed spin-state change is a fast exchange.
Here the nature of the spin-state change is very well documented,
using a great variety of methods. At high pH there is a discontinuous change in the NMR spectrum as the methionine is replaced
by a lysine showing that these two ligands, both of which give
low-spin complexes, are in intermediate exchange.

Before turning to the significance of these observations in
the reactions of cytochrome-*c*, it is worth noting that the high-
spin ⇄ low-spin equilibria of heme-proteins are usually fast.
For some time it has been known that many of the iron porphyrin
proteins occur in such equilibrium mixtures of spin states. More
recently it has been demonstrated that in certain of them, *e.g.*,
cytochromes-*c'*, a quantum mechanical mixing of the spin-states
occurs so that the ground state is not a simple identified spin-
state. It is important that these two situations are recognised

to be quite different for in quantum mechanical mixing there is
but one structure and there is no problem of conformational mobility between two states, while in spin-state equilibria there are
two states of different structure. Even those proteins which
contain quantum mechanical admixed spin states, *e.g.*, cytochrome-
c', also exhibit classical spin-state equilibria over some pH
regions. Proof of this spin equilibrium for many of the proteins
rests on the fact that NMR shows but one spectroscopic signal for
the mixtures, while UV/visible, EPR and Mössbauer spectroscopies
reveal two or more species. Thus the rate of jumping between
structures is between 10^4 and 10^8 times a second. We consider
that for cytochrome-*c* there must be a fast motion of small portions of the protein chain close to the heme and not of the whole
protein for the following reasons:

(1) Proton relaxation time constants for all the different types
     of proton in the protein give a value close to 10^9. There
     is undoubtedly some fast breathing of the protein apart from
     the rotational tumbling which has a rate constant close to
     10^8.
(2) The flipping of tyrosine and phenylalanine is slower than
     10^4 per sec in many parts of the molecule but is faster in
     others.
(3) From our measurements on lysozyme *(2)* there is also certain
     to be motion of other residues such as valine and leucine in
     which rapid flipping between conformers of the side chains
     takes place.
(4) Surface mobility of lysine side chains is high and motion
     here is faster than the electron transfer rate.
(5) In cytochrome *c* there is no suggestion of any segmental
     motion as we have proposed in haemoglobin.

     In summary, cytochrome *c* is a relatively rigid protein with
an outline structure which is constant in the two oxidation
states in solution. It is against the background of this structure and its dynamic fluctuations that we must view the mechanism
of electron transfer. While making this examination we must remember that the structural comparison is of the ground states for
two oxidation states at temperatures below 50°C. Above that temperature the Fe(III) state is in much larger part high-spin, a
change which does involve a breaking of the methionine-iron heme
bond. Thus there is a readily accessible excited state. We shall
contend that it is not the actual breaking of the coordinate link
that is important for fast electron transfer but its easy extension through stretching, and the presence of the low-lying excited state is but an indication of the ease of bond stretching.
Low-lying excited states of Fe(III) and of high spin character
are observed in many cytochromes, *e.g.*, cytochromes *c'*, cyto-

```
    TABLE 1:   AROMATIC AMINO-ACIDS VISIBLE IN NMR
SPECTRUM OF HORSE-HEART CYTOCHROME-c AT 30°C, pH = 7
```

Residue	No. in Protein	No. seen by NMR reduced	oxidised
Tryptophan	1	1	1
Phenylalanine	4	4^a	2
Tyrosine	4	2	2
Histidine	3	3	2

^{a}*All protons of the four phenylalanines are only seen clearly at 360°K. One of the two tyrosines is seen clearly only below 310°K and above 360°K.*

2.4 SOLUTION STRUCTURE AND DYNAMICS - A remarkable finding after fairly extensive assignment of the spectrum of cytochrome-c was that at 30°C only about one half of the expected number of resonances of the aromatic residues could be resolved (Table 1). Of those that can be seen the magnetic equivalence of the 2' and 6', and the 3' and 5' protons of tyrosines and phenylalanines shows that a proportion of the aromatic residues are flipping at rates > 10^4 times per sec while others, those that can not be seen readily, must be much less mobile and probably cover a range of mobilities with flipping times of 10^3 or less per second. Our assignments, which are virtually complete for the aromatic amino-acids, place the mobile amino-acids as in Table 2. ·*Thus the protein is mobile to different extents in different regions.* These observations, which were prompted by the study of lysozyme *(2)*, have led us to change our attitude to the nature of the conformation of a protein in solution and we have been able to describe a variety of dynamic features of these molecules (lysozyme and cytochrome-c) which were previously unsuspected through the real and apparent success of X-ray crystal structure studies *(2)*.

There is a very considerable difference in the NMR spectrum of the Fe(III) and Fe(II) forms of cytochrome-c and separate assignment of the different states was needed, again aided by reference to the outline crystal structures. The assignments were confirmed in many cases using cross-saturation techniques. The NMR study shows a previously unrecognised difference between the oxidation states, for while the oxidised protein exists in a dynamic equilibrium mixture of spin states at temperatures above 40°C (see Fig. 2), we have failed to find any evidence for spin-state equilibria in the reduced form. Although the mobilities of the aromatic residues in the two oxidation states are very

TABLE 2: MOBILITIES OF AROMATIC AMINO-ACIDS IN CYTOCHROME-*c*		
Aromatic Amino Acid	Mobility	
	Fe(III) State	Fe(II) State
Tryptophan 59	Not Mobile	Not Mobile
Phenylalanine 10	Mobile	Mobile
Phenylalanine 36	Mobile	Mobile
Phenylalanine 46	?	Only Mobile
Phenylalanine 82	?	Above 350° K
Tyrosine 97	Mobile	Mobile
Tyrosine 48	?	Only Mobile Above 350° K
Tyrosine 74	Not Observed	Not Observed
Tyrosine 67		

NOTES: (a) An assignment is only complete when all protons are
           assigned and therefore the assignment of some of the
           above groups is not yet certain.
       (b) The amino-acids are placed in relation to certain
           channels in the protein structure in Fig. 4.
       (c) Mobile implies a flip rate of greater than 10^4 per
           sec at 300°K and immobile a flip rate of less than
           10^3 per sec at the same temperature.

similar, there is a marked increase in overall stability to denaturation in the reduced form. This form is stable at 90°C and
over a wide pH range, 2-12, at room temperature, while the oxidised form is only stable to about 50°C and in the pH range 3.5
to 9.0. Thus the change of unit charge on the iron atom has
brought about a remarkable change in the cooperativity of the
protein structure.

We have examined the NMR spectra of cytochromes-*c* from horse
heart, bonito, yeast, pigeon, rabbit, cow, and chicken. All the
cytochromes have the same fold in solution. We state this with
confidence as the NMR spectrum of a protein is a very good diagnostic fingerprint of its conformation and this is especially
true of paramagnetic proteins (see under peroxidases). The most
important use of the different species has been therefore in
assisting assignment. It should be observed that there is really
no functional difference between the different cytochromes-*c* from
the different eucaryotic species. However, we do observe that
changes in the sequence alter the mobility of aromatic residues
to some extent. Thus in horse heart cytochrome-*c* the relative
mobilities of tyr 48 and phe 46 are different from the mobilities
of tyr 48 and tyr 46 of tuna cytochrome-*c*.

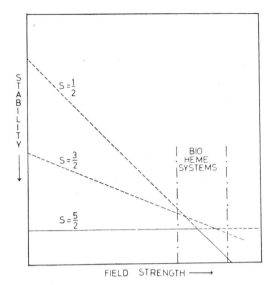

FIELD STRENGTH ⟶

Fig. 2. The spin-states of iron(III) in different complexes of increasing ligand field strength. Note that at low field strength the high-spin state is stable. The intermediate spin-state is rarely of greatest stability. The ligands, porphyrin plus protein groups, place iron in biological systems close to the crossover points. A similar diagram can be drawn for the three spin-states of iron(II).

A major problem which has to be overcome in the study of heme-proteins is the correct alignment of the g-tensor of the low-spin Fe(III) state, for once this is achieved structural work is much easier. We have proceeded through the assignment of the porphyrin ring resonances of the heme-group relying upon external and internal probes. For example, some of the heme methyl groups can be assigned to their appropriate pyrrole rings, in part through observations of the effects of externally added probes such as $[Cr(CN)_6]^{3-}$. The full details of the assignments will be published later. It suffices to say here that, together with an outline knowledge of the disposition of resonances of the aromatic groups in the structure from crystallographic data, even limited knowledge of the Fe(III) heme g-tensor permits the following conclusions:

1. The fold of the protein as seen in crystals is the same as that seen in solution.
2. The positions of many side chains are not well defined due to internal motion.
3. The exterior of the protein can not be given a defined shape as motion of surface groups is very extensive.

These properties are common to some five or six proteins which we have looked at in detail. Further details of the structure can now be tackled using isomorphous replacement.

2.5 METAL-REPLACEMENT IN CYTOCHROMES-c - Recently cytochromes-*c* with different metal ions in the centre of the porphyrin have been prepared *(3)*. This development permits the use of the method of isomorphous metal replacement in cytochrome-*c*, as has been used previously in other heme proteins and by Vallee and Williams, and Malmstrom and coworkers in a long series of studies on various metalloenzymes *(4)*. We note that the use of this method in heme proteins was attempted very early by biochemists who were not aware of the physical-chemical studies which are permitted by such replacement. Table 3 lists some of the heme-probes that can be used.

TABLE 3: SOME METALLO-PORPHYRIN PROBES[a]		
Metal Ion	Spin State	Use in NMR
Fe(II)	4/2	Shift and Relaxation Probe
	zero	Diamagnetic Blank
Fe(III)	5/2	Shift and Relaxation Probe
	1/2	Shift Probe (largely)
Mn(II)	5/2	Relaxation Probe
Co(II)	1/2	Shift Probe
Ln(III)	zero to 7/2	All types of Probe
Cu(II)	1/2	Relaxation Probe
VO(II)	1/2	Relaxation Probe
Co(III)	zero	Diamagnetic Blank

[a]*It is feasible that a much greater variety of metals can be incorporated but this has not yet been attempted.*

Of particular interest in cytochrome-*c* is the use of the diamagnetic Co(III) porphyrin in place of heme for it can serve as a diamagnetic reference state for the Fe(III) low-spin state of the protein, allowing by difference a direct determination of the paramagnetic shifts of the NMR lines in the Fe(III) state, which are essential data in structure determination. It is also possible to compare diamagnetic Fe(II) and Co(III) proteins looking for evidence of a conformation change due to a change of oxidation state. Again cobalt(II) porphyrin is an excellent shift reagent for it has strictly axial symmetry and thus the methods used in the study of molecular complexes of metal porphyrins by Hill and Williams *(5)* will become possible. Relaxation studies can be used if Mn(II) high-spin or, probably better, Cr(III) high-spin (d^3) porphyrins can be inserted, but note that high-spin Fe(III) is both a relaxation and shift probe. If the heme protein does not contain a six-coordinated low-spin central iron atom to which access is blocked, then shift and relaxation probes may be developed in an alternative way by using different ligands to the heme. This procedure can not be used in cytochrome-*c* without affecting its native structure and we leave its discussion until later (see peroxidases). The last method we call heme-tickling.

In collaboration with Professor Chien, who has prepared several metallocytochromes-c_3, we have now examined the NMR spectrum of cobalt(III) cytochromes-*c* in detail. The isomorphous replacement is very precise. In fact a comparison of the diamagnetic Co(III) and the diamagnetic Fe(II) enzymes shows that the perturbations due to the intrinsic ring current probes within the two molecules, including the effect of the metal-porphyrin, are almost precisely the same (Table 4).

TABLE 4: CONFORMATIONAL PARAMETERS FOR TRYPTOPHAN

Cytochrome *c*	Position of Resonance (ppm)					
	d	*d*	*t*	*t*	*s*	shifts
Fe(II)	7.07	7.60	6.68	5.74	(6.99)	ring current
Co(III)	7.07	7.59	6.62	5.77	(7.06)	ring current
Fe(III)	7.57	7.37	6.54	6.31	(6.86)	ring current + pseudo-contact
Fe(III)-Co(III)	+0.50	-0.22	-0.08	+0.54	(-0.20)	pseudo-contact
d = doublet, t = triplet, s = singlet						

Thus within the limits of NMR studies Co(III) and Fe(II) enzymes show no conformational changes in the heme pocket or elsewhere, although there is probably a 5% change in bond lengths to the ligands. Apart from the structural implication - that there is little or no conformation change in the heme region of cytochrome-*c* on change of oxidation state from Fe(II) to Co(III), and pre-sumable therefore to Fe(III), there is the mechanistic implication that the electron-transfer changes only the metal-ligand bond dis-tances by a small amount. (Note: the discussion applies to low temperatures only, *i.e.*, <50°C). This result is in direct con-tradiction with deductions based on general physical properties and on the early crystal structure of the two oxidation states. There may be small conformation changes elsewhere in cytochrome *c* on oxidation/reduction, but the changes in solution are indeed very small internally. (Very recently we have been informed by Prof. Dickerson, who was unaware of our NMR results, that the refined accurate crystal structure analyses of Tuna cytochrome *c* in the two oxidation states shows no really detectable conforma-tion change (June, 1976)).

2.6 TEMPERATURE AND pH CHANGES - Monitoring the NMR spectrum of cytochrome-*c* (Fe(III) form) with changing temperature shows that there are different states of Fe(III) cytochrome-*c* in fast ex-change. We find, for example, that some of the resonances with large paramagnetic shifts do not follow the expected 1/T depend-ence, and indeed some of them show an inverted dependence on temperature which implies that the magnetic field experienced by these protons, *i.e.*, the local susceptibility, increases with temperature. The simplest explanation is that there is a fast spin-equilibrium

$$\text{low spin state} \underset{\longleftarrow}{\overset{\longrightarrow}{\rule{2cm}{0pt}}} \text{higher spin state}$$
$$(\text{low } T^{\circ}C) \hspace{3cm} (\text{high } T^{\circ}C)$$

These two states must have somewhat different geometries with the average bond distances, Fe to ligand atoms, somewhat longer in the high-spin than in the low-spin state (Fig. 3). As the above equilibration is fast on the NMR time scale, ligand motions of at least 0.1 Å (and perhaps as much as five times this) must occur at rates exceeding 10^4 times per second. The same temperature ef-fects are not seen in the low-spin Fe(II) state, but note that the low-spin Fe(II)-ligand bond lengths are expected to be somewhat longer than those of low-spin Fe(III) complexes.

On changing the pH of the Fe(III) form of cytochrome-*c* from 4 to 3 there is again a continuous change in the NMR spectrum which shows that the dissociation of the ligand(s) from the heme

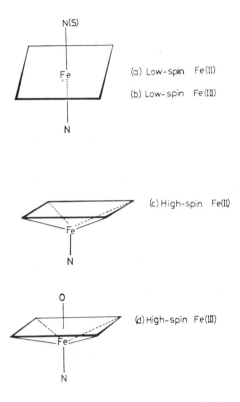

Fig. 3. Some of the geometric arrangements found for iron in porphyrin complexes and in different spin-states.

which produces the observed spin-state change is a fast exchange. Here the nature of the spin-state change is very well documented, using a great variety of methods. At high pH there is a discontinuous change in the NMR spectrum as the methionine is replaced by a lysine showing that these two ligands, both of which give low-spin complexes, are in intermediate exchange.

Before turning to the significance of these observations in the reactions of cytochrome-c, it is worth noting that the high-spin $\rightleftarrows$ low-spin equilibria of heme-proteins are usually fast. For some time it has been known that many of the iron porphyrin proteins occur in such equilibrium mixtures of spin states. More recently it has been demonstrated that in certain of them, *e.g.*, cytochromes-c', a quantum mechanical mixing of the spin-states occurs so that the ground state is not a simple identified spin-state. It is important that these two situations are recognised to be quite different for in quantum mechanical mixing there is but one structure and there is no problem of conformational mobility between two states, while in spin-state equilibria there are two states of different structure. Even those proteins which contain quantum mechanical admixed spin states, *e.g.*, cytochrome-*c'*, also exhibit classical spin-state equilibria over some pH regions. Proof of this spin equilibrium for many of the proteins rests on the fact that NMR shows but one spectroscopic signal for the mixtures, while UV/visible, EPR and Mössbauer spectroscopies reveal two or more species. Thus the rate of jumping between structures is between 10^4 and 10^8 times a second. We consider that for cytochrome-*c* there must be a fast motion of small portions of the protein chain close to the heme and not of the whole protein for the following reasons:

(1) Proton relaxation time constants for all the different types of proton in the protein give a value close to 10^9. There is undoubtedly some fast breathing of the protein apart from the rotational tumbling which has a rate constant close to 10^8.

(2) The flipping of tyrosine and phenylalanine is slower than 10^4 per sec in many parts of the molecule but is faster in others.

(3) From our measurements on lysozyme *(2)* there is also certain to be motion of other residues such as valine and leucine in which rapid flipping between conformers of the side chains takes place.

(4) Surface mobility of lysine side chains is high and motion here is faster than the electron transfer rate.

(5) In cytochrome *c* there is no suggestion of any segmental motion as we have proposed in haemoglobin.

In summary, cytochrome *c* is a relatively rigid protein with an outline structure which is constant in the two oxidation states in solution. It is against the background of this structure and its dynamic fluctuations that we must view the mechanism of electron transfer. While making this examination we must remember that the structural comparison is of the ground states for two oxidation states at temperatures below 50°C. Above that temperature the Fe(III) state is in much larger part high-spin, a change which does involve a breaking of the methionine-iron heme bond. Thus there is a readily accessible excited state. We shall contend that it is not the actual breaking of the coordinate link that is important for fast electron transfer but its easy extension through stretching, and the presence of the low-lying excited state is but an indication of the ease of bond stretching. Low-lying excited states of Fe(III) and of high spin character are observed in many cytochromes, *e.g.*, cytochromes *c'*, cytochromes b, and cytochromes P-450 (see Fig. 2).

2.7 MECHANISM OF ELECTRON TRANSFER - As far as the mechanism of electron transfer is concerned the overall conclusions are that the constraints imposed by the protein fold have placed the ligands to the iron of cytochrome-c at such bond distances or angles as to affect grossly:
(1) The static structure such that it is held very close to the cross-over point of spin states in oxidation state III (entatic state);
(2) The dynamics of ligand motions are such that the heme pocket is of some flexibility and there are some motions of groups at greater distances from the iron.
    This study of cytochrome-c thus shows it to be a protein with a relatively rigid structure within which the side chains possess a range of mobilities, and these fluctuations of structure are quite probably vital in the actual mechanism of electron transfer. We know that electron transfer cannot be an inner-sphere reaction for the iron ligands cannot be displaced by an electron transfer agent. Clearly it is possible for outer sphere electron-transfer to occur through one edge of the heme, but in fact it is not well exposed and in electron-transfer reactions with some biological ligands, *e.g.*, the Cu(II)/Cu(I) proteins of the azurin type where the copper ion is buried within the protein fold, the distance between the metal ions of the two reactants must be equal to or exceed 1 nm. For this reason it has been postulated by others that electron transfer occurs via a hopping mechanism in which the electron goes from the metal ion in the porphyrin to an aromatic residue such as tyrosine, tryptophan or phenylalanine of the protein. The objections to this mechanism can be put most clearly after a discussion of peroxidases, and it is sufficient to say here that a tunnelling through preferred solvent (hydrophobic) channels is more probable (Fig. 4). A full review of the work leading to this conclusion has been published *(1)*.
    The mobility of the proteins may be vital in function for it allows:
(1) Fluctuation of the redox potential so as to match the donor or acceptor potential - an essential condition for tunnelling.
(2) Lowering of the Franck-Condon activation energy for electron transfer.
    In some regards it is as well to look at cytochrome-c in the same way as one inspects other proteins such as lysozyme, treating the heme group as the bound substrate (for it undergoes the reaction) within a reaction groove set up within the protein. As with lysozyme, the "groove" must be mobile to allow the

reaction (of bound substrate $\rightleftharpoons$ product) to proceed. This must be true of all proteins. Now the groove is adjustable by movements of amino-acid side-chains, which we see in cytochrome-c, or by small adjustments of segments or larger sections of the protein as a whole. In most proteins we believe now that the size of the whole groove can be varied and that this is not only an important component of the reaction mechanism but is also a requirement if the protein is to relay information to its neighbours (see cyto-chrome P-450).

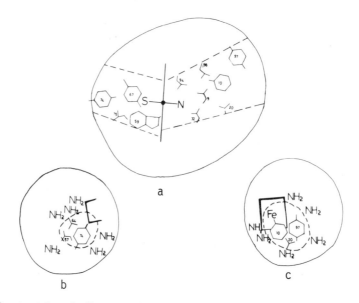

Fig. 4. Schematic diagram of horse ferricytochrome c: (a) left and right channels viewed from the front of the molecule; (b) left channel viewed from the left side of the molecule; (c) right channel viewed from the right side of the molecule (-----), boundaries of channels. S is methionine, N is histidine.

We can return now to the anion binding sites of cytochrome-c, asking about directions of electron-transfer within the protein.

2.8 BINDING SITES OF CYTOCHROME-c - In order to follow electron transfer from the iron atom of cytochrome-c to some added redox reagent we need to know how the added reagent binds to the pro-tein. With this knowledge the study of electron transfer becomes an experiment in a matrix and as such will be much easier to appreciate than electron transfer in solution where many differ-ent combinations of reactant are present at once. Fortunately, the use of paramagnetic redox reagents (see Table 5) enables us

TABLE 5:	SITES OF BINDING OF PARAMAGNETIC REAGENTS TO CYTOCHROME-c	
Reagent	Oxidised State	Reduced State

Lanthanides	1. strong: His 33 (Pr, Gd)
	2. (strong): Trp 59 (Pr)
	1/3. strong: Ileu 9(95) (Pr, Gd)
	    Tyr 97
		1. strong: Tyr 97 (Gd)
		    Phe 10

$K_3[Cr(oxalate)_3]$	1. strong: Heme crevice, His 26 2. weak: His 33	1. weak: His 33, His 26
$K_3[Cr(CN)_6]$	1. strong: His 26 2. weak: Heme crevice 3. weak: Lysines	1. weak: Lysines

$[Cr(biguanidine)_3]$
    Cl_3
	1. strong: N-acetyl terminus
	    (negative charge region)
		Not attempted

$CuSO_4$	1. strong: His 33 2. weak: His 26 3. weak: Lysines	1. strong: His 33 2. weak: His 26 3. weak: Lysines
$Gd(tiron)_3$	1. Tyr 97	Not attempted
Ferredoxin II (D. gigas)	1. Tyr 97; Ala 15	Smaller effect, if any

Note: Differential binding of two oxidation states at Heme crevice.

to find where the electron transfer compound is binding. As the table shows, we have studied several such reagents and we note that whereas anions bind to specific positively charged regions, there are differential preferences even between anions such as an oxalato-complex and a cyanide complex. Again the binding of the anion is different in the oxidised and reduced forms of the enzyme. Similar selectivities are shown in the binding data for cations. The copper(II) ion binds to histidines with considerable strength, but lanthanides seek out the carboxylate clumps on the protein surface. The positively charged chromium(III) biguanidine complex goes to a quite different site at the N-terminus where there is also some negative charge. Now none of the anions or cations in the table themselves undergo redox reactions, but it is readily seen that we can change the metal, say Cr(III) to Fe(III), and so obtain an isomorphous anion which is a reactant. Such experiments have been done by others in part but without sufficient knowledge of the protein.

At the bottom of Table 5 we have added a protein reagent

which will bind to cytochrome c. We note here only the preferential binding of one oxidation state [Fe(III)] just below the heme pocket region. This type of combination of proteins becomes of more direct interest when we describe protein-protein combinations such as occur in the cytochromes P-450.

Cytochrome c has mobile surface residues, much as we find for other proteins. In particular, surface lysines are freely mobile with side-chain rotational rate constants of $>10^9$ sec^{-1}. This means that structure of the surface as provided by crystallography could be quite misleading as to the conformations of side-chains required to bind substrate or small molecules. Thus, although we now have excellent markers for the study of the association of cytochrome-c with other electron-transfer agents, great care will have to be taken in the examination of the structures of the complexes. Change of redox state alters the surface while it has little effect on internal conformation.

3. PEROXIDASES

3.1 INTRODUCTION - Most peroxidases are water soluble monomeric heme-proteins which catalyse the oxidation of a wide variety of organic molecules by H_2O_2 and/or O_2. There are peroxidases in all plants, many bacteria, and in most animals. In a vegetable such as a turnip there are at least seven peroxidase iso-enzymes, some properties of which are given in Table 6 *(6)*.

TABLE 6: AMINO-ACID COMPOSITION OF TURNIP PEROXIDASES P_1 AND P_7					
Amino Acid	P_1	P_7	Amino Acid	P_1	P_7
	mol/mol	mol/mol		mol/mol	mol/mol
Lysine	7	12	Alanine	34	32
Histidine	3	3	½ Cysteine	11	8
Arginine	11	16	Valine	22	20
Asp-X[a]	58	40	Methionine	3	6
Threonine	33	18	Isoleucine	19	15
Serine	44	40	Leucine	43	22
Glu-X[b]	29	16	Tyrosine	3	5
Proline	21	13	Phenylalanine	20	15
Glycine	33	25			

[a]*Aspartic acid plus asparagine*   [b]*Glutamic acid plus glutamine*

Note the low content of tyrosine and absence of tryptophan.

The following properties are common to most of these enzymes:
1. They contain heme, usually iron protoporphyrin IX;
2. The iron is bound by a single nitrogen ligand, probably histidine, to the protein, while the other side of the heme is open or, in the high-spin Fe(III) state, probably has an iron-bound water molecule (see Fig. 3);
3. The proteins are large, usually of molecular weight $\sim$40,000, and they are highly soluble, partly due to surface sugar residues;
4. The protein chains are cross-linked by -S-S- groups.

The enzymes have two substates, H_2O_2 (or O_2) and usually some aromatic organic group such as a phenol or an indole (but see below). Their reactions, one electron oxidation of the phenol etc., are of *low selectivity* and many substituted aromatic compounds can be attacked. The reaction intermediates at the iron are well described and starting from Fe(II) or Fe(III) the states are: FeO(IV), so called compound II, $Fe^{II}O_2$ (compound III), and (compound I), probably FeO^{IV} plus a radical (Fig. 5).

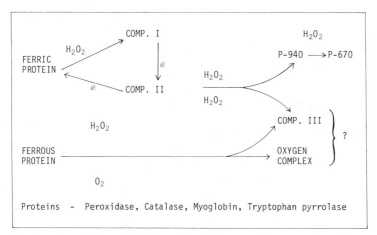

Fig. 5. The relationship between the different oxidation states of iron in peroxidases and catalases. P-940 and P-670 are decomposition products. Compounds I, II, and III are two, one and three oxidizing equivalents above Fe(III).

In plants the peroxidases probably regulate hormone activity, the plant auxins are indole derivatives, and they may also be used to make a protective scab after injury, while in animals they may well be part of the defence mechanism (see below). In seaweeds the peroxidases on the surface of the frond produce iodine, perhaps as a simple antiseptic, and in bacteria and myelocytes there are peroxidases which are sufficiently powerful that they can oxidise chloride to chlorine. This halogen activation exposes life to grave risk while providing protection. We return to this point later.

3.2 GENERAL STRUCTURE - Today, an understanding of any enzyme should start from the outline "rigid" structure provided by a combination of sequence and X-ray diffraction data. In the case of the peroxidases the complete or partial sequences of several peroxidases are available, but there are little crystal structure data. Although it may be a wise course to wait for this information, we have become increasingly impressed by the ability of NMR methods to handle structure problems and thus we have been tempted to make a general comparative study of the isoenzymes of turnip and horse radish peroxidases *(7)*. My comments on peroxidases arise from the totality of the efforts of three independent groups, two groups of colleagues from Marseille and Copenhagen, having merged to complete the peroxidase sequences (see the introduction).

Spectroscopic and chemical studies by several groups have shown that the heme iron is bound to one histidine (proximal) of the three that occur in horse radish peroxidase. Furthermore, the chemical studies suggest strongly that there is a second histidine very close to the heme - a so-called distal histidine. If we accept these chemical studies, then the surround of the heme becomes very similar to that in myoglobin or hemoglobin (Fig. 6), and using this comparison and the close sequence analogies between myoglobins, hemoglobins and peroxidases, Welinder and ourselves have proposed the following structure for the active site (Fig. 7). The proposed proximal and distal histidine peptide sequences of the different isoenzymes can be aligned (Table 7). It will be noticed that, with the exception of P_7, three histidines are always near to the iron. In partial confirmation of this structure the NMR spectra of horseradish peroxidase and the acidic turnip peroxidase isoenzymes show no histidine resonances, which could well be broadened by paramagnetic broadening. We note that the histidine resonances close to the heme in metmyoglobin cannot be seen in its NMR spectrum, while more remote histidines are seen. The suggested position of the S-S bridge in this peroxidase structure is close to that of the pair of R-S heme links in cytochrome-*c*. The cysteine closer to the heme in the peroxidase is reminiscent of cysteine-93 in the β-chain of hemoglobin (Fig. 6). It also has possible analogies with the cysteine of cytochrome P-450 which binds directly to the iron.

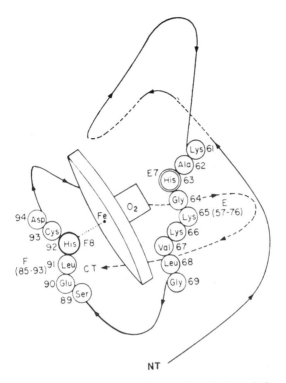

Fig. 6. A semi-schematic representation of the β chains of hemoglobin. Note the position of the cysteine (93) in the β chain and compare structures of cytochrome c and peroxidase.

TABLE 7: HORSE RADISH PEROXIDASE SEQUENCE (K.G. WELINDER)				
40				49
Leu - His - Phe - His - Asp - Cys - Phe - Val - Asn - Gly - Cys				
E.5	E.7	E.9	E.11	E.14
	170			
Ser - Gly - Gly - His - Thr - Phe - Gly - Lys - Asn - Gln				
	F.8	F.10	F.12	

Note: The His sequences contain no tyr, or Tryp amino-acids. The top sequence is that for the distal histidine while that at the bottom is the proximal histidine. One of the Phe residues becomes iso-leucine in turnip peroxidase P_7 and is thought to be at the binding site of the substrate. The letterings E.x and F.y refer to corresponding residues in hemoglobins.

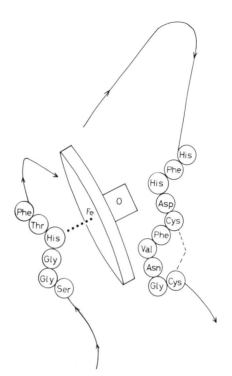

Fig. 7. *The outline structure of the heme pocket of peroxidase based on the sequence of Table 7 and the structure of hemoglobin (Fig. 6). The positions of the residues can also be seen in Fig. 8. The NMR spectra on the different isoenzymes and the dependence of these spectra on temperature and substrate binding, together with the NMR data on the substrate resonances, do not yet establish an unequivocal positioning of the substrate, the sequence, and the heme, but the figure does indicate correctly the broad general features of the active site.*

It would not require a large change in the protein to provide an RS⁻ ligand for coordination to the iron of peroxidase and this may occur in chloroperoxidase. It is relatively easy to denature hemoglobin slightly so that cysteine-93 becomes bound to the iron. There is then a possible phylogenetic link between these groups of heme proteins.

There is also good evidence for a distal histidine in the heme pocket of peroxidase from pH titrations, which generate similar spectral perturbations to those seen in myoglobin. The differences between myoglobin and peroxidase are striking, however, and T_1 and T_2 relaxation time studies of the bulk water of peroxidase solutions lead to the suggestion that the iron water ligand is much more strongly hydrogen-bonded to the proposed distal histidine of peroxidase. The finding that the ferro peroxidase reaction with oxygen requires the uptake of a proton, which is not the case for myoglobins, suggests that the reaction is as follows:

$$\begin{bmatrix} \text{Imid} \cdot H^+ \cdot {}^-O_2C \\ - Fe(II) - \end{bmatrix} + X \longrightarrow \begin{bmatrix} \text{Imid } H^+ \ldots CO_2H \\ - Fe(II) \ X - \end{bmatrix}$$

Note that uptake of anions to Fe(III) in peroxidases also requires the uptake of a proton. The closer association of the distal histidine with the iron in peroxidases could lead to the formation of a low-spin bis imidazole iron(III) complex at high pH, and not a hydroxide complex, for this reaction is quite unlike that of myoglobin.

Although peroxidases have the same or a very similar iron coordination sphere to that of myoglobin, the physical and chemical properties of the proteins indicate that there must be other subtle differences. The physical properties, both resonance Raman and *low temperature* EPR measurements, suggest that the heme in peroxidases must be more like the heme of cytochrome-c'. Now cytochrome-c' heme probably has the same coordination sphere as myoglobin, but in its ground state it can occupy several different quantum mechanical mixtures of spin-states and it has been proposed that, as there is admixture of a low-lying $S = 3/2$ quartet and $S = 1/2$ Kramers doublet into the ground state $S = 5/2$ sextet of the iron, the iron atom lies more closely in plane than in myoglobin. It is possible that this is also the case for peroxidase, though to a lesser degree. Table 8 relates the various structures and properties of these heme proteins. The differences in the coordination sphere are also indicated by the slow exchange rate of water from the Fe(III) state of peroxidase. We are led to conclude that the various heme proteins adjust the histidine-iron, the porphyrin-iron and the water-iron links differentially from protein to protein and we may suppose that this has a functional consequence. These proteins are then excellent examples of the *entatic* state modification of the structure of a metal coordination sphere by a protein so as to optimalise function *(4)*.

We turn next to the redox potentials of peroxidase *(6)*, Table 8, and note a gross variation from myoglobin which cannot be due to the heme ligands. Rather it is due to the side-chains of other amino acid residues in the heme pocket or just possibly to constraints on the porphyrin ring. The known residues in the

TABLE 8:   REDOX POTENTIALS OF SOME HEME-PROTEINS

	pH = 6	pH = 7	pH = 8	pH = 9	pK
Sperm Whale Myoglobin	+0.06	+0.05	+0.04	+0.02	8.7
+ imidazole			-0.10		
Chironomus thumni[a]	+0.15	+0.12	+0.06	+0.01	7.4
Hemoglobin A α chain	+0.05	+0.05	+0.05	< 0.0	~8.0
β chain	+0.11	+0.11	+0.06	< 0.0	(~7.0)
Cytochrome P-450		-0.30			
Cytochrome-b	±0.05	0.05	±0.05		
Cytochrome-c		0.25	0.25		
Peroxidase P_1	-0.18	-0.22	-0.22	-0.22	10.2
Peroxidase P_7	-0.10	-0.13	-0.25		8.4
Catalase		< -0.50			
Hemoglobin $(-CO_2^-)$		~-0.40			
Hemoglobin $(-\phi O^-)$		< -0.50			

[a]*No distal histidine. There are a series of such cytochromes with E^o values (pH = 7) from 0.0 to +0.2 volts.*

heme pockets of myoglobin and hemoglobin are largely non-polar *e.g.*, phe, leu, val (Fig. 8), and it is very probable that this is also true of cytochrome-c, but it is not likely to be so markedly the case for peroxidases (see above picture, Fig. 7). The redox potentials and absorption spectra of turnip peroxidases have been measured over a wide pH range and the results show that there must be several ionising groups in the vicinity of the heme which can influence it (Table 8). The redox properties of myoglobin and cytochrome-c do not show such pH effects. Thus we see that the ground state properties of peroxidase are under control of several parts of the polypeptide to a much more marked degree than was observed for cytochrome-c or myoglobin.

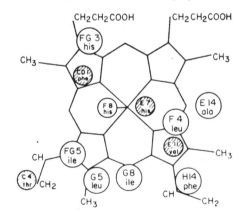

Fig. 8. The structure of myoglobin. The different shading indicates different sides of the heme group.

3.3 NMR STUDIES OF PEROXIDASES (7) - The first use of the NMR
method in the examination of peroxidases was to find the resonances of the methyl groups of the heme in the Fe(III) complexes.
Reference 7 gives a comparison of these resonance positions with
those found for other heme proteins. The peroxidases are obviously high-spin S = 5/2 Fe(III) species by this criterion and absorption spectra, although at very low temperatures EPR indicates
that other spin-states may be involved. Addition of cyanide produces a low-spin ferric state (S = $\frac{1}{2}$), whilst addition of azide
gives an equilibrium mixture of spin states. *There is only one
set of NMR lines for the azide complex* so that the environment of
the heme must undergo rapid conformational switching between low-
spin and high-spin species at a rate which is equal to or exceeds
10^4 per sec. This rate is of the same order as some steps of the
reactions of peroxidases, so that the protein-linked heme adjustments (spin-states changes) could be rate limiting; *i.e.*, as the
iron goes through the steps:

Fe(III) → Compound I → Compound II → Fe(III)
high-spin   Fe(IV) low-spin high-spin
              ↑

Fe(II) → Fe(II)O$_2$, Compound III
high-spin low-spin

We note, however, that in catalase the turn-over rate is very
fast indeed, $\sim 10^9$ sec^{-1}. This means that the changes in iron
oxidation states are at this rate at least, and it is not really
conceivable that major parts of the protein are responding at
these rates. However, we know that there are internal motions in
lysozyme and cytochrome-*c* with such rate constants and these are
probably local rotational/vibrational changes of amino acid side-
chains with respect to one another. It would appear, however,
that despite the changes

Fe(III) ⟶ FeO(IV) ⟶ Compound I
high-spin ⟵ low-spin ⟵ low-spin

there can be no real geometric switches of any magnitude except
those associated with the organic substrate.
    The next step in the analysis will be to assign the methyl
resonances specifically to the four different corners of the heme,
as was done using deuterium substitution in myoglobin or as we
have done in part for cytochrome-*c*. Once this assignment has
been accomplished, the *g*-tensor of the Fe(III) states will be
known and the internal heme probe becomes much more useful in the
mapping of the protein. It is possible to substitute various

other metal porphyrins for heme in peroxidases, *e.g.*, Co(II), Mn(II), Mn(III), and these substitutions will permit a great variety of probe studies, as has been done in myoglobin and cytochrome-*c*.

The NMR spectrum shows no resonances to very high field which could be due to -S-CH_3 or NH_2-CH_2-, so confirming that the iron is not bound by met or lys ligands. However, we are not able to see histidine bound directly to the heme. The heme methyl resonances are sensitive to a variety of changes in the heme pocket, including a titrating group with a pK = 8.5 and the binding of substrates. By comparison with myoglobin, heme-linked pK = 8.5 most likely reflects the titration of a distal group, probably histidine.

We can also examine the resonances of the protein side-chains of peroxidase *(7)*. A general point here is that the NMR spectrum is a finger-print of structure and not just of sequence. The acidic turnip peroxidases are rather similar, whilst there are both similarities and differences between the P_7 isoenzyme (basic) and the P_1 isoenzyme (acidic). By studying the temperature dependence of the resonances, we can find many which arise from groups close to the heme, and by heme-tickling methods (see below) and using difference spectroscopy, we can confirm that these resonances belong to atoms near the heme. Now although it is not yet possible to assign these resonances with any certainty to individual amino acids, a number of points can be made. In turnip P_1, P_2 and P_3, and in horse radish peroxidase spectra, there is an aromatic resonance with an unusual chemical shift which is notably temperature dependent, *i.e.*, near the heme. This resonance is not present in P_7. In the turnip P_7 NMR spectrum there is a methyl resonance which is temperature dependent, *i.e.*, near the heme, but is absent in the spectra of the other enzymes. In the sequences of these enzymes a particular residue near the heme is phenylalanine for turnip P_1 and P_3 and horseradish peroxidase, but the corresponding position is isoleucine for turnip P_7 peroxidase (see Fig. 7). These resonances are sensitive to the binding of substrate in the isoenzymes and they have similar temperature dependencies. Using the substrate site as a reference point (see below) we can put the groups into the picture of Welinder and the heme pocket now becomes as in Fig. 7. Thus the NMR and the sequence data begin to provide a picture of the peroxidase active site.

We have already mentioned a new technique for the study of the heme-pocket which we call heme-tickling (Table 9). In order to see its value, we must refer to the work on the structure of lysozyme where we could *exchange* different metal ions of the lanthanide series at the active site, and so obtain at this site

		Ligand	Probe
	TABLE 9:	HEME-TICKLING PROBES	
Native enzyme		Imidazole, H_2O	(Shift)/Relaxation
	$+F^-$	1% F^-	Relaxation
	$+N_3^-$	1-10% N_3^-	Shift
Azide protein	$+H_2O$	90% N_3^-	Relaxation
Native enzyme	+ Acetate	1-10% Acetate	Shift

a variety of NMR relaxation and shift probes. These probes permitted the clear definition of the conformation of lysozyme in
solution. In a heme protein we can achieve the same variety of
probes by changing the metal (see above), or the ligand on the
Fe(III) of the heme, provided that ligands are in fast exchange
on the NMR time scale and provided that the ligands adjust the
spin-state equilibrium. As pointed out above, this proves to be
the case. Thus if to native peroxidases (95% high-spin) we add
fluoride in small aliquots, then we alter the unpaired electron
spin relaxation rate of the iron and it is driven toward 100%
high-spin. By taking difference spectra in the presence of different fluoride concentrations added to peroxidase, we obtain the
same information as from the difference spectroscopy of lysozyme
with and without Gd(III). The result is that the regions of the
protein at certain distances from the iron (gadolinium) are revealed in the difference NMR spectrum in turn as the fluoride
(gadolinium) concentration is raised. In order to make a shift
probe from heme comparable with lanthanide shift probes used to
determine structural features of lysozyme, we titrate the native
protein with azide which continuously alters the g-tensor at the
iron. The results are more difficult to interpret in the case of
the azide titration of peroxidase than those from the addition
of, for example, praesodymium(III) to lysozyme, as lower symmetry
is involved. The methods have been tested on myoglobin, a protein of known crystal structure, and we now have confidence in
their value.

The mapping of the binding sites of some substrates, namely
phenols and indoles, has been achieved by the same, now standard,
methods using shift (diamagnetic ring current and paramagnetic
probes) and relaxation (paramagnetic) probe procedures where the
central heme acts as the probe *(9)*. Note that the method must
be used with great care and appropriate blanks are essential.
The structure derived for peroxidase/indole complexes is shown in
Fig. 9 and that for peroxidase/phenol complexes is rather similar.
All our data show that different substrates are held in different,

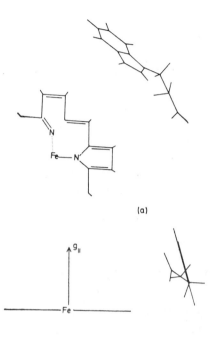

(a)

(b)

*Fig. 9.   Structure of the indolepropionic acid-horseradish per-
oxidase complex: (a) projection onto the heme plane, and (b)
projection onto the plane containing the symmetry axis (z) and
perpendicular to the indole ring.   The outline of the porphyrin
ring has been included in an arbitrary orientation in the xy-plane.*

though related, regions of the protein, suggesting that, although
there is no well-defined substrate binding *site,* there is a
hydrophobic region into which substrates partition readily.   The
hydrophobic character of the pocket has been confirmed by an examination of proton and fluorine shifts in fluoro-phenols.   We
suspect that changes of amino-acids from turnip peroxidases P_1 to
P_7 (from phenylalanine to leucine as described above) alters the
binding by the pocket to a small degree, for the pocket is probably not highly structured but rather mobile.   The binding site
is some 0.8 - 1.0 nm from the iron in all cases, which raises the
question of the mechanism of electron transfer to give the substrate radical.   Some peroxidases are more efficient than horseradish peroxidase in the oxidation of halides, and we conjecture
that these peroxidases must have a site similar to the other peroxidases, but we speculate that there is now a single basic side-
chain substituted into the sequence in the heme region and that

this becomes a hydrophobic anion binding site. Such sites are
known to bind halides except fluoride very well, and can be made
to have the selectivity sequence $I^- > Br^- > Cl^- > F^-$.

It is of considerable interest that when the substrate binds
to the protein, the ESR signal of the iron is changed. The substrate binding site is far from the iron and we suppose that
binding induces a small conformation change in the heme pocket
which is relayed to the iron through minor conformational changes.
Evidence for a minor conformational change is also obtained from
NMR, small shifts occurring upon substrate binding for a small
number of amino acid resonances (including a phenylalanine in HRP
and the acidic turnip isoenzymes). Small shifts in porphyrin
ring resonances reflect a change in the electronic structure of
the heme. Thus, once again, the heme pocket is shown to possess
mobility.

We return now to the problem of the electron-transfer path to
the iron higher oxidation states from substrate raised by the
structure of the substrate/protein complex (Fig. 9). There is
clearly no direct contact between the oxygen of the FeO(IV) and
the substrate for there is no direct oxygen atom transfer. Moreover, only relatively small ligands can replace H_2O_2 or O_2 at the
iron, and we imagine that the side-chains of the protein leave
only a rather small pocket for the O_2 molecule to bind to the
iron. This is the same situation as in hemoglobins where again
only relatively small ligands can bind to the iron atom. Thus,
we conjecture that electrons must be able to hop from the indole
or phenol to the iron porphyrin π^* states. It must be a function
of compound I and compound II that the π^* states are of high
electron affinity. The π^* orbitals will spread some distance
from the iron and there is probably little need to invoke any
other mechanism than long-range outer sphere electron transfer,
but note that we have every reason to believe that the pocket for
binding is hydrophobic, assisting the rather long-distance electron hop *(1)*.

There are additional features of interest in these structures.
Although there are phenylalanine groups in the heme pockets of
peroxidases, hemoglobins and myoglobins, there are no tryptophan
or tyrosine residues, but these residues are very close to the
heme of cytochrome-*c*. Elsewhere *(1)* we have pointed out that a
protein which shuttles between Fe(II) and Fe(III) states cannot
oxidise tyrosine or tryptophan if the redox potential of the iron
couple is lower than +0.5 volt, for the redox potential required
for oxidation of phenols or indoles is about +1.0 volt. Cyto-
chrome-*c* has both a tryptophan and a tyrosine in the heme pocket
and its redox potential is +0.25 volts. However, the redox potentials of the FeO_2 or FeO(IV) complexes, *i.e.*, higher oxidation

states of iron are about +0.9 to +1.2 volts and are able to attack either phenol (tyrosine), or indole (tryptophan) (Table 10).

TABLE 10: INSPECTION OF POSSIBLE ELECTRON-TRANSFER CENTRES		
Oxidised Centre	Reduced Centre	Observation
Fe^{III} heme peroxidase	Phenols	No reaction
Fe^{III} heme peroxidase	Indoles	No reaction
Fe^{III} in oxygenases	Phenols	No reaction
Cu^{II} in oxidases	Phenols	Reaction
Phenol radicals	Indoles	Little reaction
Fe^{IV} in peroxidases	{ Phenol	Reaction
	{ Indole	
Fe^{V} in peroxidases	{ Phenol	Reaction
	{ Indole	
Fe^{IV} or Fe^{V} in peroxidases	Benzene	No reaction
Fe^{IV} (or Fe^{VI}?) in P-450 cytochromes	{ Alkyl chains	Reaction
	{ Benzene	

It would appear to be essential, in general, to build a pocket
without these groups for those heme proteins in which the iron
binds O_2 or H_2O_2. Interestingly, and by way of contrast, in the
case of cytochrome-*c* peroxidase, the initial product of H_2O_2
attack produces compound II [Fe(IV)O] and a radical (probably
phenolate) in the protein. Here a tyrosine is attacked. Perhaps
this phenolate sits roughly in the position of the substrate in
other peroxidases and in the case of cytochrome-*c* peroxidase
electron transfer to cytochrome-*c* is by hopping through this
tyrosine. Of course, this peroxidase has no small molecule substrate at all, but interacts with cytochrome-*c*. If the phenolate
radical is brought close to another phenolate group in a second
protein, electron exchange can occur. Thus the electron transfer

$$[Fe \rightarrow phenol] \longrightarrow [phenol \rightarrow Fe]$$
cytochrome-*c* peroxidase   cytochrome-*c*

is possible, but only if one of the Fe atoms is raised to a very
high redox potential. It is for this reason that we excluded
this path in our discussion of electron transfer in cytochromes-
c. We draw attention to this possibility in copper proteins
which have a very high Cu(II)/Cu(I) potential, >0.7 volts.
Models for this type of reaction have been examined *(1)*.

Perhaps the most important results from this study of peroxidases are:
1. The substrate binding site is rather differently disposed
    with respect to the heme for phenols and indoles. The nature
    of the substrates, and the fact that they do not bind to the
    heme iron, suggests that in fact they are just solubilised in
    an "oily droplet" part of the protein about 0.8-1 nm from the
    heme/iron and on the open side of the protein.
2. Substrate binding does affect the iron to some degree which
    indicates that substrate binding modifies a region of the
    protein of length about 1 nm.
3. The substrates phenols, indoles, or iodide are not oxidised
    by the Fe(III) state but only by the higher states, compound
    I and compound II, *i.e.*, Fe(IV) or higher states.
4. In the different peroxidase enzymes, *e.g.*, P_7 and P_1, the
    substrate binding pocket is modified but there is still quite
    general binding.
    The plant peroxidases use hydrogen peroxide to oxidise *organic*
substrates, but there is also a special function of some peroxidases which is to oxidise the simple inorganic anions of the
halogens to halogen radicals or halogen cations, which then act
as very aggressive reagents. So aggressive are these reagents,
in fact, that they can be used to destroy bacteria, as for example
in myelocytes where peroxidative oxidation of Cl^- by myelo-per-
oxidase is employed in the immune system. This action is directly comparable to the use by man of bleach (NaClO) as a disinfect-
ant. Now conventional plant peroxidases will *not* oxidise either
Br^- or Cl^-, although they can readily oxidise I^-. We suspect
that the redox potential required for oxidation of Cl^- is too
high for oxidation by compound I of conventional forms of peroxidases. There are two ways of raising the redox potential of iron
in a protein, no matter what the oxidation state; the first is to
change the ligands on the iron so that they are better electron
acceptors when lower oxidation states are stabilised and E^0 becomes more positive; the second is to make adjustments in the
protein surrounding the iron, *e.g.*, of charge distribution, so as
to favour high potentials. The first solution is found in the
myelo- and lacto-peroxidases which have redox potentials considerably higher than normal peroxidases, as the heme in them has an
aldehyde substituent making it a good electron acceptor (see
Table 11). The second solution is to be found in chloro-peroxidases. Chloro-peroxidases, unlike normal peroxidases, in the
Fe(II) carbon monoxide state have a Soret absorption band at 450
nm, which is otherwise seen only in P-450 cytochromes in these
complexes. In the case of the P-450 cytochromes, this absorption
arises in part from the presence of the thiolate iron ligand.

TABLE 11: PEROXIDASES		
Source	Heme	Substrates
Plants: Horse Radish Turnip	Protoporphyrin IX	Phenols, Indoles
Yeast	Protoporphyrin IX	Cytochrome-*c*
Mould Chloro-peroxidase	Protoporphyrin IX	Chloride (Iodide)
Myelocytes	Unsaturated*	" "
Lactating Glands	Unsaturated*	" "
Blood Stream	None (selenium)	Glutathione

These hemes are more unsaturated than protoporphyrin IX.

We are left with the question as to whether chloro-peroxidase has different ligands from normal peroxidases, *i.e.*, it has a sulphide donor. Chloro-peroxidase like cytochrome P-450 (see below) shows anomalous physical properties such as:
1. Highly rhombic g-values of high-spin Fe(III) species.
2. The Fe(III) species in the native protein is largely low-spin.
3. The absorption spectra of the Fe(II), Fe(II)CO, Fe(III)CN$^-$, and Fe(III)N$_3^-$ complexes are anomalous and are usually shifted to longer wavelengths than expected.

Our conclusion is that as in P-450 complexes (see below), the protein also adjusts the heme unit so as to alter the metal ion properties (the entatic state hypothesis). It is unfortunately the case that only a very detailed structural study can reveal the nature of this strain and such a study is exceedingly difficult for proteins of very high molecular weight.

While describing these chloro-peroxidases, we wish to draw attention to the fact that they could well be a source of weakness as well as of strength for biological systems, just as the P-450 enzymes are both detoxifying agents and potential sources of carcinogens, due to the epoxides they produce. The fact that peroxidases can produce halogen radicals, superoxide radicals, chlorinated organic compounds, and even epoxides, makes us believe that they are not only a source of protection, but they are also a potential source of carcinogenic compounds. Thus this enzyme could be responsible for damage to DNA through production of halogenated compounds or oxidised hormones.

4. THE CYTOCHROMES P-450

Detoxification in many cells and organelles is assisted by the reactions catalysed by the cytochromes P-450. Their overall reaction is

$$2H^+ + O_2 + 2e + RH \rightarrow R\text{-}OH + H_2O$$

These enzymes require a supply of electrons, either from an Fe/S protein or a simple cytochrome such as cytochrome b_5. Unlike the reactions of peroxidases, these reactions occur in membranes. The central functional unit of P-450 is the same iron protoporphyrin as in conventional peroxidases, but the iron is bound to a thiolate ligand and not a histidine. Note that as with peroxidases, these are very catholic enzymes. Their function is to activate oxygen (or hydrogen peroxide). The substrates which can be attacked by P-450 enzymes are, however, much more resistant to oxidation than those which are attacked by peroxidases. Clearly the ancillary enzymes, the Fe/S *or* the cytochrome b_5, assists by reduction of the enzyme-O_2 complex to give a powerful oxidizing species. Another difference from the peroxidases is that P-450 inserts one atom of the O_2 molecule into the organic substrate. Effectively, this action is brought about by the simultaneous action of three substrates at the iron atom

$$2\bar{e} \longrightarrow Fe \begin{array}{c} \nearrow O_2 \\ \searrow RH \end{array}$$

All three must therefore be present together and in close juxtaposition, for one atom of the oxygen must leave the iron to be inserted in RH while one leaves as H_2O. There is now evidence that the special intermediate is of the compound I type found in peroxidases. Finally, it must be the case that the RH group, though probably bound in a similar hydrophobic pocket to that in peroxidases (*i.e.*, a low selectivity site) must be nearer to the O_2 of the FeO complex for the oxygen atom to be transferred.

We return now to the states of the iron in the P-450 enzymes. All the evidence suggests that there is generated an exposed high-spin Fe(II) in cytochrome P-450 which picks up oxygen and then becomes low-spin. This step is unusual in normal peroxidases. Thus the active site must have a similar flexibility to that observed in myoglobin. Further evidence for this flexibility comes from the spin-state changes of the Fe(III) form of the protein in the presence or absence of substrates. The substrates can be sterols, camphors, amines or a vast number of other

molecules, including many drugs. These substrates usually cause
a partial switch in the high-spin/low-spin equilibrium balance on
binding to the protein, showing that the surrounds of the iron
are adjustable for many of the substrates cannot bind to the
iron. Thus, a conformational change is inevitable and this conformational change could be propagated to distant parts of the
protein as in hemoglobin. The spin-state change can act, too, as
a switch of structure in neighbour proteins. This is a cooperativity akin to that found in hemoglobin, except that here the
cooperativity is between functions of different kinds, while in
hemoglobin functions of the same kind are cooperative. Our first
conclusions are that binding O_2 and binding substrate (RH) will
give a spin-state and/or an oxidation state change and that both
will give a conformation change.

It is of interest to ask next what amino-acid residues could
be used to form the heme pocket and bind the substrate of these
enzymes. In the reactions of peroxidases and especially of the
P-450 enzymes, there is the apparent risk that the activated H_2O_2
or O_2 will attack the protein side-chains, for the substrates of
these enzymes are the very groups which are common amino acid
side-chains. In functioning hemoglobins it is known that tyrosine and tryptophan are absent from the heme pocket, and surely
this must be true of peroxidases and P-450 enzymes, for these
residues are readily oxidised. The P-450 series presents an
extreme problem, for this enzyme can even attack *aliphatic* chains
so that there would appear to be no safe side-chain of an amino-
acid for the heme pocket. A possible solution is as follows:

1. After combination with oxygen (before substrate binding),
   there must be a cavity remaining around the heme O_2 complex
   on the O_2 side.
2. The substrate should either be added before oxygen uptake or
   should induce a conformation change in such a way that following the conformation change (induced by ligand binding),
   the carbon to be attacked is sterically compressed upon the
   FeO_2 unit, or the FeO unit, whichever is the attacking group.
3. This allows the attack on the saturated chain to occur once
   the FeO_2 or FeO unit is activated by electron transfer from
   the electron sites (Fe/S or cytochrome b_5 proteins).

Now we turn to the connection between the electron donor,
cytochrome b_5, and the cytochrome P-450. As in the case of cytochrome c, the cytochrome b_5 has a structure in which the heme
comes to the surface, and it is undoubtedly true that outer
sphere electron transfer is one likely route of reduction of this
protein. However, the P-450 lies in a membrane, while the cytochrome b_5 lies exposed on the membrane surface, and it appears to
be highly unlikely that the electron can pass from cytochrome b_5

to cytochrome P-450 through an outer sphere mechanism of the
normal kind. This is all the more the case when it is remembered
that the cytochrome P-450 only takes the electron from cytochrome
b_5 after binding oxygen and a very bulky substrate in the heme
pocket. A more probable model is that the disposition of heme
groups is such that the two hemes are quite far apart in space.
Electron transfer may now be through a hydrophobic channel as
detailed by Moore and Williams *(1)*. The transfer will be trig-
gered by conformational changes.

Recently, with Dr. A. Xavier, we have started a study of
protein/protein interaction between electron transfer proteins.
We have observed that an oxidised ferredoxin from sulphur bacteria binds to cytochrome *c* in its oxidised state and we can make
a direct NMR study of this binding. The ferredoxin binds the
cytochrome *c* close to the heme exposed edge. It may well be possible to study electron transfer reactions between such proteins
as those by NMR methods once we have established the surface of
interaction between the two proteins.

ACKNOWLEDGEMENT. The authors wish to thank Dr. R. Ambler, Dr. C.
Greenwood, Dr. G. Pettigrew and Professor R.E. Dickerson for considerable help with the biochemistry and structure of cytochrome
c.

REFERENCES

*Note: A limited number of references are supplied here, as we
have given a very extensive list in a recent review (Ref. 1).*

1. G.R. Moore and R.J.P. Williams, *Coordination Chemistry Reviews*,**18,** 125 (1976).
2. I.D. Campbell, C.M. Dobson and R.J.P. Williams, *Proc. Royal
   Society (London)*, A**345,** 41 (1975).
3. L.C. Dickinson and J.C.W. Chien, *Biochemistry*,**14,** 3526 (1975).
4. See B.L. Vallee and R.J.P. Williams, *Proc. Nat. Acad. Sci.
   U.S.A.*,**59,** 498 (1968).
5. L. Ford, H.A.O. Hill, B.E. Mann, P.J. Sadler and R.J.P.
   Williams, *Biochim. Biophys. Acta*,**430,** 413 (1976).
6. J. Ricard, G. Mazza and R.J.P. Williams, *Eur. J. Biochem.*,**28,**
   566 (1972).
7. R.J.P. Williams, P.E. Wright, G. Mazza and J. Ricard, *Biochim.
   Biophys. Acta*,**412,** 127 (1975).
8. Y. Hayashi, H. Yamadi and I. Yamazcki, *Biochim. Biophys. Acta*,
   427, 608 (1976).
9. P.S. Burns, R.J.P. Williams and P.E. Wright, *J. Chem. Soc.,
   Chem. Commun.*, 795 (1975).

index